Aero Detail 23 # DE HAVILLAND MOSQUITO

カバー・表紙イラスト／カルミネ・ディ・ナポリ
Cover Illustrations / Carmine di Napoli

Photo / Raffaele Mancini

CONTENTS

モスキート原型１号機（モスキート・エアクラフト・ミュージアム）……2
Mosquito Aircraft Museum

モスキートF.B.Mk.VI シリーズ II（モスキート・エアクラフト・ミュージアム）……3
Mosquito Aircraft Museum

モスキートN.F.Mk.30（ベルギー王立軍事博物館）……4
Musée Royal de l'Armée et d'Histoire Militaire

モスキートB.Mk.35 s/n TA634（モスキート・エアクラフト・ミュージアム）……5
Mosquito Aircraft Museum

モスキートB.Mk.35 s/n TJ138（イギリス空軍博物館）……6
Royal Air Force Museum

モスキートB.Mk.35 s/n RS709（アメリカ空軍博物館）……7
United States Air Force Museum

機首……8
NOSE

キャノピー ＆ コクピット……16
CANOPY & COCKPIT

胴体 ＆ 爆弾倉……30
FUSELAGE & BOMB BAY

主翼……42
WINGS

エンジン……54
ENGINES

主脚……68
MAIN GEAR & GEAR BAY

尾部……72
TAIL

武装……78
ARMAMENT

モスキートの基本塗装 ＆ マーキング……80
Basic Camouflage & Markings of the MOSQUITO

デ・ハヴィランド・モスキート実機写真集……94
DE HAVILLAND MOSQUITO Photo Album

モスキート ディテール・イラスト……104
Detail Illustrations

モスキート各型変遷……78
Variations of the Mosquito

D.H. 98 Mosquito Prototype
s/n E0234 / W4050

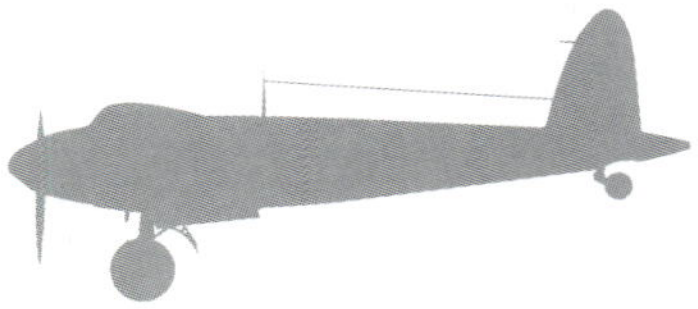

Mosquito Aircraft Museum

モスキート・エアクラフト・ミュージアム

Address 住所 PO Box 107, Salisbury Hall, London Colney, Hertfordshire AL2 1BU, UK
TEL (1727) 822-051
Open from March to the end of October,Tuesday, Thursday and Saturday 14:00 to 17:30, Sunday and Bank Holidays 10:30 to 17:30
5月から10月末まで公開
火、木、土曜： 14:00~17:30
日曜日と銀行休日： 10:30~17:30

モスキートの原型第1号機は現在、ソールズベリーホールのモスキート・エアクラフト・ミュージアムに保存、展示されている。この、世界に現存する歴史的な航空機のなかで、もっとも貴重な機体のひとつは、初飛行以来、空力的試験やエンジンのテスト機として数々のフライトを行ない、モスキートの最高速度も記録して、現在まで生きのびてきた。第二次大戦機のなかにあって、原型第1号機が現存するというのはきわめてまれな例であろう。

原型第1号機はソールズベリー・ホールで完成。デ・ハヴィランド社のBクラス登録番号E0234を与えられ、1940年11月25日に操縦ジェフリー・デ・ハヴィランドJr.、航法士にエンジン関係艤装設計主任ジョン・E・ウォーカーが搭乗し、初飛行に成功した。原型機は左右エンジン・ナセルに1,298hpのロールスロイス・マーリンRM.3SMを装備。機体全面はトレーナーイエロー、スピナーは黒であった。

翌月、空軍省から正式なシリアルナンバーが与えられると、登録番号E0234は、W4050に描きかえられた。1941年2月、ボスコムダウンのA& AEE(Aeroplane and Armament Experimental Establishment)で、実用試験を開始するにあたり、主翼上面と胴体はダークアースとダークグリーンに塗りかえられ、胴体両側面のラウンデル直後に大きな"P"の字が記入された。これは、当時イギリスで、すべての試作機に施されていた基本迷彩であった。

1941年2月24日、この原型機は着陸後のタキシング中に、胴体後部右側の乗降／アクセス用ハッチ周辺に亀裂が入るという深刻なアクシデントにみまわれた。このことから、胴体の各部を改修し、新型の尾輪を採用、同時に胴体右側の乗降／アクセス用ハッチ上部に、胴体補強用のストレーキが追加された。原型機は同年5月にも無理な着陸によって胴体にふたたびダメージを受けたが、すみやかに修理が行われた。この修理の際に貼られた合板のパッチは、現在も展示機にみることができる。

1942年6月、ロールスロイス・マーリン61、2段2速過給器付エンジンに換装された原型機は、この新しいエンジンでの初飛行に成功。その後、原型機は、エンジンをロールスロイス・マーリン77に換装し1943年8月8日に最高速度706km／hを記録。これは水平飛行におけるモスキートの最高速度記録となった。同年12月ごろ、原型機を使用した各種試験は終了。同機はハットフィールドに保存され、デ・ハヴィランド社で実習用の機体として用いられた。1945年、ロールスロイス・マーリン77エンジンにかわって、1段過給器付のマーリン25エンジンが装備され、現在にいたっている。

本機は1958年にソールズベリーホールのハンガー内に収められ、翌年5月15日から、正式に公開されている。原型機は初飛行後にさまざまな改修が加えられ、特にエンジン・ナセルの長さは、原型機のオリジナルでは尾翼にバフェッティッングが生じることから、後端が延長されると同時に、それまで1枚式だったフラップが内翼／外翼に2分割され、本機はその延長ナセルに2分割フラップを装備したタイプに改修されている。しかし、それも改良の歴史であり、ここでは貴重な原型機の姿を、今日もみることができる。

The Mosquito Aircraft Museum has on display what is perhaps one of the most historic aircraft existing in the world today: The original Mosquito prototype. This plane not only survived extensive aerodynamic and engine tests during its brief but intense flying career, but it can also claim the title of the fastest Mosquito ever flown.

The prototype, with the Company Class "B" registration E0234, was completed at Salisbury Hall in November 1940 and made its maiden flight on November 25 with Geoffrey de Havilland Jr. at the controls and John Walker, the Chief Engine Installation Designer, as navigator. The airplane, powered by two 1,298hp Rolls-Royce RM.3SM engines, sported an overall Trainer Yellow paint scheme with Black spinners. In December 1940 the registration E0234 was replaced by the official Air Ministry serial number W4050. In February 1941, before starting Official Service Trials at the A&AEE (Aeroplane and Armament Experimental Establishment) at Boscombe Down, the upper surface of the wings and the fuselage were repainted in Dark Earth and Dark Green camouflage, with large "P" markings applied to both sides of the fuselage, just behind the roundels. This was the standard marking in the U.K. that identified all prototype service aircraft of that era.

On February 24, 1941, the plane was involved in a serious taxiing accident following landing resulting in the fuselage fracturing near the crew access hatch on the right side. An entire new fuselage was subsequently fitted to the aircraft and this incorporated a new tail wheel and several other minor modifications. A reinforcing strake was also added to the right side of the fuselage just above the entry/access door, and this fitting became standard for all of the production Mosquitoes. In May 1941, after a heavy landing, W4050's fuselage was again damaged, but this time it was quickly repaired by the application of a large plywood patch, still visible today.

In June 1942, the prototype made its first flight in its new configuration with Rolls-Royce Merlin 61 two-speed, two-stage supercharger engines. Subsequently powered with the Merlin 77 engines, W4050 attained on August 8, 1943, a top speed of 706 km/h, the highest speed ever achieved by a Mosquito in level flight. The aircraft was retired from test flying, and probably had its last flight sometime December 1943.

It remained at Hatfield, where it was used as an instructive airframe by de Havilland apprentices, and in 1945 its two-stage Merlin 77s were replaced with single-stage Merlin 25s, this being the configuration in which it is preserved today. After a long period of storage at Salisbury Hall, Hatfield and Chester, W4050 was eventually preserved inside a little hangar at Salisbury Hall in October 1958, and on May 15, 1959, it was officially exhibited, becoming the nucleus of Britain's oldest aircraft museum. Because of the numerous modifications and mishaps that took place during its service career, the prototype as it stands today is significantly different from when it made its maiden flight. However, its superb condition and role in history make it a true aviation treasure to be preserved for all time.

D.H. 98 Mosquito F.B.Mk.VI Series II
s/n TA122

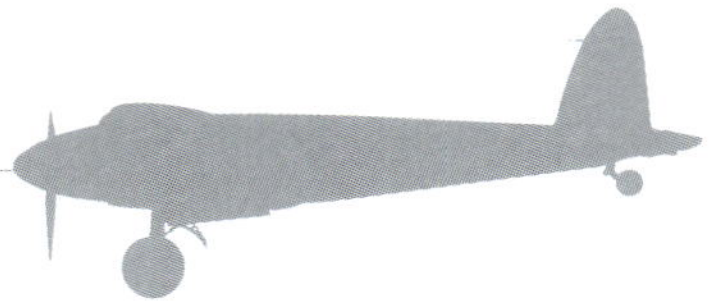

Mosquito Aircraft Museum

モスキート・エアクラフト・ミュージアム

Address　PO Box 107, Salisbury Hall, London Colney,
住所　Hertfordshire AL2 1BU, UK
TEL　(1727) 822-051
Open from March to the end of October,Tuesday, Thursday and Saturday 14:00 to 17:30, Sunday and Bank Holidays 10:30 to 17:30
５月から10月末まで公開
火、木、土曜：　14:00~17:30
日曜日と銀行休日：　10:30~17:30

現在ソールズベリーホールのモスキート・エアクラフト・ミュージアムでレストア中の戦闘爆撃機型モスキート、F.B.Mk.VIシリーズII（S/N TA122）は、1945年初頭にハットフィールドのデ・ハヴィランド社工場で完成した機体である。本機は まず、エドゼルのNo.44MU（メンテナンス・ユニット）に配属され、のちにNo.605Sqnに割り当てられ、同年4月3日、ベルギーに到着。ヨーロッパのドイツ占領地域において、補給施設などへの夜間爆撃任務についている。この、第2戦術空軍のNo.605Sqnは、第二次大戦中、もっとも有名になったモスキート部隊のひとつであった。4月25日、No.605Sqnは、オランダのフォルケルへ移動、同じくモスキート部隊のNo.418Sqnと第2グループのNo.136ウイングを編成した。

終戦後、本機TA122はNo.４Sqnで練習機や連絡機として使用され、1948年11月にNo.1 BRISD予備機保管所に送られた。1949年1月13日、本機は機体コードUP◎Gを与えられてふたたび現役に復帰。旧西ドイツに基地をおくNo.4Sqnに配備されて、最初にヴァーンのRAF基地、のちにツェーレに移動している。夜間攻撃作戦用の練習機として使用された本機は、1950年、現役を引退した。

翌年、機体はオランダのデルフト工科大学に売却され、数年のあいだ、航空工学科の教材として用いられた。この間に、左右両エンジンと装備品のいくつかが失われ、主翼も胴体から取り外されている。分解された機体はその後、長い間オランダ王立空軍博物館で保管され、はじめはヴーンスドレヒトに、のちにはギルゼ=ラインに保存されたのち、モスキート・エアクラフト・ミュージアムに寄贈されることに決定。1978年2月26日、ハットフィールドに到着した。

到着後、本機の胴体に部分的な補修を施したミュージアムのスタッフは、壮大で意欲的なレストア計画を発案した。それは本機をもとに完全なF.B.Mk.VIを再現するというものである。主翼をはじめとする使用可能な部品は、イスラエルに残っていたシーモスキートTR33のものが調達され、計画は1985年に主翼の再製作から始められた。計器や搭乗員乗降ハッチなどの多くの部品も、海外にあったストック、または回収されたものが送られてきており、レストアは国際的なプロジェクトになっていると、いえるかもしれない。。

資金はブリティッシュ・エアロスペース社の援助をうけ、作業は7、8人のボランティアによって続けられている。完成した際には、モスキートによるもっとも有名な作戦として知られる、アミアン監獄爆撃作戦"オペレーション・ジェリコ"の隊長ピカード大佐の乗機、機体コードEG◎Fの塗装になる予定。まだ手に入らないパーツもあるが、計画が終了し、TA122号機が完全な姿をあらわしたとき、ソールズベリーホールのボランティアたちにとって、誇るべきものになるであろう。

This fighter-bomber Mosquito F.B.Mk.VI Series II, serialled TA122, was built at the de Havilland factory at Hatfield in early 1945. It was initially delivered to No.44 MU (Maintenance Unit) at Edzell. After moving to No.417 ARF on March 30, 1945, it was assigned to No.605 Squadron of Warwick County, joining the squadron at Coxyde in Belgium on April 3, 1945. Operating in the night interdiction role over occupied European territories, No.605 Squadron, 2nd Tactical Air Force, was one of the most famous of all the Mosquito squadrons in the course of the Second World War. In April, the squadron moved on to Volkel, Holland, where, with the Mosquito-equipped No.418 Squadron they formed No.136 Wing of No.2 Group.

In the post-war years, No.4 Squadron operated TA122 as training and courier aircraft, and in November 1948, it was transferred to the No.1 BRISD Reserve Pool. On January 13, 1949, wearing squadron code UP-G, it was officially reissued to No.4 Squadron based in the former West Germany, first at RAF Wahn and then at RAF Celle. Operated as a trainer in the night interdiction role, the Mosquito was eventually retired from operational service in June 1950. One year later it was bought by Delft Technical University in Holland, and missing both engines, some components and fittings and having the wing separated from the fuselage, it was used as a teaching aid for several years by the Department of Aerospace Engineering. After its dismantling, TA122 was put into storage for a number of years with the RNLAF Museum, first at Woensdrecht AFB and then to Gilze-Rijen AFB, and finally the fuselage was donated to the Mosquito Aircraft Museum, where it was transferred on February 26, 1978. After having partly restored the fuselage, the Museum's staff undertook a great and ambitious restoration project devoted to rebuilding a complete Mosquito F.B.Mk.VI. A wing and other parts in acceptable condition were recovered in Israel, and in the spring of 1985, the wing rebuild was started. Many other components, such as instruments and the cockpit entry door, were located and recovered from other countries, making this a truly international project.

Though the airplane has lost some of its original parts, when the long restoration program is completed TA122 will be the pride of the hard-working team of Salisbury Hall's volunteers.

D.H. 98 Mosquito N.F.Mk.30
s/n RK952 / MB-24

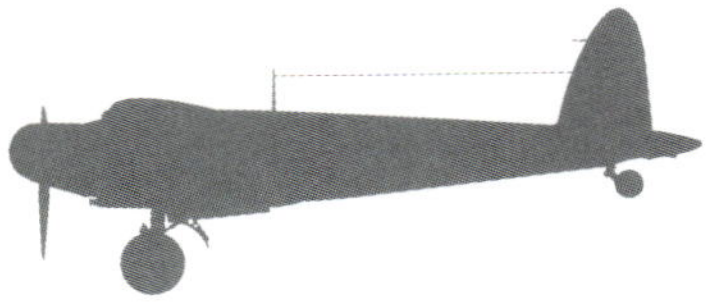

Musée Royal de l'Armée et d'Histoire Militaire

ベルギー王立軍事博物館

Address　Parc du Cinquantenaire, 3
住所　B-1040 Bruxelles - Belgique
TEL　(02) 734-5252
Open 9:00 to 12:00 and 13:00 to 16:45 (Closed on Mondays, Christmas Day, January 1, May 1, Election Days)

公開時間　9:00~12:00、13:00~16:45
月曜日、12月25日、1月1日、5月1日と議員選挙日は休日

ベルギー王立軍事歴史博物館が所蔵、展示するモスキートN.F.Mk.30は、リーヴスデンのデ・ハヴィランド社秘密工場で製造された夜間戦闘機型モスキートである。エンジンは2段2速過給器付ロールスロイス・マーリン76エンジンを装備している。オリジナル・シリアルはRK952。初飛行は1945年5月22日、ヨーロッパでの戦争が終結した数週間後のことであった。

本機ははじめRAFのNo.218MU（メンテナンス・ユニット）に送られたが、作戦部隊に配備されることのないまま、同年7月11日、NO.10MUに移され、長い間保管されたのち、1951年10月30日にベルギー政府に売却された。ベルギーに輸送する前にリングウェイのフェアリー航空機で整備を受けている。

1953年9月4日、ベルギー空軍の戦力となったRK952は、新しいシリアル・ナンバーMB-24を与えられ、ベルギー空軍向けのモスキートN.F.Mk.30最後の24機の1機として、新しい任務についた。機首上面に透明ガラスの新型レドームが設けられた本機は、機体コードND-Nをつけ、ボーヴァシャンのNo.1ウイングNo.10（夜間戦闘機）Sqnに配備されている。本機はここで、もっぱらレーダー／ビーコン測定飛行と、標的機として使用されている。

1955年8月18日、MB-24最後の公式フライトが行われた。搭乗していたのは、パイロットがアンリ・ユベール中佐、航法士がヨ・ヴェイネンであった。1956年10月17日、本機はボーヴァシャンで退役し、翌年3月17日に、ブリュッセルの王立軍事博物館へ移送され、保存されることになった。1968年には一時的な修復が行われている。

1979年、金属部分と装備品の修復を目的としたレストアが行われ、1984年にはベルギー空軍の基本的迷彩である全面ミディアムシーグレイに上面ダークグリーンの雲形迷彩を施され、機体コードND-Nを記入された。NDはスコードロンのコード、Nは機体の個別コードをあらわしている。また、近年、数名のボランティアによりコクピットに関するレストア計画が実行されている。レーダーのオリジナル部品の一部、20mm機関砲まわりなどに失われた部品はあるが、本機は夜間戦闘機型モスキートとして、たいへん良好なコンディションで保管されている。

The Mosquito N.F.Mk.30 owned and exhibited by the Musée Royal de l'Armée et d'Histoire Militaire was built at de Havilland's Leavesden shadow factory, powered by Rolls-Royce Merlin 76 two-stage supercharger engines. This aircraft, originally serialled RK952, made its first flight on May 22, 1945, just a few weeks after VE-Day. Initially delivered to No.218 MU (Maintenance Unit) of the Royal Air Force, this Mosquito never saw wartime service. On July 11, 1945, it was assigned to No.10 MU at Hullavington, where it was stored for a number of years until it was sold to the Belgian Government on October 31, 1951. Before delivery to Belgium, the aircraft was transferred to the Fairey Aviation facilities at Ringway for general maintenance.

RK952 was finally put into service by the Belgian Air Force on September 4, 1953, receiving the new serial number MB-24, being the last of twenty-four BAF N.F.Mk.30s to be delivered. The aircraft, sporting a new glazed radome fitted on the upper side of the nose, was coded ND-N and assigned to No.10 (Night Fighter) Squadron, No.1 Wing, at Beauvechain, but it was not used operationally, being operated for radar and beacon calibration flights and as a target aircraft. MB-24 made its last official flight on August 18, 1955, flown by pilots Kpt. Henri Boels and Kpt. Jos Wijnen. It was eventually retired from active service on October 17, 1956, at Beauvechain and was transported to the Royal Army Museum (Air Section) at Brussels on March 17, 1957, where it was preserved. In 1968, it was temporarily transferred to Koksijde for repainting.

In 1979, a restoration project began focussing mainly on repairing metal components and equipment, and in 1984, the Mosquito was completely repainted at the Royal Army Museum to a more representative camouflage scheme of Medium Sea Grey overall and disruptive pattern of Dark Green on the upper surfaces. This was exactly as it was marked in BAF service. It was coded ND-N, ND being the squadron mark and N the individual aircraft letter. Recently, a small team of volunteers has undertaken a new restoration program centering on the cockpit. Though some of the airplane's original radar equipment, some fittings and the 20mm cannon installation are missing, this night fighter Mosquito is in excellent, authentic condition.

D.H. 98 Mosquito B.Mk.35
s/n TA634

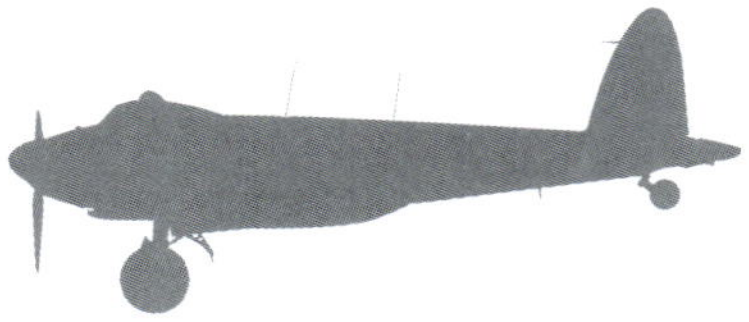

Mosquito Aircraft Museum

モスキート・エアクラフト・ミュージアム

Address 住所 PO Box 107, Salisbury Hall, London Colney, Hertfordshire AL2 1BU, UK
TEL (1727) 822-051
Open from March to the end of October,Tuesday, Thursday and Saturday 14:00 to 17:30, Sunday and Bank Holidays 10:30 to 17:30
５月から10月末まで公開
火、木、土曜： 14:00~17:30
日曜日と銀行休日： 10:30~17:30

爆撃機型モスキート最後の生産型であるB.Mk.35は、実戦には間に合わなかったが、1948年春まで、計274機生産されたバリエーションである。

モスキート・エアクラフト・ミュージアムが所有するB.Mk.35（S/N TA634）は1945年春にデ・ハヴィランド社のハットフィールド工場で生産され、実戦部隊に配備されることもなく、1945年4月14日、ハットフィールドから直接シャウビュリーのNo.27MU（メンテナンス・ユニット）に向けて飛行している。本機はここでおよそ7年のあいだ保管され、1952年2月22日、サイウェルに所在するブルックランド・アヴィエーション社に送られて、標的曳航機に改造されることになった。機体は、1953年12月31日にランドウのNo.4CAACU（Civirian Anti-Aircraft Cooperation Unit／民間協同防空隊）に送られ、標的曳航任務に就いた。

本機はのちに第2戦術空軍の司令部に配備され、武装訓練機として用いられた。その後、しばらく保管されていた本機は、1959年にエクセターのNo.4CAACUに送られ、1963年には、RAFが行ったモスキート公式最終飛行にその1機として参加している。そして、同年6月14日、B.Mk.35（S/N TA634）は正式に退役した。

TA634はのちにリヴァプール・コーポレーションに売却され、1968年、グレイ／グリーン系迷彩に塗装に、HT◎G(HJ896)という架空の機体コードをつけた本機は、映画「モスキート爆撃隊」に出演。同年７月16日に最後のフライトを終えた。

1970年、リヴァプール・コーポレーションは本機をモスキート・エアクラフト・ミュージアムに寄贈。シリアルはオリジナルのTA634にもどされ、新しい機体コードEG◎Fが与えられた。機体はミュージアムのスタッフによって注意深く大切に保管されてきたが、10年にわたって屋外展示を続けてきた結果、機体と装備品に腐食が進んだため、1981年に新しい格納庫が作られ、本機はこのなかで屋内展示されるようになった。そして、機体構造のダメージと木製部分の綿布に関する重要な修復が開始された。また、同時に標的曳航仕様に改修されていた爆弾倉後方ドアを、もとどおりに修復することになった。作業はミュージアムのスタッフと、モスキート修復を決意した人々による特別チームによって行われていった。部品はほとんど完備しているが、内翼前縁内部のラジエターと滑油冷却器は、失われたままである。

現在、B.Mk.35（S/N TA634）は、1944年から1945年にかけて夜間爆撃に活躍したNo.571Sqn所属のB.Mk.XVI、8K◎K機として、爆撃機型の基本迷彩である上面ダークグリーン／オーシャングレイ、下面ミディアムシーグレイが施され、ソールズベリーホールのモスキート・エアクラフト・ミュージアムに展示されている。

The B.Mk.35 of the Mosquito Aircraft Museum was built at de Havilland's Hatfield factory in the spring of 1945, bearing military serial number TA634. The aircraft was never flown by an operational squadron, being flown direct from Hatfield to No.27 MU (Maintenance Unit) at Shawbury on April 14, 1945, where it was stored for about seven years. On February 22, 1952, TA634 was delivered to Brooklands Aviation Ltd. at Sywell for conversion work to target tug standards, and then on December 31, 1953, it was assigned to No.4 CAACU (Civilian Anti-Aircraft Co-operation Unit) at Llandow to be operated as a target tug. Later it was delivered to HQ 2nd Tactical Air Force, flown by the Armament Practice Station. In September 1959, after a brief period of storage, TA634 was delivered to No.3 CAACU at Exeter, taking part in the "Official Last Fly-Past" of RAF Mosquitoes on May 9, 1963, and then, on June 14, it was officially retired from active duty.

The aircraft was later bought by Liverpool Corporation, and in 1968, camouflaged in a gray/green paint scheme with the code letters HT-G and the false serial HJ896, it played a starring role in the *Mosquito Squadron* movie, flying for the last time on July 16, 1968. In 1970, the airplane was donated by Liverpool Corporation to the Mosquito Aircraft Museum, and it received the new code EG-F, retaining the serial TA634. The Mosquito was admirably preserved by Museum's staff with great care, but ten years of exposure to the elements virtually destroyed many of the airframe components. In 1981, the Mosquito was finally housed inside a new hangar built with the assistance of de Havilland enthusiasts, and an important restoration program began to repair the structural damages and the fabric on the wooden structure. In the course of work on the bomb doors, they were converted back from target tug to bomber specifications. The special group that undertook the restoration work included members of the Museum and people who had dedicated their careers to working on Mosquito airplanes. Only a few components are missing today, among which are the oil and coolant radiators inside the inboard wing leading edges.

B.Mk.35 is currently exhibited at Salisbury Hall and is undergoing further cockpit restoration work. It sports a standard finish of Dark Green and Ocean Grey upper surfaces and Medium Sea Grey undersurfaces with the insignia of No.571 Squadron, coded 8K-K. This marks it as a B.Mk.XVI of the Light Night Striking Force, one of Bomber Command's most potent weapons during 1944/45.

D.H. 98 Mosquito B.Mk.35
s/n TJ138

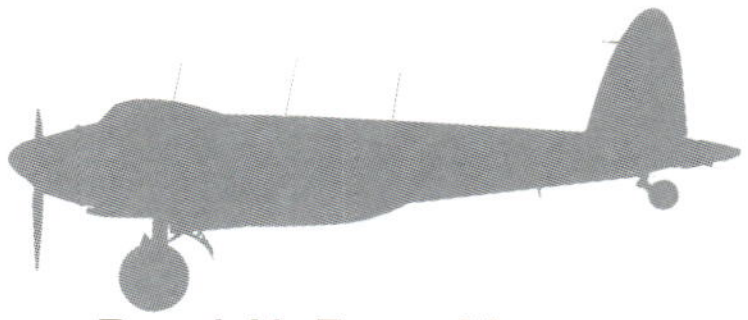

Royal Air Force Museum

イギリス空軍博物館

Address 住所 Grahame Park Way, Hendon, London NW9 5LL, UK
TEL (181) 205-2266
Open Monday-Saturday 10:00 to 18:00, Sunday 14:00 to 18:00 (Closed Bank Holidays)
公開時間
月～土： 10:00~18:00
日曜日： 14:00~18:00
銀行休日は休館。

イギリス空軍博物館のTJ138は、現存するB.Mk.35のなかで、唯一爆撃機として部隊配備された機体である。2段2速過給器付ロールスロイス・マーリン113／114エンジンを装備した本機は、1945年夏にデ・ハヴィランド社のハットフィールド工場で生産され、8月にシャウビュリーのNo.27MU（メンテナンス・ユニット）へ、保管のため送られている。1950年10月31日、本機は旧西ドイツのツェーレへ移動し、短期間のあいだNo.98Sqnで部隊配備についた。機体コードはVO◎L機であった。同年11月初旬に部隊は旧西ドイツのファスベルクへ移動。翌年2月20日、本機は新しい配備計画により余剰機となってイギリスへもどり、ランドウのNo.38MUに保管された。

1953年7月15日、本機は標的曳航機に改造するためサイウェルのブルックランド・アヴィエーション社に送られて、翌年1月に作業が終了。はじめにNo.22MUへ移動、1954年3月8日からは、ランベルのNo.5CAACU（Civirian Anti-Aircraft Cooperation Unut／民間協同防空隊）で新しい標的曳航機として、ミサイル攻撃シミュレーションのためにレーダー照準の目標機となったり、対空砲火訓練の標的機に使われた。

1959年6月、現役を退くことになったTJ138は、処分のためにNo.27MUへふたたび送りだされた。翌月3日、本機は兵力外とされ、8日には新しい機体取扱番号7607Mを与えられている。7月29日、機体は当時イギリス国内で徴募のための機体展示を熱心に行っていたバイスターのNo.71MUに送られた。1960年、近衛騎兵連隊のパレードに展示された本機は、No.98Sqnの塗装に塗り直され、最初のシリアルTJ138をつけられていた。

1967年1月、本機はイギリス空軍博物館に移されて、しばらくのあいだ展示されたのち、1975年12月にフィニングリーへ、1977年2月にはRAFの新兵教育学校所在地であるスウィンダーバイに移送された。この時点でエンジンはかろうじて始動可能することができたが、機体の状態は悪く、羽布張りの部分も欠落していた。

1986年11月、機体はセント・アサン基地に保存される歴史的航空機として展示するために、移送され、B.Mk.35として大々的な修復が加えられた。しかし、この修復では標的曳航機に改修された胴体下面後部フェアリングやそのまわりまで、手がつけられていない。

1991年、本機は戦後旧西ドイツ駐留していた時の塗装であるNo.98SqnのVO◎L機に塗り直され、1992年にRAFアヴィントンのスタッフによって分解、イギリス空軍博物館へふたたび搬送されている。1992年2月7日、TJ138はヘンドンに到着し、現在まで同地のイギリス空軍博物館ボマーホールに展示されている。

The Royal Air Force Museum's Mosquito TJ138 is the only surviving B.Mk.35 that was actually operated in squadron service as a bomber. It was manufactured at de Havilland's Hatfield factory during the summer of 1945 and powered by 1,690hp Rolls-Royce Merlin 113/114 two-stage supercharger engines. It was delivered in August to No.27 MU (Maintenance Unit) at Shawbury for storage. On October 13, 1950, this Mosquito moved to Celle, in the former West Germany, where it started its brief squadron service operated by the No.98 Squadron as VO-L. In early November 1950, the Squadron moved to Fassberg, and then on February 20, 1951, the aircraft was flown back home to the UK as re-equipment plans made it surplus. It was then held in storage with No.38 MU at Llandow.

On July 15, 1953, it was delivered to Brooklands Aviation Ltd. at Sywell for conversion work as a target tug, and it was ready for operational service in January 1954. Initially allocated to No.22 MU at Silloth, it was delivered to No.5 CAACU (Civilian Anti-Aircraft Co-operation Unit) at Llanbedr on March 8, 1954, to be operated in its new target towing role out of Woodvale, acting as targets for radar plotting in connection with simulated missile attack and for anti-aircraft gun practice. TJ138 ended operational service in June 1959, when it was retired and was assigned for disposal to No.27 MU once again. On July 3, 1959, it was declared non-effective, and on July 8, 1959, it received instructional airframe number 7607M. On July 29, 1959, the aircraft was delivered to No.71 MU at Bicester. This Maintenance Unit was at the time the RAF's prime salvage and transportation unit devoted to enlisting and exhibition initiatives around UK. In 1960 the aircraft was displayed at Horseguards Parade, by now returned to No.98 Squadron colors and re-serialled TJ138.

In January 1967 the Mosquito was handed over to the Royal Air Force Museum and was part of Colerne's collection of historic aircraft until the station's closure, after which it was transferred first to Finningly on December 31, 1975, and later in February 1977 to Swinderby, home of the RAF's School of Recruit training. At this time, the engines were still run from time to time, although the aircraft needed refurbishment and lacked the fabric covering on the fuselage.

In November 1986 it was transferred to RAF St. Athan to join the regional collection of historic aircraft, and eventually refurbished and restored as a B.Mk.35. Nevertheless, it retains some target tug modifications such as the rear bomb bay fairing and some fittings on the underside of the fuselage. In 1991 it was repainted again in No.98 Squadron colors and coded VO-L, representing the actual post-war period when the bomber was deployed in the former West Germany. Dismantled and loaded at St. Athan for transport to the RAF Museum by a team from RAF Abingdon, TJ138 arrived at Hendon on February 7, 1992, where it is now on permanent display in the Bomber Hall.

D.H. 98 Mosquito B.Mk.35
s/n RS709

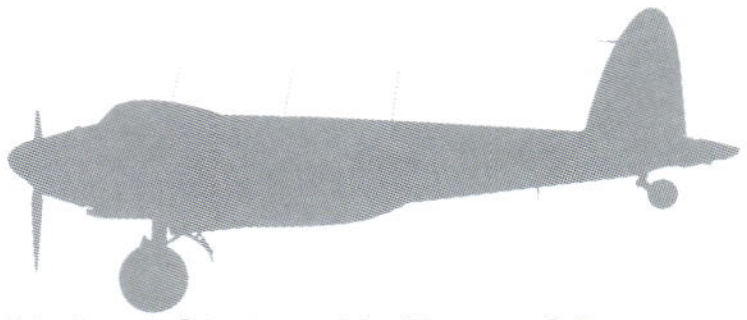

United States Air Force Museum

アメリカ空軍博物館

Address　Wright-Patterson Air Force Base
住所　Dayton, Ohio 45433-6518, U.S.A.
TEL　(513) 255-3286
Open daily 9:00 to 17:00 (Closed on Thanksgiving Day, Christmas Day, January 1)
公開時間　毎日 9:00〜17:00
1月1日、感謝祭、12月25日は休館。

アメリカ空軍博物館に展示されているモスキート（S/N RS709）は、1946年にイギリスのエアスピード・エアクラフト社によって、クライストチャーチで製造されたB.Mk.35である。本機は1946年初頭にまず、ケンブリッジのマーシャル・フライング・サービス社に送られ、その後、ロートンのNo.15MU（メンテナンス・ユニット）で保管され、標的曳航機に改修するため、1952年にはブルックランド・アヴィエーション社に送られている。

改修を終えた本機は、1952年10月31日に、No.236OCU（Operational Conversion Unit／作戦行動転換訓練部隊）へ送られたのち、シャウビュリーのNo.27MUで保管され、最終的にエクセターのNo.3CAACU（Civirian Anti-Aircraft Cooperation Unut／民間協同防空隊）の標的曳航機となった。本機は1963年5月16日に現役を引退。民間機G-ASKA機となったのち、映画「633爆撃隊」に出演、活躍し、アメリカの個人オーナーに売却された。1972年1月2日、機体はテキサス州ハーリンゲンの南部空軍に送られ、可動しない展示用エンジンを装備され、静態展示されていた。アメリカでの登録番号はN9797であった。その数年後、本機はイエスタディズ・エアクラフト・ミュージアムに売却され、1976年にはフォーブス・フィールドでコンバット・エア・ミュージアムに展示されていた。

1979年、RS709はイギリスの個人オーナーに売却されて、同年11月、ふたたび英国の地を踏み、大がかりな修復を行われた。金属部分の多くは交換され、オリジナルでは木製だったテイルコーンが、グラスファイバー製のものにおきかえられた。また、プロペラとエンジンはアメリカでオーバーホールされている。機体はイギリスの新しい民間機登録番号G-MOSIを与えられ、塗装は、1944年当時のM.Mk.XVIをモデルに、スピナーが赤い、グレイ／グリーンのカモフラージュにされた。1983年、本機は大修復後の初飛行を済ませると、アメリカ空軍博物館に売却され、2年後の1985年2月4日に、ライト・パターソン飛行場に到着した。

現在、本機はバルジ型の爆弾層ドアを、フォト・フラッシュ爆弾用ドアに改修して、PR.Mk.XVIとして展示されており、1944年から1945年にかけてイギリスで活動したアメリカ陸軍第25爆撃航空群653爆撃（偵察）飛行隊所属の機体として、全面PRUブルーで塗装されている。

第二次大戦中、アメリカ空軍はカナダ製モスキート40機を使用したほか、レンドリースの見返りにイギリスから100機以上のモスキートを受け取っている。第25爆撃航空群のモスキートは尾部を赤く塗装していたのが特徴で、写真の展示機に塗られているDデイ・インヴェイジョンストライプは、上面部が消された1944年７月当時を再現している。

The Mosquito on display at the United States Air Force Museum, s/n RS709, is a British-built B.Mk.35 manufactured in 1946 by Airspeed Aircraft at Christchurch. It was initially delivered to Marshalls Flying Services Ltd. at Cambridge early in 1946, after which it went to No.15 MU (Maintenance Unit) at Wroughton for storage. In 1952, RS709 was delivered to Brooklands Aviation Ltd. at Sywell for conversion work to the target-towing role. Delivered to No.236 OCU (Operational Conversion unit), on October 31, 1952, it was again put into storage with No.27 MU at Shawbury, and finally operated as a target tug by No.3 CAACU (Civilian Anti-Aircraft Co-operation Unit) at Exeter. It retired from active service on May 16, 1963. The aircraft played a major role in the *633 Squadron* and *Mosquito Squadron* movies with the civil registration G-ASKA, though it never wore this code. In 1969, an American private owner bought this Mosquito, and on January 2, 1972, it was transferred to the Confederate Air Force at Harlingen, Texas, where, with the American registration N9797, it was put on static display due to an unserviceable engine. After a few years, it was sold to Yesterdays Air Force, and in 1976, it was exhibited at Forbes Field with the Combat Air Museum. RS709 was again sold in 1979 to a British private owner, and in November, it returned home to the U.K. There, the Mosquito was completely refurbished, many of the metal components were replaced, the original wooden part of the tail cone was replaced with a fiberglass fitting, and both of the engines and propellers were overhauled in the U.S. The plane received a new British civil registration G-MOSI, wearing a gray/green camouflage paint scheme, with red spinners, representing a B.Mk.XVI of the 1944 period. After this overhaul, the Mosquito flew again in September 1983. Finally, RS709 was sold to the USAF Museum, landing on February 4, 1985, at Wright-Patterson.

This Mosquito has been restored to represent a P.R.Mk.XVI, with the bulged bomb bay doors replaced by flush-fitting doors, sporting a PRU Blue paint scheme as a weather reconnaissance aircraft of the USAAF 653rd Bomb Squadron, 25th Bomb Group, based in England in 1944-45. In fact, during the war, in addition to the 40 Canadian Mosquitoes acquired by the USAAF and flown under the American F-8 (photo reconnaissance) designation, the British turned over more than 100 Mosquitoes to the USAAF under Reverse Lend-Lease, and these aircraft retained their British designations as P.R.Mk.XVIs. The 25th Bomb Group adopted a red tail for their planes in August 1944, and removed the D-Day invasion stripes from the upper wing and upper fuselage surfaces in September 1944.

機首 NOSE

1

2

3

写真 1-3: 数あるWWⅡ機のなかにあってその原型第１号機が現存しているというのはきわめて希有であろう。生誕の地、ソールズベリー・ホールのモスキート・ミュージアムに現存するこのモスキート原型、s/nW4050はその代表でもある。写真はその原型の機首爆撃照準窓周囲。のちに戦闘機、偵察機などあらゆる任務についたモスキートだが、空軍発注仕様B1/40や、この爆撃照準窓などからも明らかなように、本来は爆撃機として開発された機体であった。武骨なフォルムが一般的だった当時の英空軍機のなかにあって、スマートなモスキートの姿態は完成当初からその卓抜した高性能ぶりをうかがわせる。機体全面をイエローに塗っているのは、テスト飛行時に味方対空砲火部隊から誤射されないための措置である。この当時、ソールズベリー・ホールのあったハットフィールドはドイツ空軍機の爆撃にさらされ、敵味方が入り乱れる第一線でもあったのである。

Photos 1-3: There are very few WWII aircraft for which the original prototype still exists, but the Mosquito is one of those few. S/N W4050, the original Mosquito, can be found in Salisbury Hall of the Mosquito Museum. The photos show the area around the bombardier's sighting window in the nose. Though eventually used for nearly every purpose imaginable, including fighter and reconnaissance plane, these photos testify that the Mosquito was originally developed primarily as a bomber. The all-yellow markings of the plane were to make identification simple and prevent its being shot at by anti-aircraft gunners in the vicinity during test flights. At the time, Hatfield, the location of Salisbury Hall, had been bombed by the Luftwaffe, making for nervous trigger fingers all around.

写真 4: レストア中の戦闘／爆撃機型F.B.Mk.VIの機首部。塗装が剝がされた状態のため胴体の合板部や金属製のキャノピー枠や、機体各部の構造、形状などの詳細がよく分かる。

写真 5: ブローニングMk.II 7.7mm機銃４挺(装弾数各500発)を装備する、機首ソリッドノーズ部。強度や耐熱の観点からプレス成形された金属製で仕上げられている。

写真 6: 同じく、ソリッドノーズ部を裏側からみる。４個並んでいるのが7.7mm機銃口で中央の穴は航法灯取付穴。左側上部の穴は射撃カメラ用のもので、G42またはG45射撃カメラが取り付けられた。

Photo 4: The nose of the F.B. Mk.VI undergoing restoration. Thanks to the condition of the fuselage paint, many of the details of the fuselage's structure that would normally be obscured are well shown.

Photo 5: The solid nose, which housed four Browning Mk.II 7.7mm MGs, carrying 500 rounds each. This area is a pressed metal construction, given the needs for strength and heat-resistance.

Photo 6: The same nose seem from the inside. The four holes in a row would accommodate the gun muzzles, while the central opening was for the nose navigation light. At the upper left is where either a G42 or G45 gun camera would be mounted.

5

4

6

機首 NOSE

7

8

写真 7-8: その特異な形状からブルノーズ(雄牛の鼻)とも呼ばれた、A.I.Mk.Xレーダーを収容するN.F.Mk.30の機首レドーム。通常は機体塗色によって塗られている。レーダーの電波障害を低減させるため金属製ソリッドノーズは廃され、ファイバー製のレドームに改められている。N.F.Mk.30は大戦に参加した最後のモスキートで、リーブスデン工場で530機すべてが生産された。機首右側のバルジは射撃カメラ収容部。

写真 9-10: レドーム後方の機器装備部のパネルハッチ。写真10はパネル裏側を示す。

Photos 7-8: The nose radome of the N.F. Mk.30, which housed an A.I. Mk.X radar, the shape of which caused it to be known affectionately as the "bull nose." To prevent interference with the radar system, the metal construction was dropped in favor of a fiber one. It was normally painted in the same colors as the rest of the aircraft. The N.F. Mk.30 was the last Mosquito type to be introduced into combat in WWII. All 530 examples produced were built at Ribsden. The bulge on the right side of the nose is for a gun camera.

Photos 9-10: The panel hatch to the equipment mounting area at the rear of the radome. Photo 10 is the interior side of the panel.

9

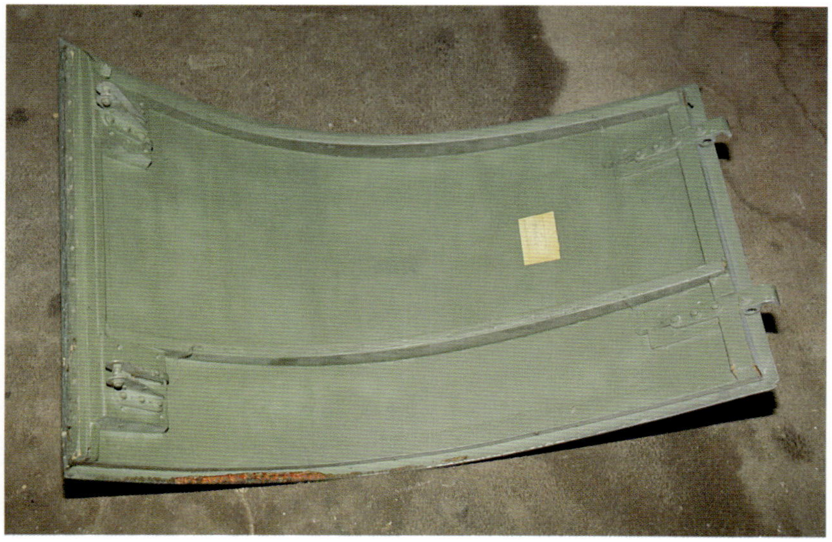

10

写真11： 機首レドームを後方からみる。A.I.(Airborne Interception:空中要撃)Mk.Xレーダーを取り付けた状態を示す。右端に一部みえるくぼみは射撃カメラの収容部。

写真12： A.I.Mk.Xレーダー設置フロア。内側はコクピットと同じくグレイグリーンで塗装されている。

写真13-14： A.I.Mk.Xレーダー本体。A.I.Mk.Xレーダーは英国のレーダーをもとに米国で開発された当時の最新鋭航空機搭載用レーダー、SCR720の英軍呼称で、10センチ波長を使用するマイクロ波要撃レーダーである。従来のA.I.Mk.Ⅷが捜索距離10.46km、左右各3.22kmだったのに対し、前方および左右各16.1kmという性能で、米陸軍のP-61ブラックウィドウの搭載レーダーとしても採用されている。

Photo 11: The nose radome seen from the rear, as it would appear with the A.I. (airborne interception) Mk.X radar installed. The bulge for installation of the gun camera can be seen to the far right edge of the nose.

Photo 12: The A.I. Mk.X mounting floor. It's painted the same gray/green as the cockpit.

Photos 13-14: The A.I. Mk.X radar. The Mk.X was the most advanced air search radar of its day, developed in the United States from a British design. Known as the SCR270, it used the 10cm microwave band. The A.I. Mk.VIII that it replaced could see only 10.46km forward and 3.22km to either side of the aircraft, whereas the Mk.X could search 16.1km forward and to either side. It was also mounted in the U.S. P-61 Black Widow.

11

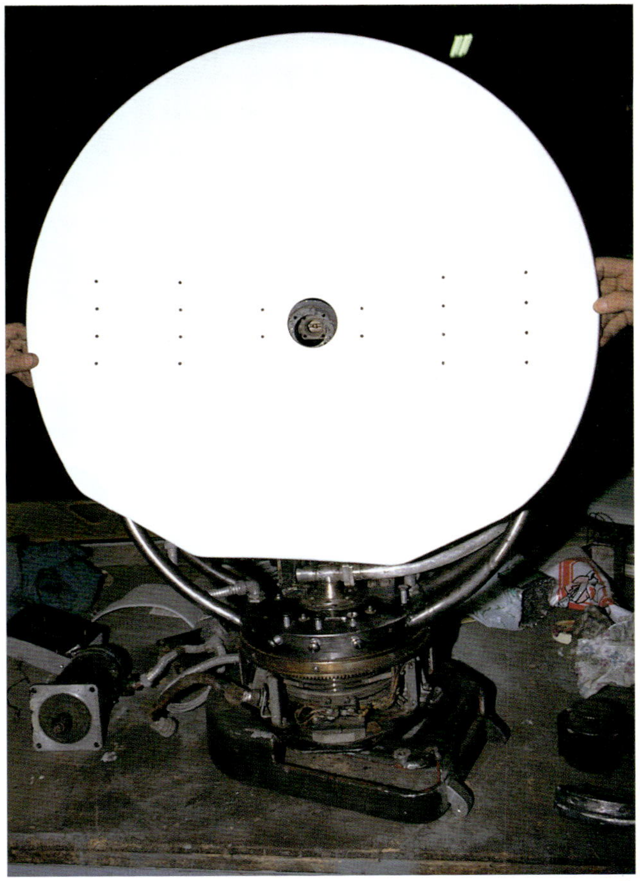
13

15

写真15： 機首右側に設けられた射撃カメラ・ハウジング。機首レドーム設置のため、機首武装は廃され、射撃カメラもこの位置への移動を余儀なくされてしまった。

Photo 15: The gun camera housing mounted on the right side of the nose. The presence of the radar forced the removal of nose armament, and the relocation of the gun camera to this position.

12

14

機首 NOSE

16

17

18

19

写真16-17: 爆撃機型の最終量産型、B.Mk.35の機首部(S/N TA634)。初飛行は終戦まぎわの1945年3月12日だったため、実戦でその高性能を発揮できずに終わってしまった。

写真18: 内側からみた爆撃照準窓。爆撃照準器は取外されている。

写真19: 爆撃照準窓先端に装着されている機首航法灯。電球自体は取外されている。

写真20: 機首左右には爆撃士用の側方窓が設けられている。機内の明りとりの役割も果たした。

写真21: 機首右側に設けられた機体アースの小ハッチ。

Photos 16-17: The nose of the B.Mk.35(S/N TA634), the final Mosquito bomber version. It made its first flight on March 12, 1945 and was unable to demonstrate its potential in combat before the war ended.

Photo 18: The bombardier's window seen from the inside. The bombsight has been removed.

Photo 19: The nose navigation light at the tip of thebombardier's window. The bulb itself is not present.

Photo 20: Windows for the bombardier are present on both sides of the nose. These helped bring light into the aircraft as well as aid visibility.

Photo 21: The external groud on the right side of the nose.

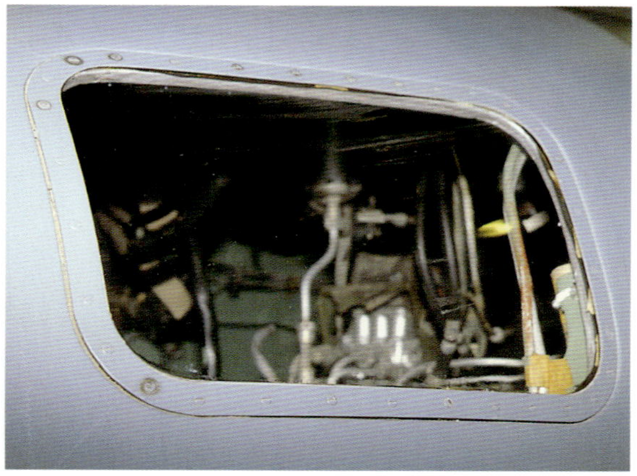
20

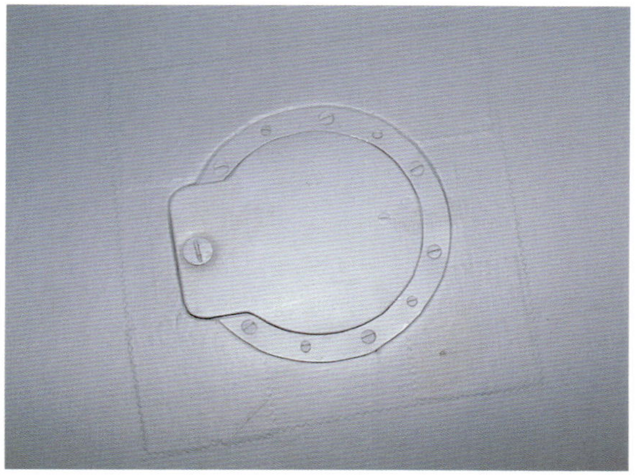
21

22

23

24

写真22-24: B.Mk.35(S/N TJ138)の爆撃照準窓。装備している爆撃照準器はMk.XⅣとよばれ、従来のMk.Ⅱ(オートマチック)、Mk.Ⅸ(コースセッティング)にかわって採用された、米陸軍のノルデン爆撃照準器と同じ方向ジャイロ式の新型爆撃照準器。ランカスターなどにも装備された大戦中期以降の英空軍の代表的爆撃照準器である。昼間爆撃やパスファインダー(目標指示／先導機)任務のモスキートの照準器として広く使われたが、のちに夜間爆撃やターゲット・マーカー任務の機体では、従来の爆撃照準器を外して爆撃士席にオーボエと呼ばれる全天候爆撃／航法システム、またはH2Sレーダーを搭載してレーダーによる爆撃照準を採用した機体も多い。これらの機体ではこの爆撃照準窓／側方窓の部分を機体塗装によって塗りつぶす場合が一般的である。

Photos 22-24: The bombardier's sighting window on the B. Mk.35(S/N TJ138). The Mk.XIV bombsight that was normally installed was based on a directional gyro, like the famed U.S. Norden bombsight. Used in the Lancaster as well, it was the standard U.K. bombsight from the middle of the conflict onward. Though commonly used for daylight bombing and pathfinder missions, later night bombing and target marker aircraft frequently used an H2S radar bombing system, or the Oboe all-weather bombing/navigation system in place of the standard bombsight. Most such aircraft had the nose window and side windows painted in the same scheme as the rest of the aircraft.

25

26

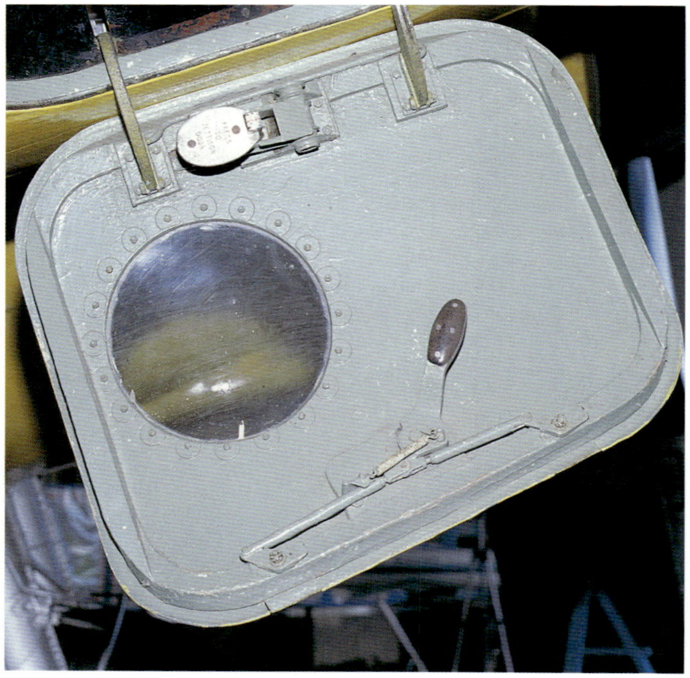

27

写真25-27: モスキート原型機の搭乗員乗降ハッチ。機首下面右側よりに設置されたこのタイプのハッチは爆撃機型、偵察機型で採用されている。ドアには下方をのぞける丸型の窓が設けられており、開閉の際に下に障害物や地上員の有無を確認できるようになっていた。ドア内側のレバーはロック用のものだが、ヒンジ近くにあるノブは緊急時のドア投棄用のものでロックを解除状態にしたあと、このノブを押すことによりヒンジ部が外れ、ドア自体が投棄されるようになっている。

写真28,29: 同じくB.Mk.35(S/N TA364)の乗降用ハッチ。ドア中央にステーが追加されているのが異なるが、構造的には大きな差異はない。写真ではみえないが、コクピット床板(爆撃／航法士席手前)にも同じく丸窓付のドアが設けられている。もちろんこれは飛行中に爆撃照準や航法、観測にと前後にあわただしく動き回る爆撃／航法士への配慮からである。写真28に写っているのは昇降用ラダーでこれを使用して機内に出入りする。なお、このラダーは折り畳まれて入口前部内側にある収容部に格納される。

Photos 25-27: The crew boarding hatch of the Mosquito prototype. Located on the underside of the right side of the nose, a similar configuration was also used in bomber and recon. versions of the plane. A small, round window allows crew to confirm the presence of obstacles or ground personnel before opening. The lever on the inside of the door is to lock it, while the knob close to the hinges is the emergency release. After releasing the normal lock, this knob would unlock the hinge and release the hatch from the aircraft completely.

Photos 28-29: The crew boarding hatch of the B. Mk.35(S/N TA634). Except for the presence of a stay in the center of the door, there is little difference from that of the prototype. Though not visible in the photo, a similar hatch is present in the cockpit floor, just ahead of the bombardier/navigator's seat, as he would frequently change position in the aircraft while in flight. Photo 28 shows the boarding ladder. The ladder folds up for storage in a special compartment just forward of the hatch.

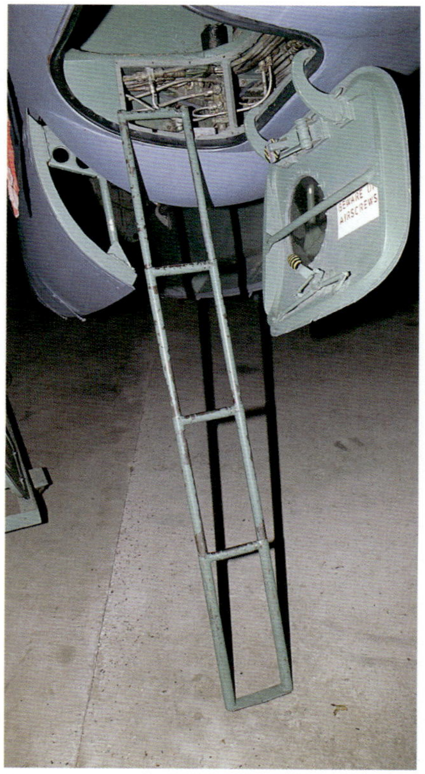

28

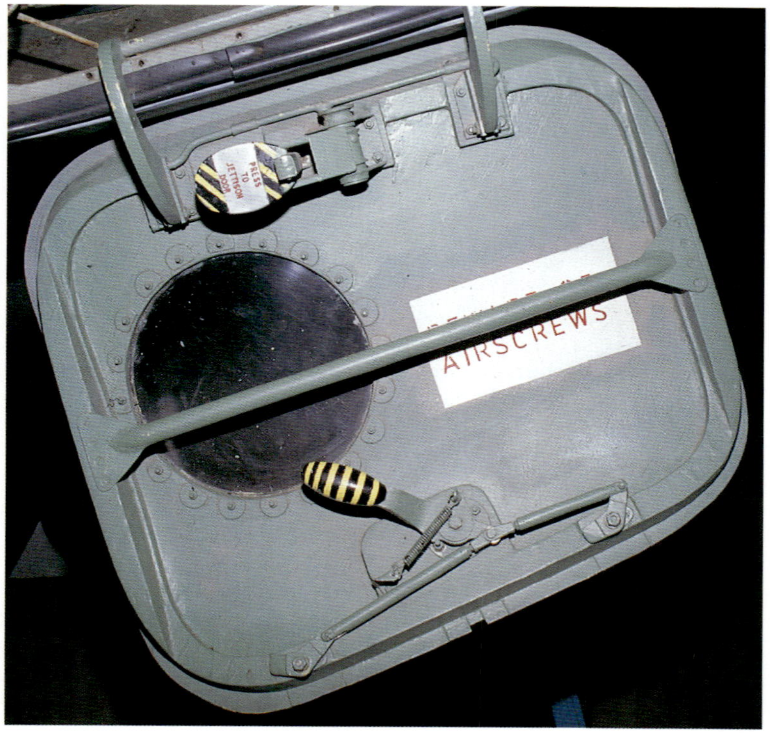

29

30

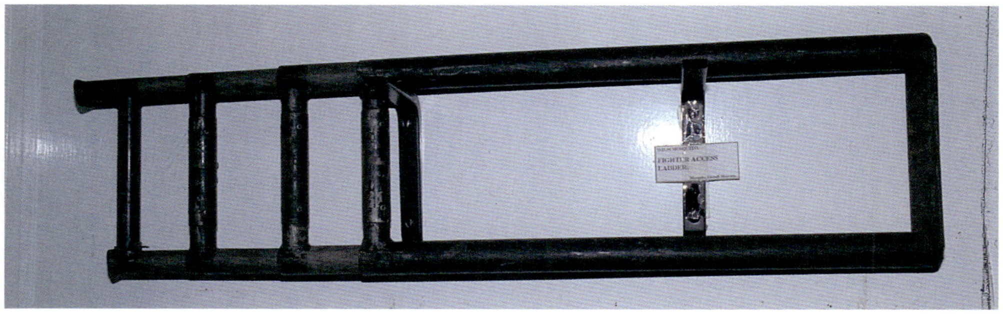

31

写真30: F.B.Mk.Ⅵの搭乗員乗降口。胴体下面にイスパノ20mm機関砲を装備した夜間戦闘機型／戦闘爆撃機型では下面乗降口のかわりに機首右側に乗降口が設けられた。機体外板が赤く見えるのはプライマーの色が残っているため。モスキートでは機体外板にあたる合板の上、全面にマダポラムと呼ばれる綿布が張られたあと、このプライマーが3層塗られ、さらにアルミニウム・ドープが2層塗られたあとに、機体迷彩塗装が施されている。

写真31: 昇降用のラダー。機種にかかわらず、モスキート全般に使われている。

写真32-33: 同じくN.F.Mk.30の乗降口およびドアの詳細。戦闘爆撃機型とほぼ同一だが、ドア投棄用レバーが大型化されているのに注意。下端の黄色いステーはドア固定用のバー。

Photo 30: The crew boarding hatch of the F.B. Mk.VI. Night fighters and fighter-bombers with the 20mm Hispano cannons on the underside had the crew boarding hatch relocated to the right side of the nose. The red coloring visible here is primer. The Mosquito took three coats of primer and two coats of aluminum dope before the final camouflage scheme was painted on the aircraft.

Photo 31: The boarding ladder, present on every model of the Mosquito.

Photos 32-33: The crew boarding hatch on the N.F. Mk.30. It's quite similar to that of the bomber, but note the enlarged emergency door release lever. The yellow bar is for operating the latch mechanism of the hatch.

32

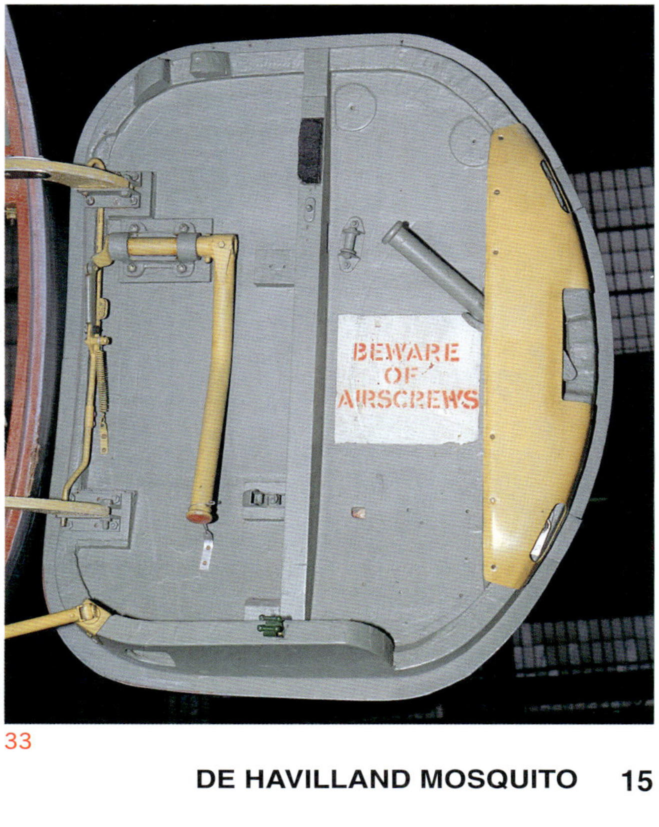

33

キャノピー&コクピット　CANOPY & COCKPIT

34

35

36

写真34-37: 原型第１号機のコクピット・キャノピー。形状的にはのちの量産型と大きな相違はないが、窓枠の一部形状および数に原型機だけの特徴がみられる。操縦士横に位置するガラスはフラットなタイプがはめこまれているが、のちの爆撃機型などで多用された側下／後方の視界をかせぐためのブリスター付きを装着した時期もある。フレームの数は比較的多いが、モスキートの視界性は非常に高く、搭乗員からは歓迎された。操縦席真上の部分は緊急脱出用ハッチとして分離／投棄できるようになっていたが、飛行時は搭乗員の身体が垂直尾翼にたたきつけられるおそれがあるため、地上不時着時での機外脱出などに限られていた。

Photos 34-37: The canopy of the prototype. Though essentially the same shape and design as that which appeared on the production versions of the plane, there are some differences in the frame and window divisions. The glass to the side of the pilot is flat, but many Mosquitoes have blister-type panes there to boost visibility. Despite the relatively large number of canopy panes, the Mosquito had excellent visibility and this point was praised by its crews. The area directly above the pilot functions as an emergency escape hatch and is designed to be able to be removed, but bailing out of the aircraft thorough this hatch while in flight risked the crew member being hit by the vertical stabilizer. As such, it was primarily limited to use as an escape hatch following belly landings.

37

写真38-42: 同じくコクピット内側よりみた原型機のキャノピー。流麗な外側にくらべるとパイプ・フレームに並んだネジ山の列など、かなりごつい印象を受ける。キャノピー頂部(座席背当ての上にあたる)は前述のように脱出用ハッチを兼ねており、フレームについているレバーを下に引けばその矩形のキャノピー部を外すことができる。操縦士席寄りの三角窓は内側に開閉できる。写真39右上にみえる円形ハッチ付きの四角形パネルは信号弾ピストルの発射口で、無線封止時に僚機や後続機に合図を送るために用いられる。ちなみに信号弾、発射ピストルは爆撃／航法士席下に収納されている。写真42の太いパイプはラジエターの余熱を利用してコクピットの暖房を図るためのダクトと思われる。

Photos 38-42: The prototype's canopy seen from the interior. Compared to the smooth exterior, the canopy frames jut out significantly on this side. The lever on the inside of the frame above the pilot's seat is for operating the emergency escape hatch, described above. The triangular window near the pilot could be opened to the inside. The square panel with the round hatch visible in photo 39 is for launching signal flares from a flare pistol. The pistol itself was stored under the bombardier/navigator's seat. The large pipe in photo 42 is believed to have been a cockpit-heating device that employed radiator heat.

38

39

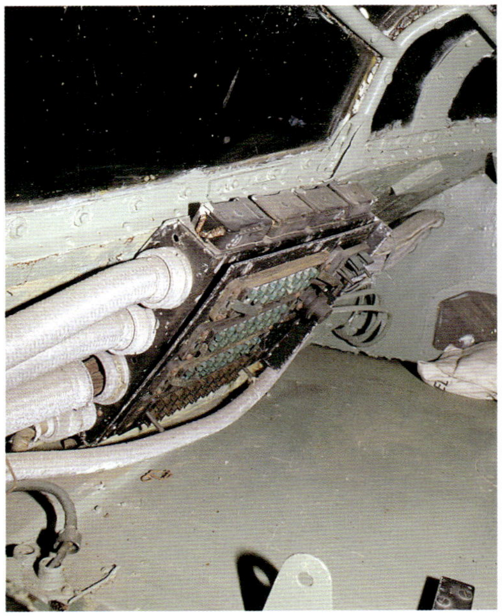

40

41

42

キャノピー&コクピット CANOPY & COCKPIT

43

45

46

44

写真43-46: モスキートB.Mk.35(S/N TA634)のキャノピー。原型機のそれと大差ないが 、爆撃／航法士席寄り頂部に天測ドームが追加されているのが目に付く。これは星座などの位置を確認することにより、目視で自機の位置確認を行うためのもの。一方、操縦士席横の窓はそれまでのブリスター付きのかわりにガラス全体が外側にふくらんだタイプに改められている。写真45で機首上部に見られる小さな突起は、風防に凍結した氷を除去するためにグリコール液を風防表面に噴射するためのノズル。写真44は頂部の天測ドーム部分の窓を外した状態を示している。

Photos 43-46: The canopy of the B. Mk.35(S/N TA634). There are no major differences compared to the prototype, but the navigational (star) sighting dome added above the navigator/bombardier's seat does attract attention. In addition, the glass to the side of the pilot has been changed from the blister type to one where the entire assembly is bulged outboard of the aircraft. The small projection visible on the top of the nose in photo 45 is the nozzle that was used to spray de-icing glycol on the windscreen. Photo 44 shows the navigation dome window removed.

47

写真47-50: 内側からみたB.Mk.35(S/N TA634)のキャノピー周囲。キャノピー頂部の天測ドームは円形ではなく、空気抵抗の少ない涙滴型であることが分かる。写真49は前部風防で、フレーム中央のダイヤルは方向舵トリムタブ修正用。その上の黄色いレバーは頂部キャノピー解除レバー。前部風防から導かれているコードは合わせガラス内の湿気を除去するためでこのコードの先にシリカゲルの入ったカプセルが接続している。前部風防は厚さ6.4mm、3.2mmのガラスで3.2mm厚のガラスを挟んだ３層ガラスが使われている。写真48は取り外した状態の頂部キャノピーである。この天測ドームは爆撃機型だけでなく、偵察機型でも装備したものが多い。また旧型のB/Mk.Ⅳや B.Mk.Ⅸなどにフィードバックされている例もあり、かならずしも本型の識別点とはいえない。

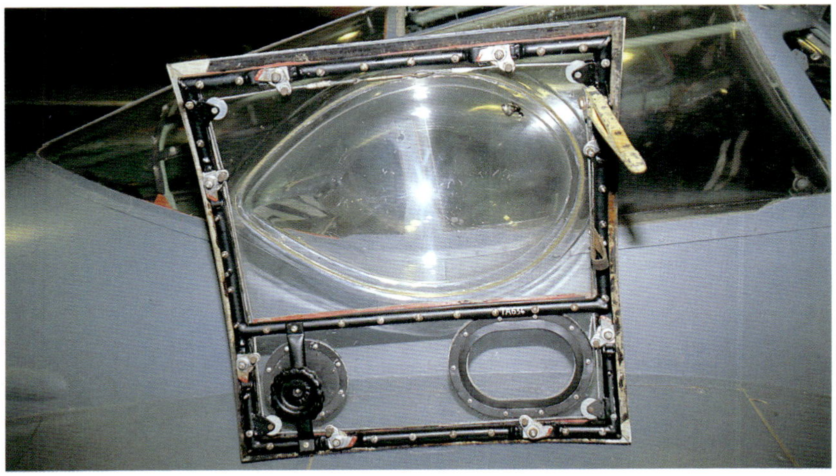

48

49

50

Photos 47-50: The canopy of the B. Mk.35(S/N TA634) seen from the inside. Note that the navigational dome is not a perfect circle, but an elliptical shape to lower air resistance. Photo 49 is of the forward canopy. The dial visible in the center of the frame is the rudder trim control, while the yellow lever above it is the emergency forward canopy release control. The cords visible running through the window were part of a dehumidifying system and were connected to a capsule containing silicone gel. The forward windscreen was built from three layers of glass, measuring 6.4, 3.2 and 3.2mm in thickness. Photo 48 shows the blister section of the canopy removed. Note that this navigational dome was not unique to the bomber version of the aircraft, being found on reconnaissance versions and also retrofit to older B. Mk.IV and B. Mk.IX planes as well, so it cannot be used as an identification point.

キャノピー&コクピット　CANOPY & COCKPIT

51

52

53

54

55

写真51-55: 夜間戦闘機型および戦闘爆撃機型モスキートのキャノピーの大きな特長は、前部風防が一枚板の耐弾ガラスになっていること。これは操縦席前方のグレアシールドに射撃照準器を設けるため風防ガラスのゆがみをきらったことや、耐弾ガラス採用のための措置と思われる。また、ワイパーを標準装備しているのも異なる点。写真ではワイパーブレード自体は外されているが、基部は残っているのがわかる。写真55では右側、航法士席のキャノピーが外側に張り出したタイプが使用されているが、これはN.F.Mk.30型特有のものではないようである。

Photos 51-55: One of the major unique points of the night-fighter and fighter-bomber versions of the plane was the single-pane bulletproof glass at the front of the canopy. This was installed not only for protection, but to eliminate distortion caused by having curved glass installed in front of the gunsight. The presence of wipers is also an identification point. The example in the photos has lost the wiper blades, but the mounts can be identified. In photo 55 the right side canopy by the navigator's seat is of the bulged type, but this is not unique to the N.F. Mk.30.

56

57

58

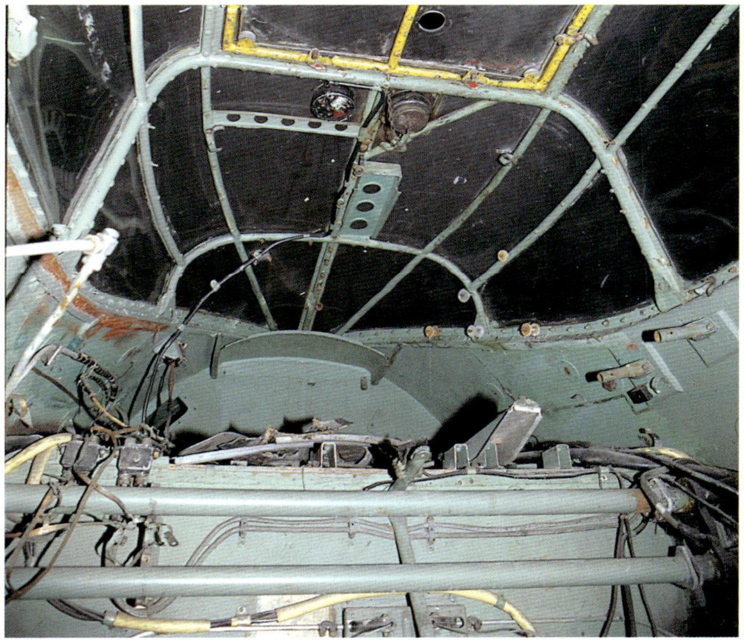

59

写真56-59: 同じく内側からみた戦闘爆撃機型F.B.MkⅥ(写真56)、および夜間戦闘機型N.F.Mk.30のキャノピー。F.B.MkⅥはガラス部分がすべて外され、艤装も省略されているので各フレームの状態が明瞭に把握できる。写真57の中央はワイパー基部にあたる。射撃照準器は取り外されているが、写真ではグレアシールド部下にボルトが2本突き出しているところが取り付け位置で、操縦士席に正対する。モスキートで主に用いられた射撃照準器はMk.Ⅱと呼ばれるジャイロ式射撃照準器(GGS)で、従来の光像式照準器GM-2にかわって大戦後半より装備がはじめられた。N.F.Mk.30のキャノピー頂部はフレームの部分が黄色く縁取られているが、これは夜間時の脱出の際などですぐ目に付くようにとの配慮であろう。

Photos 56-59: Views of the inside of the fighter-bomber F.B. Mk.VI (photo 56) and the night-fighter N.F. Mk.30. As the F.B. Mk.VI is lacking all glass and other equipment, the structure of the frame itself can be clearly studied. The wiper mount is in the center of photo 57. The gunsight has been removed, but the two bolts projecting out below the glare shield are where it would have been mounted. The Mosquito was generally equipped with the gyroscopic Mk.II gunsight, which replaced the earlier reflecting-type GM-2 beginning in the latter half of the war. The yellow markings on the N.F. Mk.30's canopy frame are to aid identification of emergency hatches at night.

キャノピー&コクピット CANOPY & COCKPIT

60

写真60: 原型機の計器盤。主計器盤は左上から、速度計、人工水平儀、昇降計、下は高度計、方向指示計、旋回計。右操縦桿把下は酸素流量計、3重圧力計。右上副計器盤は左から、ビームアプローチ指示計、プロペラ・フェザリング・ボタン、外気温度計。

写真61: 左側計器盤。上から、出力減退警告灯、回転指示計、ブースト圧計、燃圧計(左右)、油温計、油圧計(左右)、冷却温度計。

写真62: 操縦桿中央のブレーキレバー。

写真63: 左コンソールのスロットルレバー、側横はプロペラ調速レバー、細板でガードされているのは過給器ギア変換スイッチ。

写真64: 左側計器盤下のコンパス。

Photo 60: The instrument panel of the prototype. From the upper left are the air speed indicator, artificial horizon and vertical speed indicator. The lower row, from the left, contains the altimeter, direction indicator and turn indicator. The panel under the right grip of the control column includes the oxygen flow gauge and triple pressure gauge. The smaller panel to the upper right contains the beam approach indicator, propeller feathering controls and external temperature gauge.

Photo 61: The left side panel. From the upper row are the power drop warning lights, tachometers, boost pressure gauges, fuel gauges, oil temperature gauges, oil pressure gauges and coolant temperature gauges for both engines.

Photo 62: In the center of the control column is the brake lever.

Photo 63: On the left side console can be found the throttle levers with the propeller speed controls nearby, and the supercharger gear reduction control semi-covered by a thin strip of metal.

Photo 64: The compass located below the left side console.

61

62

63

64

65

66

写真65: 操縦桿後方にのびる支柱は昇降舵連結桿、その下はトルクチューブ。

写真66: 木製のフットペダル。木工、家具産業は英国が世界に誇る産業のひとつでもあり木製機が本機の要求項目とはいえ、ここまで徹底させることもあるまいにと思ってしまう。さすがに量産機では金属製のフットペダルに改められているが、その微妙な線といい、工芸品の趣さえただよう。

Photo 65: The base of the control column is connected to the control rods for the elevator. Below that are the control rods for the rudder that lead forward to the rudder pedals.

Photo 66: The wooden rudder pedals. Though one of the purposes of building the Mosquito from wood was to employ the craftsmen of England's world-class woodworking industry, one questions whether it was necessary to go this far! Still, these pedals have an air about them that is not present in the metal versions employed in production models.

写真67-70: 機首爆撃窓内部の詳細。操縦席計器盤の各メーター、配線が剝出しになっており、乱雑さは否めない。写真67の手前にならんでいるのは信号弾で、赤、緑、黄色など各色が用意されている。前述のように、各種信号弾は無線封止時に合図を送るために装備しているもので、信号弾発射ピストルを使用して、コクピット・キャノピー頂部の発射口から打ち上げた。モスキートの場合、爆撃先導機任務や、爆撃目標都市上空に滞空して後続の爆撃機群に爆撃位置を指示するターゲット・マーカー任務などに多用されており、多くの信号弾を装備していた。写真69、70で機内に取り付けられたラックは、それら信号弾を収納するためのもの。

Photos 67-70: Interior details of the nose of the plane. The scene behind the instrument panel is a jumble of cables and wires. Note the signal flares, and the racks for storing a good supply of them in photos 69 and 70. Signal flares were a crucial means of communication during periods of radio silence and were fired from a pistol stored in the plane through an opening in the canopy. They were particularly important for Mosquitoes used in pathfinder and target marker missions.

67

68

69

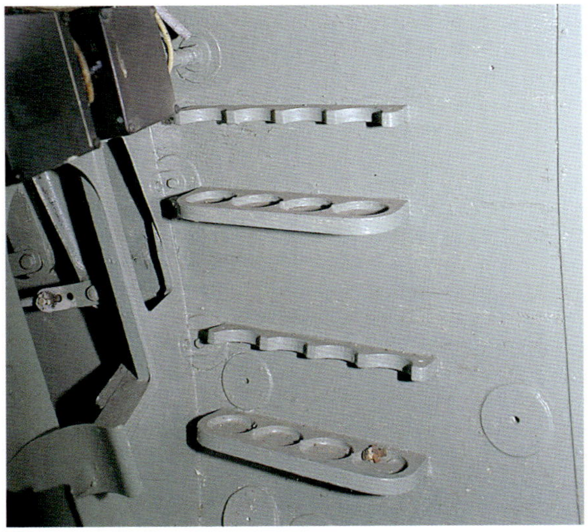
70

キャノピー&コクピット　CANOPY & COCKPIT

71

写真71：N.F.Mk.30の操縦席。計器盤は外されており、フットペダルが残るのみ。実際にはコクピットとレーダー機器室とは仕切りによって塞がられている。

写真72：N.F.Mk.30主計器盤下部。左上から、ベンチレーター(欠)、降着装置位置指示計(欠)、フラップ位置指示計、酸素流量計(欠)、点火スイッチ、酸素調整器、そして３重圧力計(欠)。

72

Photo 71: The cockpit of the N.F. Mk.30. Unfortunately, little remains save the rudder pedals. A bulkhead normally separated the radar area and cockpit as well.

Photo 72: The lower portion of the N.F. Mk.30's main instrument panel. From the upper left are the ventilator (missing), landing gear position indicator (missing), flap position indicator, oxygen flow gauge (missing), ignition switch, oxygen flow control, and the triple pressure gauge (missing).

73

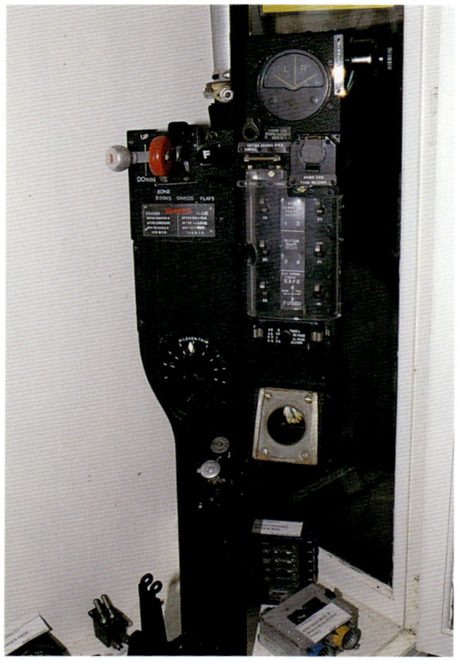

74

写真73：N.F.Mk.30左側計器盤。上から、人工水平儀、回転指示計、ブースト圧計、燃圧警告灯(左右)、油温計、油圧計(左右)、冷却温度計、一番下は左右着陸灯スイッチ。

写真74：F.B.Mk.Ⅵの計器盤の一部。ビームアプローチ指示計、下の矩形は爆弾切換え／信管セットスイッチ、その左は各種切換えレバー。

写真75：F.B.Mk.Ⅵ右コンソールの一部。左上から、燃料計(内翼タンク)、同(中央および増加タンク)、同(外翼タンク)、下は識別灯スイッチ、ワイパー加減抵抗器、外気温度計、そして航空時計。

Photo 73: The left side console of the N.F. Mk.30. From the upper row is the artificial horizon, turn indicator, boost pressure gauge, left and right fuel warning lamps, oil temperature gauge, left and right oil pressure gauges, coolant temperature gauge and at the very bottom is the left and right landing gear indicator lights.

Photo 74: Part of the instrument panel of the F.B. Mk.VI. Below the beam approach indicator is the short bomb selector/fuse control, and to that left of that are a number of different selector switches.

Photo 75: Part of the right console of the F.B. Mk.VI. From the upper left is the fuel gauge (inboard wing tanks), fuel gauge (fuselage and external tanks), fuel gauge (outboard wing tanks), recognition light control switches, wiper speed control, external temperature gauge and aviation clock.

75

76

77

写真76-79：N.F.Mk.30の計器盤。先の原型機やF.B.Mk.Ⅵのそれと比較しても基本的な違いはみられない。写真77の操縦桿先端には機関砲発射スイッチがついているのがわかる。写真79のスロットルレバー上にあるのは昇降舵トリムタブ調整器。その左横にはビームアプローチ、コンパスなどのスイッチが並んでいる。

Photos 76-79: The instrument panel of the N.F. Mk.30. It is essentially unchanged from the F.B. Mk.VI that it was derived from. Note the machine gun firing switch visible on the control column in photo 77. Above the throttle levers in photo 79 is the elevator trim control. To its left are the switches for the beam approach, compass and other equipment.

78

79

80

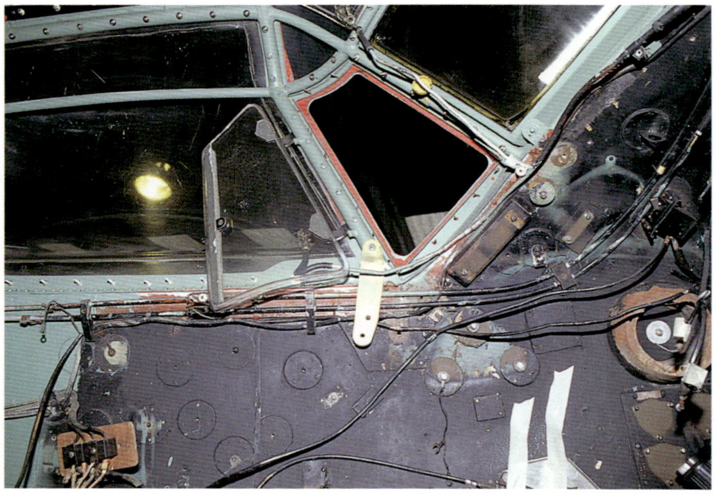

82

81

83

84

写真80-84: B.Mk.35(S/N TA634)のコクピット内部。計器盤は外されているが、基本的なレイアウトなどは原型機のそれと変わらない。写真81はコクピット後部の詳細。この手前にはT.1154送信装置が搭載される。写真84は操縦士席部分のフロア。チェーンを介するシャフトは昇降舵作動伝達用。チェーンによってその上のプーリーを回し、その動きをケーブルに導いて昇降舵を操作するものである。

Photos 80-84: The cockpit of the B. Mk.35(S/N TA634). The instrument panel is missing, but the overall configuration is quite similar to its prototype. Photo 81 is a view to the rear of the cockpit, showing the T.1154 transmitter installation. Photo 84 is a close-up of the floor around the pilot's seat area, revealing the shaft and chain mechanism that was connected to the elevator control cable.

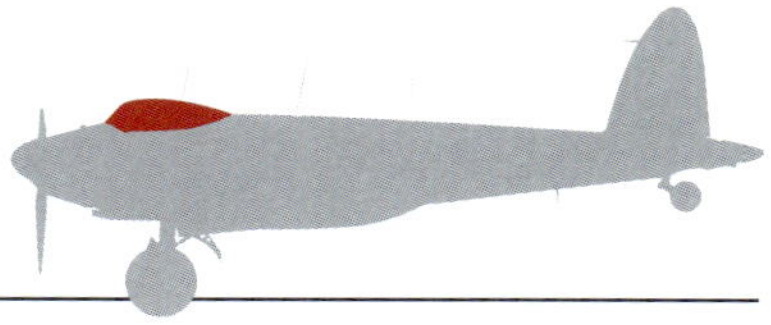

85

写真85-91：同じくB.Mk.35(s/nTJ138)のコクピット。計器盤、艤装もほぼ現役当時のままに残されており、非常に参考になろう。計器盤、各種レバーなどの配置も前述の解説と基本的に変わらない。写真86の左上の２個のレバーはラジエターフラップ調節用レバー。写真87のグレアシールド下、キャノピー枠直下にあるのは機体識別灯(コクピット後方)点灯スイッチ。その横に２個並んだダイヤルスイッチは航法／爆撃士席の室内灯用。コンソール上は左から、エンジン消火装置作動ボタン、IFF(敵味方識別装置)破壊スイッチ、航法灯点灯スイッチ、機体下面識別灯点灯スイッチ。写真90奥の矩形の機器はVHF調整装置、クッションが置かれているのは操縦士席、その横についている小さなレバーはシートベルト解除レバー、その下の箱は応急医療品を収納するボックスである。

Photos 85-91: The cockpit of the B. Mk.35 (s/n TJ138). By contrast, this example is in nearly pristine condition, providing an excellent reference. The two levers in the upper left of photo 86 control the radiator flaps. The control switch for the rear canopy tactical recognition light is visible in photo 87 directly below the canopy frame. The two dials next to it are to control the navigator's lighting. From the upper left of the console are the engine fire extinguisher controls, the IFF destruct switches, navigation light switches and the lower fuselage formation (recognition) light control. On the back wall of photo 90 is the VHF adjuster, with the pilot's seat cushion visible to the left. The seat belt release lever is also visible, with the box below that containing first-aid supplies.

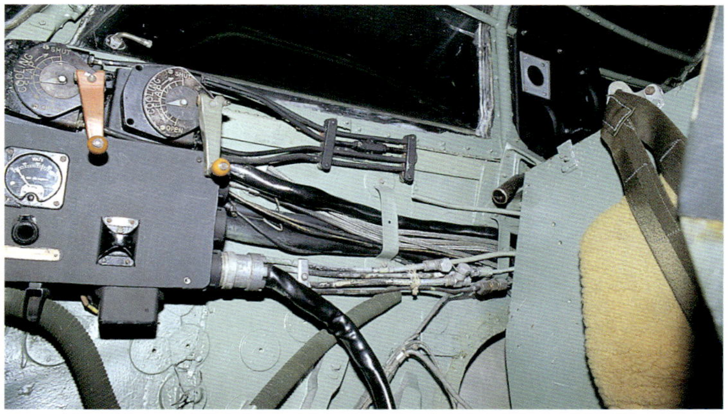

86

87

88

89

90

91

写真92-94: 原型機の操縦士席。背当てには防弾鋼板がはられており、後方からの敵の射撃より操縦士の身体を守るようになっている。本機にかぎらず、英空軍爆撃機クルーはパラシュートを身につけず手に持って機体に搭乗、飛行中は所定の位置においたままで、脱出の際にのみ身につける。もちろん、これは狭い機内では作業上支障をきたすからである。写真94は操縦士席の下側。フロアとの間にはかなりの隙間があるが、ここには漏斗(じょうご)とホースのついた「衛生用タンク」がフライトごとに設置される。これは端的にいえば、オシッコをためておくタンクで、一応暖房装置の整ったモスキートではあるが、高空を、しかも長時間の飛行を余儀なくされればどうしても生理的現象が我慢できない状況が生じる。そういう時のための装備である。もっとも漏斗はひとつしかなく、操縦士専用なのか、あるいは航法/爆撃士と仲良く共用するのかは、残念ながら機体マニュアルも黙して語ってはくれていない。

92

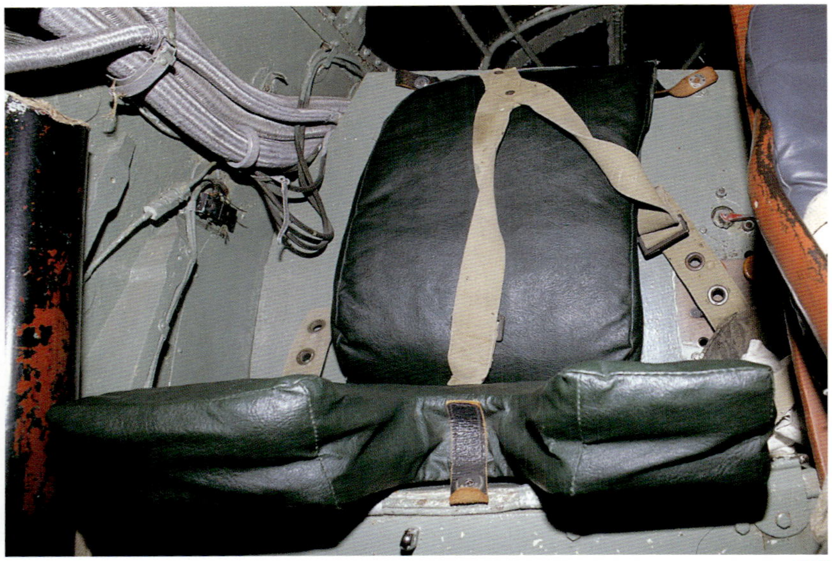

93

94

Photos 92-94: The pilot's seat. Note the armor plate on the back. British bomber crew boarded their aircraft without wearing their parachutes, storing them in a pre-determined location on board instead. They would don them only in an emergency. Of course, in the narrow spaces on board an aircraft, things did not always go smoothly when multiple crew would rush for their 'chutes in a panic. In the narrow space between the floor and the bottom of the fuselage, the Mosquito boasted a sort of septic tank system that allowed crew to relieve themselves during long flights.

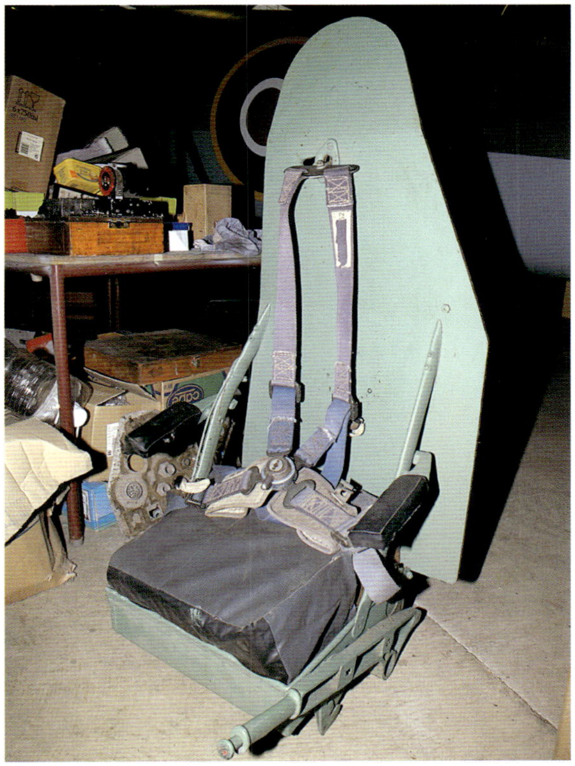

95

96

写真95-100: B.Mk.35(S/N TA634)の操縦士席全景。座席左側のレバーは座席高低調節レバー。右側の小さなレバーはシートベルト解除レバーで、これは緊急時にバックルが解除できない場合、このレバーを動かすと背当て裏側にあるシートベルト固定ラッチが解除されてシートベルトが根元から外れるようになっている。座席両側にはアームレストがついており、長時間の飛行による操縦士の疲労を軽減させる。使用しない場合には上にはねあげておく。フロア上には同じクッションのついたフットレストを備えている場合もある。シートベルトはオーソドックスな４点式で、バックル中央のソケットを回すことによってすべてのベルトがリリースされる。

Photos 95-100: Overall view's of the pilot's seat. The lever to the right is for adjusting the seat's height. The small lever to the left is to release the seat belt. If the buckle could not be operated in an emergency, this lever could be used to release the belt completely at its mounting points. Note the arm rests to help reduce pilot fatigue during long flights. These could be stored out of the way when not in use. Foot cushions were mounted on the floor in some examples. The seat belt was of an orthodox four-point design, and was released by rotating the buckle in the center.

99

97

98

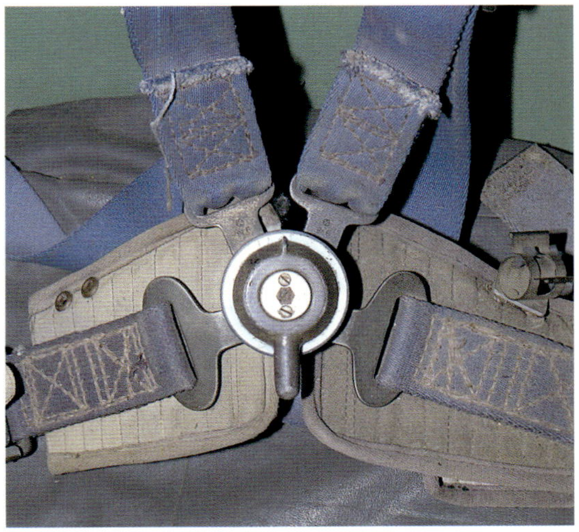

100

胴体＆爆弾倉 FUSELAGE & BOMB BAY

101

写真101：N.F.Mk.30の胴体。モスキートの胴体は機体バルクヘッドを治具に固定して合板を張って左右胴体を別々に組み立て、操縦系統のケーブルなどを組み込んでから胴体左右を張り合わせて完成させた。そして他の部分と同じく胴体全面にマダポラム綿布を張ったあと、プライマー、アルミニウム·ドープを重ね塗りして表面処理を施した。実際、胴体組み立て中の写真を見ると、まるで1/1のプラモデルを作っているようにさえ感じられる。

Photos 101: The fuselage of the N.F. Mk.30. The fuselage of the Mosquito was built in left and right halves, with the outer panels being fastened to the inner bulkheads first. Once control cables and wiring work was completed, the left and right halves of the aircraft were then attached. Like other parts of the aircraft, Madapolam fabric was then applied to the wood, and this was in turn painted with multiple coats of primer and aluminum dope.

写真102：B.Mk35(S/N TA634)の胴体。モスキートの胴体後部右側にある細長いストリンガーは、この下に設けられた機内点検ハッチをあけたための補強用に施されたもので、初期の機体ではみられない。オーストラリア製の機体、例えばMk.41などでは左側にも同様の補強材がみられるものもあるが、これも偵察カメラ窓を増設した結果、強度的に不安をもたれたためであろう。

Photos 102: The rear fuselage of the B. Mk.35(s/nTA634). The long, narrow stringer on the right, rear portion of the fuselage was attached to strengthen this area due to the inspection hatch that was opened below it, so it is not present on early models. On Australian-built Mosquitoes, such as the Mk. 41, a similar stringer is present on the left side of the plane due to the presence of the reconnaissance camera window that was added.

102

103

写真103：B.Mk.35(S/N TJ138)の胴体後部。きわめて平滑な表面処理が施されているのがよくわかる。胴体右側と比較して、搭乗員乗降ハッチや点検ハッチのたぐいもいっさいない。ただし、左側胴体真横には例のマダポラム綿布のストライプが張られているはずだが、写真ではそれすらも視認できない。

Photos 103: The rear fuselage of the B. Mk.35(s/nTJ138). The supremely smooth finish of the fuselage surface is quite evident in this shot. In contrast with the right side of the fuselage, there are no access or boarding hatches at all. However, the Madapolam fabric stripe should be present on the middle of the left side of the plane, but not even that is visible in this photo.

写真104: 後方からみたB.MK35(S/N TA634)の胴体部。機体表面の光の当たり具合で胴体断面のようすが把握できる。胴体上部にはホイップアンテナが2本取り付けられている。写真からキャノピー頂部のアストロ・ドーム(天測ドーム)が意外に大きいこともわかる。

写真105: 前述のようにモスキートの胴体は左右別々に組み立ててから胴体左右を張り合せて完成した。胴体接着面は平らではなく楔形に仕上げられており、密着性と接着面の面積確保が図られているのには感心させられる。

Photo 104: The fuselage seen from the rear. The lighting in this photo reveals the cross-section's shape rather well. Two whip antennas are present on the upper side. The canopy astrodome is surprisingly large when seen from this angle.

Photo 105: As mentioned above, the left and right halves of the fuselage were completed separately before being attached. However, rather than flat contact surfaces, one side was formed as a protruding wedge, and the other was a cut with a concave "V" shape to increase the joint's surface area and create a more secure assembly.

104

105

胴体＆爆弾倉 FUSELAGE & BOMB BAY

106

写真106: 原型機の爆弾倉全景。爆弾倉仕切板および爆弾取付架はなく(実際にも本機では装備されなかったのかもしれない)、コクピット後方に配置された胴体燃料タンク２個(各68英ガロン＝309リットル容量)がみえる。ちなみにモスキートの爆弾倉の長さは3.45mで、開発当初では各種爆弾1,000ポンドを搭載する計画であった。

Photo 106: An overall view of the bomb bay of the prototype. It lacks bomb racks and dividers, but they were probably never installed on this aircraft. The two 68 imperial gallon (309 liter) fuselage fuel tanks are visible behind the cockpit. The Mossie's bomb bay was 3.45m long and originally designed to carry up to 1,000lbs. of various types of ordnance.

107

写真107-110: 写真107は右側爆弾倉ドア内側。爆弾倉隔壁には方向舵、昇降舵、およびトリムタブを作動させるためのケーブルが走っている。写真108も右側爆弾倉ドアで、写真109は同左側。どちらも正面から撮影したもの。ともにドア前部にはドアステーが取り付けられている。また右側ドアにはフックが取り付けられており、左側のラッチにそれを引っ掛けて左右のドアをピッチリ閉じるよう設計されている。爆弾倉ドアにつけられた円形のガラス窓は偵察カメラ搭載時に使用するためのもの。

Photos 107-110: Photo 107 is a close-up of the right bomb bay door. Note the control cables for the rudder, elevator and trim tabs running through the bulkhead. Photo 108 is also a view of the right door, with photo 109 a view of the left, both seen from the front. Note the hook on the right door that goes into the catch on the left side to hold the bay securely closed. The round window in the bay is used when a reconnaissance camera is mounted.

108

109

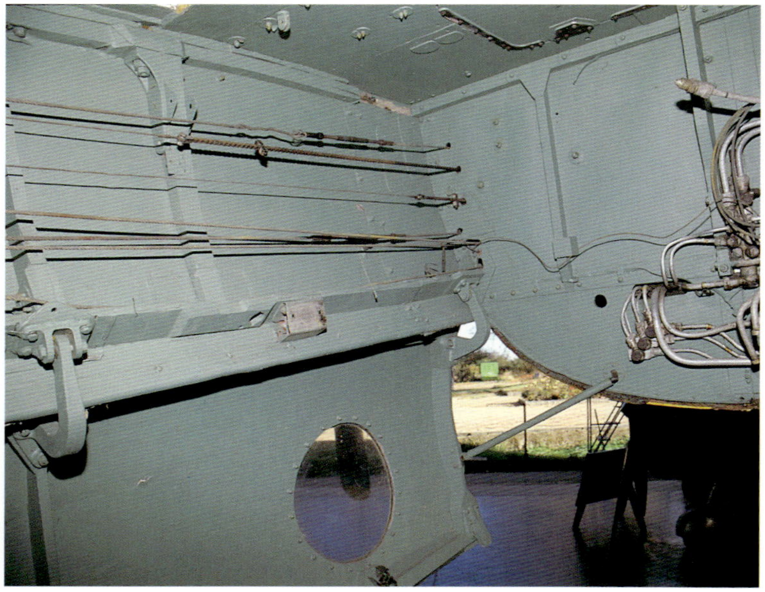
110

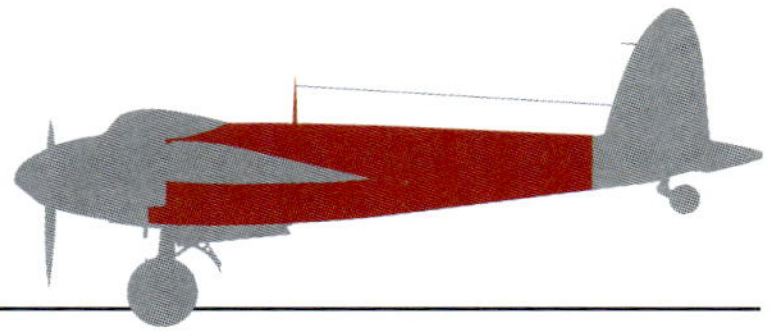

111

113

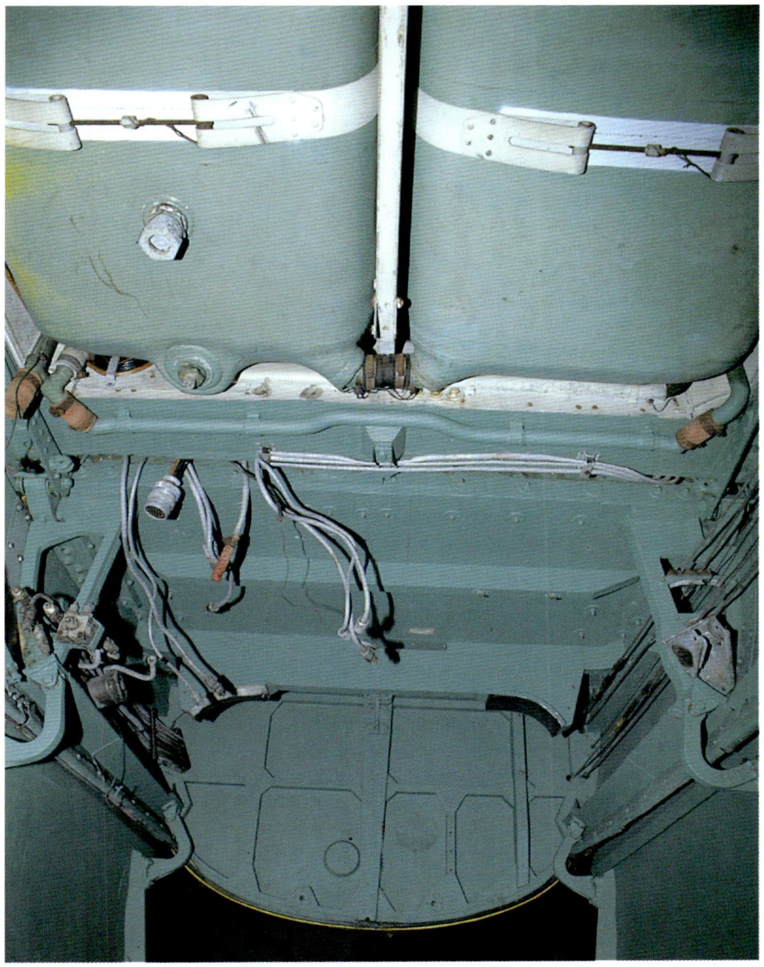

112

写真111-116: 同じく原型機の爆弾倉。写真111は爆弾倉前方隔壁を示す。写真112は爆弾倉後方隔壁の詳細。上部は68英ガロン容量の燃料タンク２個がみられるが、どちらもベルトによって固定されているのがおわかりいただけよう。写真114、115は爆弾倉開閉フックおよびラッチのクローズアップ。マニュアルによると、爆弾倉ドアが閉じて、フックが対応するラッチと向かい合ったあとに、フック部が付根のギアによって引っ張り上げられてラッチをロックし、飛行中にドアががたつかないようになっているとのことである。そういわれて写真114を見返してみると、フックの表面にミゾが切られているのがわかる。写真116は爆弾倉左側隔壁に設けられたダクト。その目的は不明だが、おそらくラジエターからの余熱を爆弾倉内に導いて爆弾倉内の温度を適正に保つためではないかと推測する。

Photos 111-116: More views of the prototype's bomb bay. Photo 111 is a view of the bay's forward bulkhead, and photo 112 is the one to the rear. The two 68 imperial gallon fuel tanks are visible above, held in place by belts. Photos 114 and 115 are close-ups of the hook and latch. According to the aircraft's manual, the hook would be pulled tight by a gear mechanism at its base after engaging the latch to avoid rattling while in flight. Photo 116 is a view of a duct on the bulkhead of the bay, unfortunately the purpose of which is unclear. This may have been to bring some radiator heat into the bomb bay for temperature maintenance purposes.

114

115

116

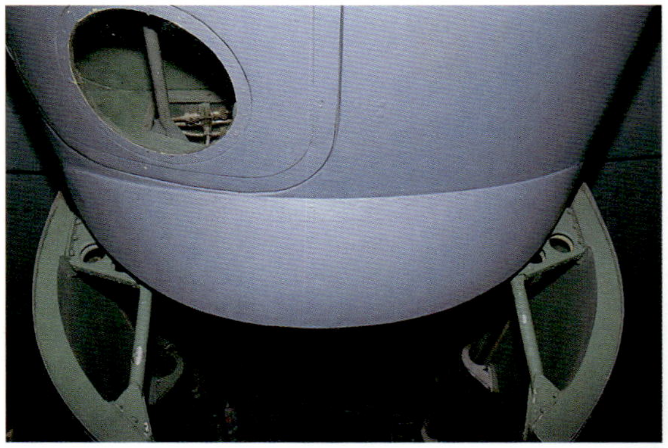

117

118

119

写真117-119: モスキートB.Mk.35(S/N TA634)の爆弾倉バルジ部。英空軍では第二次大戦中さまざまな大型爆弾を開発、使用したが、その中でも4,000ポンド(1,816kg)爆弾、いわゆる「ブロックバスター」がもっとも多用された。モスキートはこの大型爆弾が搭載できる唯一の軽爆撃機であったが、それでも従来の爆弾倉では入りきらず、バルジを設けて爆弾倉を大型化して搭載できるようにしたB.Mk. XVIが1943年7月より登場した。写真の機体、B.Mk.35はそのB.MK. XVIから発展した大型爆弾倉装備の性能向上型であったが、量産に入ったのが大戦末期だったために実戦には間に合わず、もちろん、「ブロックバスター」搭載での作戦も果たさずに終わっている。写真119は胴体バルジ後方フェアリング。

Photos 117-119: The bulge of the B. Mk.35 (S/N TA634)'s bomb bay. During the Second World War, the British developed and deployed a number of extremely large bombs. Most frequently used among these was the 4,000lb. (1,816kg) "blockbuster." The Mosquito was the only light bomber capable of carrying the blockbuster, but even it required modification to its bomb bay in order to haul this massive weapon. The B. Mk.XVI, designed to carry such ordnance, appeared in July of 1943. This B. Mk.35 was a more refined development of that Mk.XVI, but it wasn't completed until near the end of the war, and never saw any combat action. Photo119 shows the rear fairing of the bomb bay bulge.

写真120-122: 同じくB.Mk.35(S/N TA634)の爆弾倉内部の詳細。爆弾倉仕切り板には爆弾取付け架を備えており、爆弾は機内に内蔵されたウインチによって引き上げられ、吊下された。B.Mk.XVIおよびB.Mk.35(S/N TA634)ではアタッチメントの付替えにより500ポンド(227kg)通常爆弾6発を搭載することもできた。写真120は爆弾倉前方隔壁、写真122は後方隔壁を撮影している。

Photos 120-122: Views of the interior of the bomb bay of the B. Mk.35. The bombs were held in place by racks mounted on the bulkheads and raised into position by a winch installed in the plane. By changing the mounts, the B. Mk.XVI and Mk.35 could also carry six 500lb. bombs. Photo 120 is a view of the forward section of the bay, while photo 122 is of the rear.

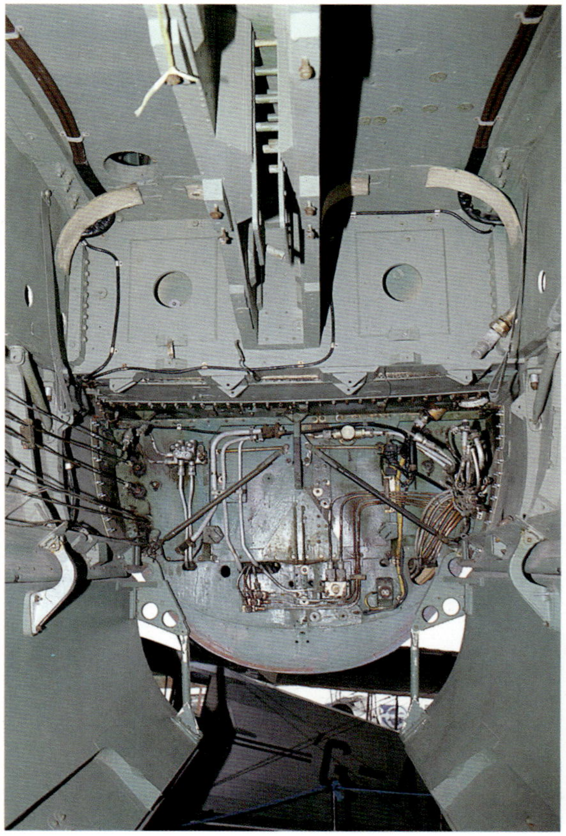

120

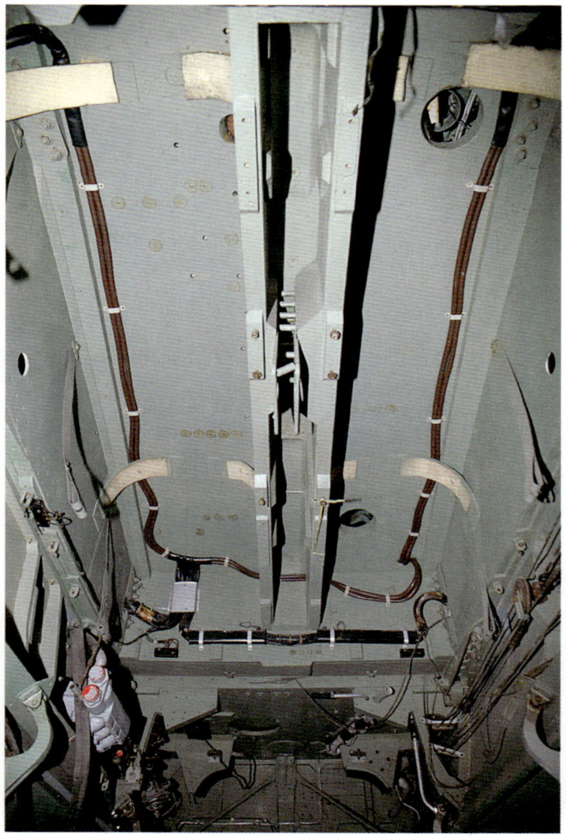

121

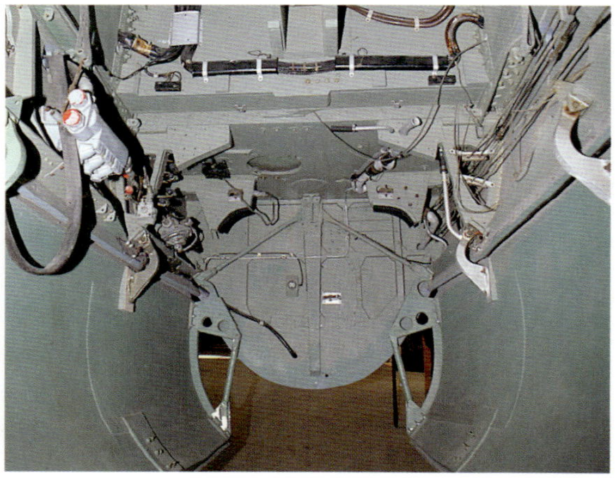

122

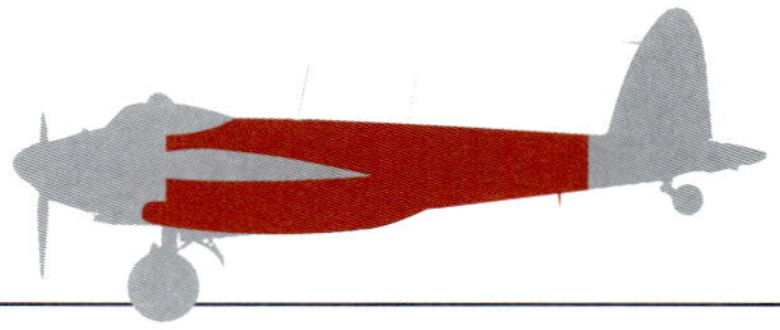

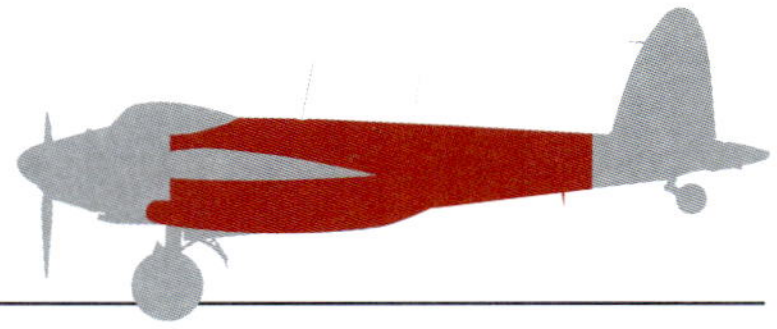

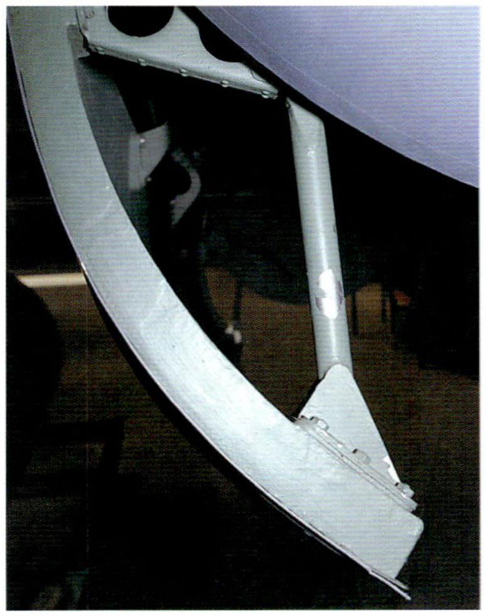

123

124

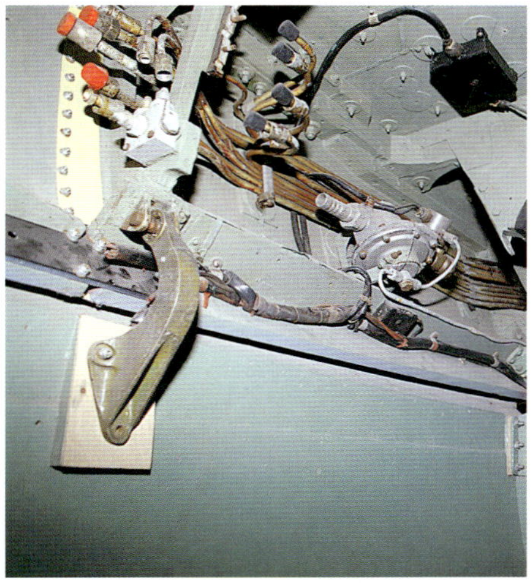

125

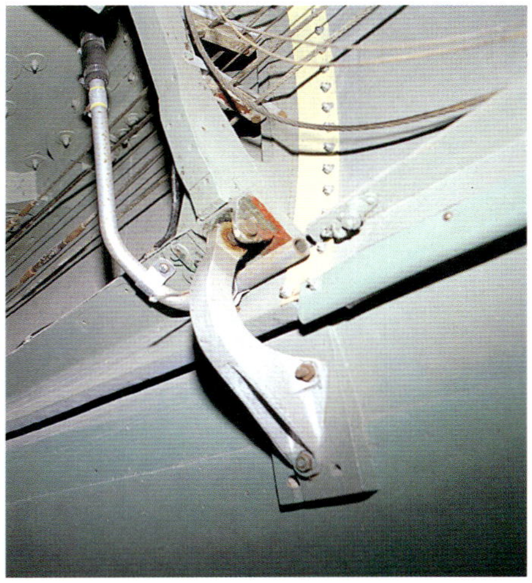

126

写真123-126: B.Mk.35(S/N TA634)爆弾倉ドアのディテール。写真123、124は前方より見た爆弾倉ドアステー。写真125、126は内側より後方隔壁方向に撮影した爆弾倉ドアヒンジ部。

写真127-128: B.Mk.35(S/N TJ138)の爆弾倉後方フェアリングと爆弾倉内部。前ページのそれと比較すると形状が異なっているのに注意。このモスキート・ミュージアムに保存されている本機は完成以降、一度も実戦に参加することなく、主として標的曳航任務に使われて現役を全うしており、その間、爆弾倉下面に空気動力式の曳航ドローグユニットを取付けていた。独自の後方フェアリングもあるいは標的曳航用のケーブルを繰り出す収容部なのかもしれない。

Photos 123-126: Close-ups of the B. Mk.35(S/N TA634) bomb bay door details. Photos 123 and 124 of the bomb bay door stays seen from the front. Photos 125 and 126 are views seen from inside the bay looking to the rear at the door hinges.

Photos 127-128: The rear fairing of the B.Mk.35(S/N TJ138) bomb bay. Note the different shape from that shown on the previous page. This particular aircraft never saw combat, spending most of its career as a target tug. During that time, an air-powered drogue-towing unit was installed in the bomb bay. This fairing's unique shape may be related to that. This particular Mosquito marked 238 flights totaling 301 hours in the air before making the final flight of any RAF Mosquito on May 9, 1963. Following this historic event, it is said to have been sold to a civilian buyer for just 720 pounds.

127

128

129

130

胴体&爆弾倉　FUSELAGE & BOMB BAY

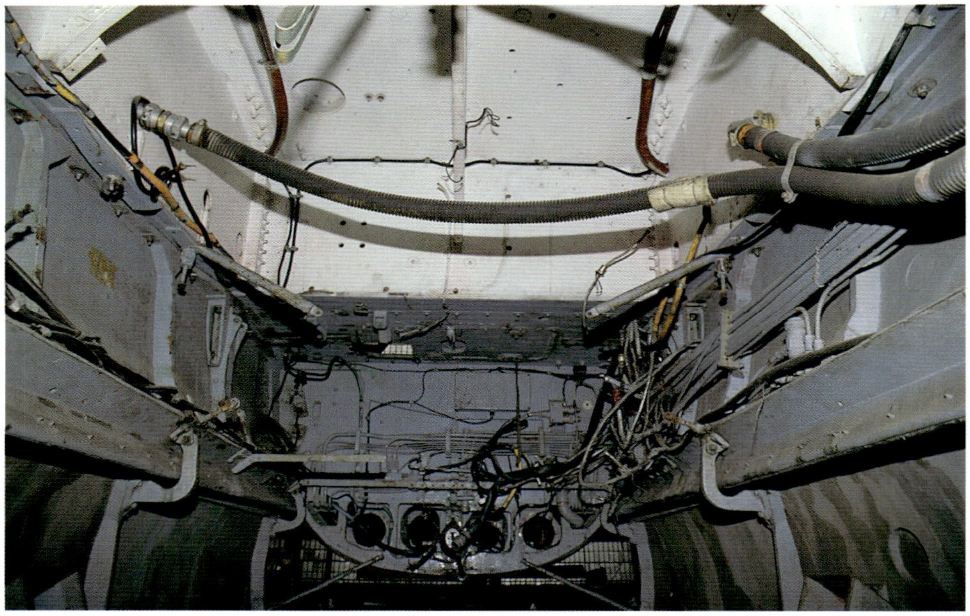

131

写真131-133: 夜間戦闘機型N.F.Mk.30の胴体下面20mm機関砲収容区画部。写真131は前方隔壁で、20mm機関砲を通す穴が4個ならんでいる。爆撃機型では爆弾倉であったこの部分は、戦闘機型ではイスパノMk. II 機関砲４門が装備された。装弾数は各150発のベルト給弾式である。機関砲は高空で凍結するのを防ぐために右主翼付根ラジエターからの余熱を送り込むようになっており、その温度の調節はコクピット内で行うことができた。またその余熱の一部はF.B.Mk.VIでは機首の7.7mm機関銃にも送り込まれているが、写真131で20mm機関砲の取付穴の右にみられるダクトを介して送られた。なお、F.B.Mk.VIではこの20mm機関砲収容部後半が爆弾倉ドアになっており、500ポンド(277kg)爆弾２発を搭載することができた。

Photos 131-133: The 20mm cannon mount area on the underside of the N.F. Mk.30's fuselage. Photo 131 shows the forward bulkhead, with the four holes to accommodate the muzzles of the cannon. In place of the bomb bay, the night-fighter mounted four 20mm cannon in this area, each holding 150 rounds in a belt-feed magazine. To prevent the guns from freezing at high altitudes, a duct brought heat from the right wing radiator back to this bay. This could be controlled from the cockpit. In addition, in the F.B. Mk.VI, some of this air was also fed to the 7.7mm MGs, and this duct can be identified in photo 131 to the right of the attachment hole for the 20mm cannon. In the F.B. Mk.VI, the rear half of the cannon bay was used as a bomb bay and could carry two 500lb. (277kg) bombs.

132

133

134

135

136

写真134：前方よりみたN.F.Mk.30(S/N RK952)の20mm機関砲周囲。砲口フェアリングは外されており、内部の構造がよく分かる。左側に見えるホース上のダクトは本来F.B.Mk.VIの機首7.7mm機関銃の凍結を防止するためにラジエター余熱を送り込むためのもの。N.F.Mk.30では機首のレーダー機器の作動を維持するため、温度調節を行なうのに使われていたのかもしれない。

写真135-136：N.F.Mk.30の機関砲点検ドア詳細。ドア左右にはそれぞれ薬莢／リンク排出口が開かれているのがわかる。写真135はドア内側であるが、薬莢排出口の後ろに見られる矩形のパネルでカバーされているのは20mm機関砲を修理する際のレンチなどの工具を収めておくところである。

Photo 134: The N.F.Mk.30 (S/N RK952) cannon bay area seen from the front. The pipe visible on the right side is for carrying heated air to the 7.7mm guns. It's also possible that it was used in the N.F. Mk.30 to keep the radar equipment warm as well.

Photos 135-136: A close-up of the N.F.Mk.30 cannon access hatch, with the shell and link ejection ports on either side. Photo 135 shows the inside of the door. The small panel behind the shell ejection port covers a tool storage area for tools commonly used when servicing the guns.

写真137：N.F.Mk30の20mm機関砲取付け前方隔壁の詳細。さまざまな配線が入り組んでいる。

写真138-139：N.F.Mk30の胴体下面。機関砲点検ドアに設けられた薬莢／リンク排出口。内側機関砲部の排出口はひとつにまとめられている。

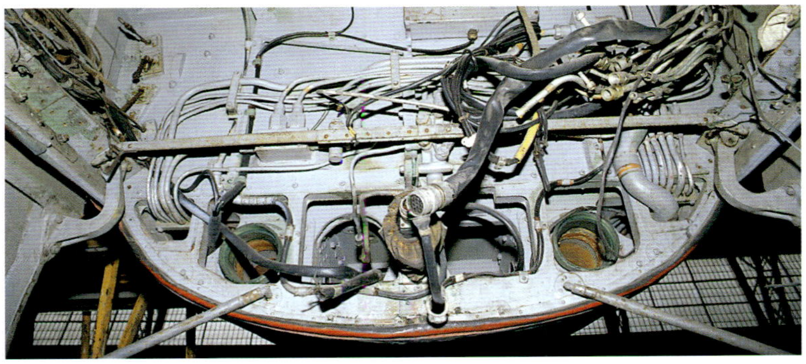

137

138

139

Photo 137: A close-up of the N.F.Mk30 forward cannon mounting bulkhead, and the complex wiring in that area.

Photos 138-139: A view of the N.F.Mk30 fuselage underside, showing the cannon access hatch with its shell and link ejection ports. The internal ejection ports were gathered together into one external port.

胴体＆爆弾倉　FUSELAGE & BOMB BAY

140

142

143

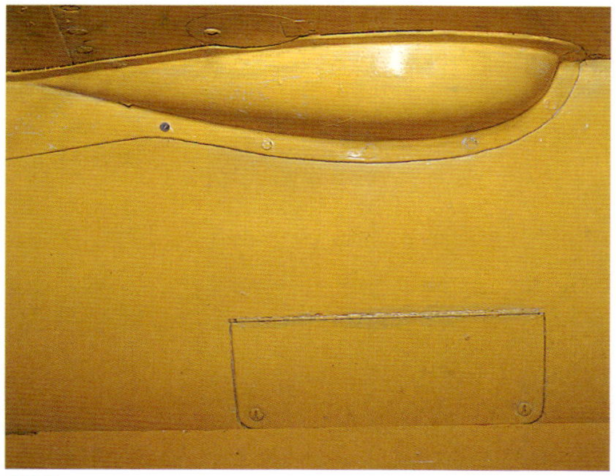
144

写真140: 原型機の爆弾倉ドア全景。前述のように本機の爆弾倉の長さは3.45mで、開発当初は250ポンド(114kg)爆弾４発搭載を計画していたが、500ポンド(227kg)爆弾でも安定フィンを切りつめれば４発収容可能で、性能的にも変わらないことがわかったため、以後の量産型では500ポンド爆弾４発が基本となった。なお、戦闘爆撃機型、F.B.Mk.Ⅵではこの爆弾倉ドアは前後で２分割されており、前半は20mm機関砲点検ドア、後半は爆弾倉ドアとして使用されている。

写真141: 右主翼付根の詳細。付根の涙滴形のフェアリングは燃料パイプをカバーするためのもの。その後方は第３隔壁にあたる部分で、細い帯金で補強されている

写真142: 主翼内側フラップ上面。機体と同じく、木製合板製である。なお、フラップ作動角は45度で、ニュートラル位置での固定はできない。

写真143: 右側爆弾倉ドア前方にあるアンテナ支柱付根フェアリング。支柱自体は失われているようだ。

写真144: 右主翼付根フェアリング下にある燃料廃棄口のハッチ。主翼および胴体燃料タンク内に残存する燃料をすべて抜き取る際には、このハッチを開いて内側の廃棄口を利用する。

Photo 140: An overall view of the bomb doors. As mentioned above, the length of the bay was 3.45m, and was originally designed to carry four 250lb. bombs. However, it was determined that if the stabilizer fins were cut back, four 500lb. bombs could be carried without significant degradation in performance, so this configuration became standard. In the F.B. Mk. VI, the bomb bay doors were split in two front to rear, with the forward doors acting as 20mm cannon access hatches, and the rear as bomb bay doors.

Photo 141: A close-up of the right wing root. The teardrop shaped fairing covers the fuel pipe. To the rear of that is the no.3 bulkhead, reinforced by a thin metal strip.

Photo 142: The upper surface of the inboard flap. Like the fuselage, it is of wooden construction. The flap could drop down 45° and could not be looked into the neutral position.

Photo 143: The antenna mount fairing forward of the right bomb bay door. The antenna itself appears to have been lost.

Photo 144: The fuel drain hatch below the right wing root fairing. This was used if it was necessary for some reason to complete drain all fuel from the wing and fuselage tanks.

141

145

146

写真145-146: 原型機右側胴体後部にある点検用ハッチ。丸型のガラス窓が取付けられているが、量産機では廃止されている。

写真147: 原型機の胴体下面の爆弾倉後方に設置されていた偵察カメラ窓。これも量産型では廃止、ないしは丸型に改められている。

Photos 145-146: A small, round inspection hatch with glass window present at the right rear of the prototype's fuselage. This is not present on production aircraft.

Photo 147: A reconnaissance camera window installed to the rear of the bomb bay on the underside of the fuselage. This is not present, or is modified to a square type on production aircraft.

147

148

写真148: レストア中のF.B.Mk.VIの主翼付根部。細い板が貼られている箇所はちょうど第3隔壁にあたる部分である。

写真149: 主翼付根部の胴体。主翼内翼前縁のラジエター部が設けられている箇所で、その断面形も把握できる。

写真150: 右側胴体後部。合板製の機体表面は塗装が剝がされていると、さながらモックアップ然としている。後部の細長い張り出しは胴体補強用のストレーキ。

写真151: 第2～3隔壁の間の胴体上部に取付けられた折り畳み救命ボート収容ボックス。

151

149

150

Photo 148: The wing root of the F.B. Mk.VI undergoing restoration. The thin metal support strip is located at the no.3 bulkhead.

Photo 149: The fuselage at the wing root. This is the location where the leading edge radiator is installed.

Photo 150: The right rear of the fuselage. The wooden construction of the aircraft is quite clear here, making it seem as if it was a mock-up if one didn't know better. Note the reinforcement strake running along the fuselage.

Photo 151: A view of the life raft storage compartment at the top of the fuselage, between the second and third bulkheads.

152

Photos 152-154: The life raft storage compartment of the B. MK.35(S/N TA634). Following a ditching, the crew was to escape through the top of the canopy and open this box. The inflatable raft was stored with a small cylinder of compressed air. What's more, the watertight compartment in the box also held marker dye, signal flares, smoke flares, drinking water, food, matches and oars. Mosquitoes frequently operated over northern Europe, where a ditching meant bitterly cold water, even in summer. The best a crew could hope for was to get into the lifeboat and hang on until spotted by a friendly rescue patrol aircraft.

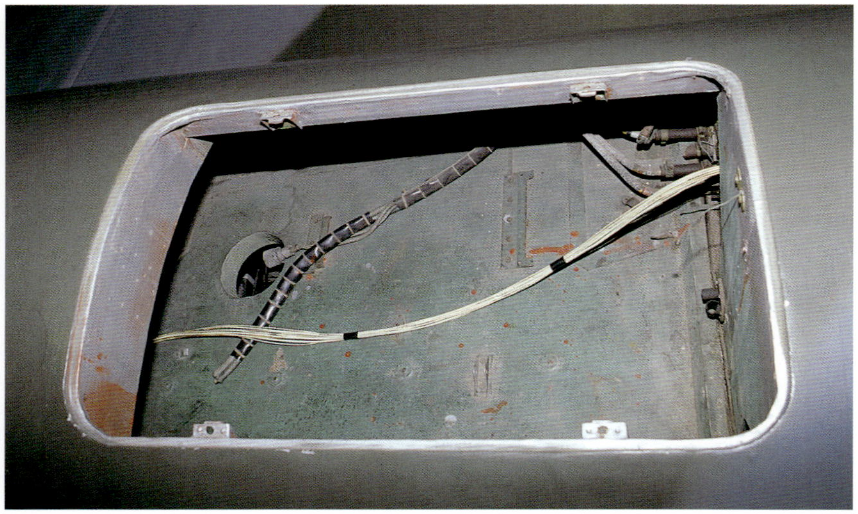

153

154

写真152-154: モスキートB.Mk.35(S/N TA634)の救命ボート収容ボックス。海上に不時着した際にキャノピー頂部から脱出した搭乗員は、この収容ボックスを開いて、なかから折り畳まれている救命ボートを引き出す。このボートには圧搾空気の入ったボトルが付属しており、瞬時に膨らませることができた。また、ボックスのふたにあたる部分には防水処理された収納袋が付属しており、ファスナーを開くと、中に海面着色マーカー、発煙筒、信号弾ピストルのほか、飲料水、食料、マッチ、そして手漕ぎ用オールなど、生存に必要な装備が収められている。モスキートの主戦場であった北ヨーロッパ戦域では、不時着水となれば夏でも水温の低い北海海域にならざるをえない。海中での漂流は即生命の危機をまねきかねないのである。とにかく、定期的に北海上空を飛行している味方救難機の到来までは、救命ボート上でじっと我慢の時間をすごさざるをえないわけである。

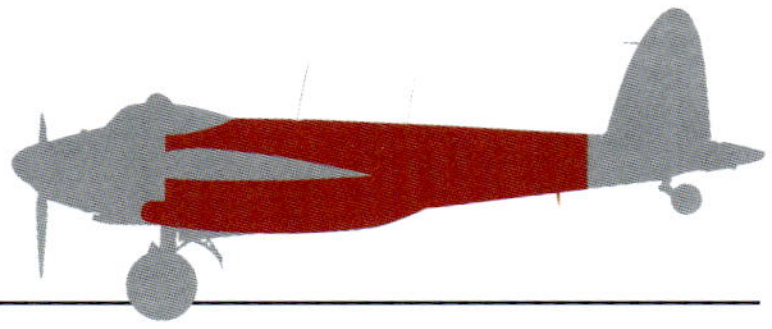

155

156

写真155: 胴体後部の点検口から胴体内側を撮影したもの。この頁の写真はすべてB.Mk.35(S/N TA634)。一番奥に救命ボート収容ボックスの一部がみられる。

写真156: T.R.9F無線送／受信器。隔壁前方の黒色のボンベは酸素ボンベで、左右計６本が装備されている。

写真157: 胴体右側下面に取り付けられた味方識別灯。本来では前方からレッド、グリーン、アンバーの３色が順番に並んでいる。

写真158-159: 胴体右側の補強ストレーキと点検口ハッチ。機体塗装は大戦当時のB.Mk.ＸⅥ(s/nML963、コード8K-K/No.571Sqn所属機)を模しているが、その機体は大戦終結間際の1945年４月10～11日のベルリン夜間空襲の際に失われている。

Photo 155: A view of the interior of the fuselage seen through the rear inspection hatch. A portion of the lifeboat storage compartment can been seen in the background.

Photo 156: The T.R. 9F radio transceiver. Black oxygen bottles are visible next to the bulkhead behind it. A total of six were carried on the left and right sides of the plane.

Photo 157: The recognition light on the lower right of the fuselage. Normally it would be colored red, green and amber from the front of the plane.

Photos 158-159: The reinforcement strake and inspection hatch on the right side of the plane. This aircraft is marked as a B. Mk.XVI (s/n ML963, code 8K-K, No. 571 Squadron) that was lost during the April 10-11, 1945 night attacks on Berlin.

157

158

159

主翼　WINGS

160

写真160-162: モスキートN.F.Mk.30の主翼外翼全景。基本構造、寸法とも原型から最終生産型にいたるまでまったく変わらなかったのは驚嘆に値する。まさしく“Wooden Wonder(木製の驚異)”の面目躍如するあたりである。表面外板に相当する合板を張った上からマダポラムとよばれる平織り綿布が全面に張られてからプライマー塗料が３層塗られ、その上からさらにアルミニウム・ドープが２層重ね塗装されて平滑な表面処理が行なわれており、その上に機体用途に応じた迷彩塗装が施された。なお、この綿布の継目部分は幅3/4インチ以上で重ね張りすることが規定されており、さらに綿布同士の継目もその上から２インチ幅の綿布のストライプが重ね張りされている。機体表面で細く盛り上がっているように見える部分がそれにあたる。

Photos 160-162: An overall view of the outer wing of the N.F. Mk.30. With very few exceptions, the wings of the Mosquito remained virtually unchanged throughout all models. Madapolam fabric was applied to the aircraft's wooden structure, and this was then painted with three coats of primer and two coats of aluminum dope, followed by the aircraft's camouflage marking scheme. Note that the seams in the fabric coating were covered by two-inch strips of the fabric and that these areas also received extra paint, creating the gentle mounds on the aircraft's surface that can be identified at certain locations.

161

162

163

164

写真163-165: 同じく主翼内翼部。木製機とはいっても、高熱を発するエンジン、ラジエター部や、構造上金属を使わざるをえない部分、あるいは点検、補給などで頻繁に開閉するハッチ部などは当然ながら金属外板が使われている。余談ながらモスキートが木製機として成功したのは、その目的が、軽量化を図るためではなかったことにあるといえる。実際、本機の空虚重量は6トン、総重量は10トンにもなり、これは日本海軍の一式陸攻11型のそれとほぼ匹敵する数値である。開発時期、要求仕様もまったく異なるのだから単純な比較は妥当ではないが、それでも最大速度で180km/h、爆弾搭載量で2倍近く凌駕するのだから、本機の高性能ぶりは歴然である。もちろん、それだけの機体を完成させた要因には長年木製機に熟知していたデ・ハヴィランド社のすぐれた設計、技術力があったためではあるが。大戦末期さまざまな木製機を開発しながらも、技術的な問題や重量の増加に悩まされて、結局ものにできなかった日独両国とはきわめて対照的である。

Photos 163-165: The inboard area of the wing. Though the Mosquito is generally described as being made of wood, the points of the plane that come into contact with the hot engines and radiators, as well as hatches, latches and other parts that simply cannot be built from wood for structural and strength reasons are naturally constructed from metal. One of the reasons for the success of the Mosquito is no doubt that fact that wood was not chosen as a building material in order to lighten the plane. The Mosquito had a dry weight of some six tons, and weighed 10 tons loaded, roughly the same as the Japanese Mitsubishi G3M "Betty" bomber. Despite this, the Mosquito flew 180km/h faster and carried twice the payload of the Betty – truly amazing performance. As the purposes for which the two planes were developed were quite different, such a direct comparison is difficult, but the remarkable design and engineering skills of the de Havilland Company are nonetheless clear. Both the German and Japanese aircraft industries attempted to produce wooden aircraft in the late stages of the war, but technical, production and weight problems prevented them from ever turning out anything of value, in sharp contrast with the British.

165

主翼　WINGS

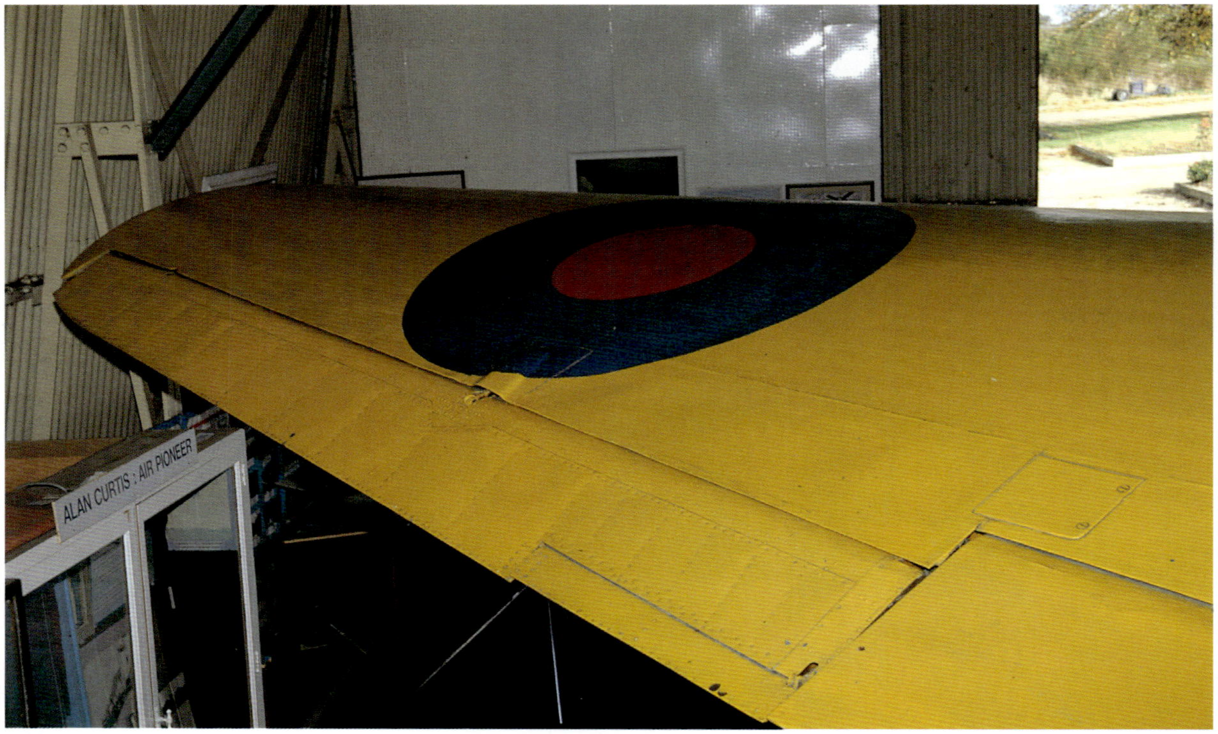

166

167

写真166-167: モスキート原型機の主翼。主翼表面はもちろん合板で仕上げられているが、補助翼は写真からもわかるように金属製外皮となっている。補助翼の作動角は＋26度30分から－11度30分の範囲で、これはモスキート最終型まで変わらなかった。写真167は主翼翼端の平面形で、のちの量産型とくらべると全幅が508mm短いため、翼端が丸みをおびている。また、当初あった翼端灯は取り外されたか、あるいは機体色によって塗りつぶされているようである。

Photos 166-167: The left wing of the prototype. The wing itself is of course of plywood construction, but the aileron is metal-skinned. The aileron could move through +26°30" and –11°30". This was unchanged throughout all models of the aircraft. Photo 167 shows the wingtip, 508mm narrower than those of the production aircraft to follow. The wingtip light appears to have been either removed or painted over.

写真168-169: 補助翼とは異なり、フラップ部は合板仕上げ。エンジンナセル上部は金属製になっているのがわかる。この原型機はその後さまざまなテストベッドとしても使われており、その間に機体外形にも改修が施された。このため初飛行当時と異なる部分も多い。エンジンナセル後端は、ナセル後流による尾翼バフェッティングを解消するため後縁から突き出したように伸長されているのが初飛行当時とは異なる箇所。

Photos 168-169: Unlike the ailerons, the flaps were of wooden construction. This prototype was subjected to extensive testing, and many modifications were made to it throughout its life. One such change was a lengthening of the trailing edge of the engine nacelles in order to reduce buffeting of the tail in flight.

168

169

170

写真170: 右主翼補助翼ヒンジ部を下面よりみる。モスキートにかぎらないが、大戦中の英国軍用機のディテールを仔細にみると、胴体、翼の下面部にいくつもの不思議な穴を発見することができる。これはドレイン・ホール(排水用の穴)で、機体の隙間や外板のツギ目などから入ったり、しみ込んだ水が機内にたまり機体構造を侵蝕しないようにするために設けられたもので、ことに機体構造の大半が木製で占めているモスキートでは他機以上に必須の設備であった。

Photo 170: The right wing aileron hinges seen from the underside. Many English warplanes, the Mosquito among them, have a number of small holes on the underside of the wings and fuselage. These are drain holes, designed to prevent fluids which have seeped into the aircraft's structure from causing corrosion. This is especially important in a plane with a wooden construction.

173

写真173-175: 主翼内翼前縁に設置されたラジエター部。写真173は左主翼ラジエター。このうち胴体に近い箇所が機内暖房用のヒーター、中央部がエンジン・ラジエター、そしてナセル寄りが滑油冷却器と、それぞれのユニットにわかれている。この配置は右側主翼でも同じである。写真174は後方よりみた右側ラジエター下面。ラジエターフラップが開いた状態になっている。写真175は右主翼ラジエター下面。3個のポッチは本来ドレイン・ホールだった部分をふさいでいるもの。

Photos 173-175: The radiator area, located on the inboard portion of the wing's leading edge. Photo 173 is the left radiator, which fulfills three roles. The most inboard portion provides fuselage heat, the central portion is the engine radiator and the outboard section acts as the oil cooler. These positions are the same on the right wing. Photo 174 is a rear view of the bottom of the right radiator, seen with the radiator flaps in the open position. Photo 175 is of the bottom of the right radiator. Note the three bulges off to the left side. These are drain holes that have been plugged.

171

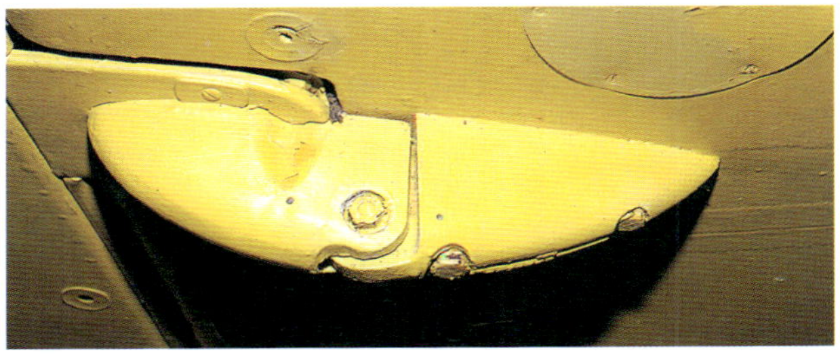

172

写真171-172: 主翼エンジン・ナセルをはさんで装備された主翼フラップのヒンジ。前述のようにモスキートのフラップの最大作動角は45度(ただし、インジケーターでは70度まで表示されている)。モスキートF.B.Mk.VIを例にすると、離陸時(機体重量17,000ポンド=7,718kg)ではフラップ15度下げで離陸速度は155ノット(287km/h)、着陸時(機体重量18,000ポンド=8,172kg)ではフラップ15度下げで着陸速度は90〜95ノット(167〜176km/h)と指示されている。

Photos 171-172: The flap hinge, located between the engine nacelle and fuselage. As mentioned before, the Mosquito's flaps could drop down to 45°, although the cockpit indicator actually went to 70°.

174

175

主翼　WINGS

176

177

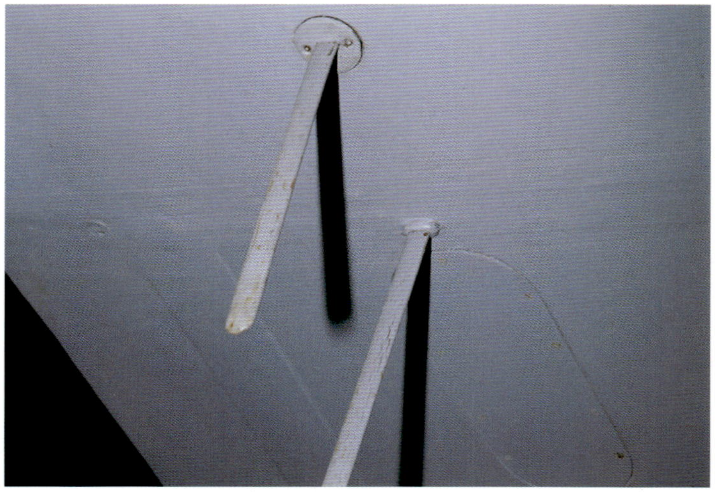

178

179

写真176-178: N.F.Mk.30の主翼。基本的に原型機をのぞいては高高度夜間戦闘機N.F.Mk.ＸＶ、および高高度写真偵察機P.R.Mk.32(ともに延長翼装備)以外の機体の主翼は寸法、形状ともにほぼ同一である。主翼翼端近くに突出しているのはパッシブ(受信)アンテナで、機首レーダーが発した電波が反射してもどってきたのを受信するものである。写真177は主翼上面、写真178は主翼下面に装着されているアンテナ。主翼左右上下面に取付けられている。

写真179: 主翼翼端後縁に取り付けられているレジンランプ。直訳すれば樹脂製灯であるが、どういう構造でどのような用途に用いられるものかは残念ながら不詳(ただし、マニュアルをみると、どうも電気が通じてはいるようなのだが)。本機のランプ本体は失われている。

Photos 176-178: The wing of the N.F. Mk.30. With the exception of the ultra high-altitude N.F. Mk.XV and P.R. Mk.32 (which both had extended wings), the dimensions and shape of the wings of the Mosquito were fundamentally unchanged throughout all models. The projection visible near the wingtip is the passive (receiving) antenna for the aircraft's radar system. This antenna picks up the reflected energy originally broadcast by the transmitter in the plane's nose. Photo 177 shows the antenna on the upper surface of the wing, while 178 is that on the lower side.

Photo 179: The so-called resin lamp on the trailing edge of the wingtip. The lamp has been lost on this example. Unfortunately, the original purpose of this lamp is not clear.

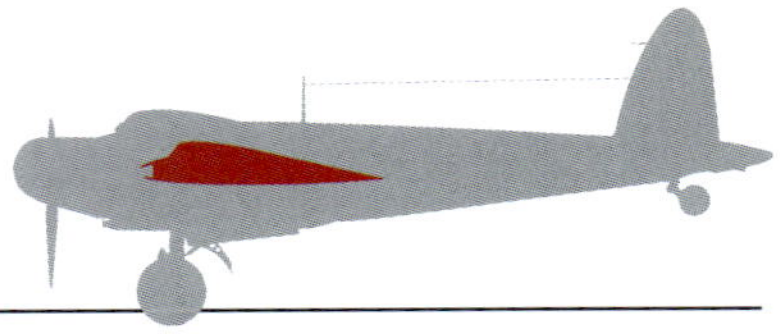

180

181

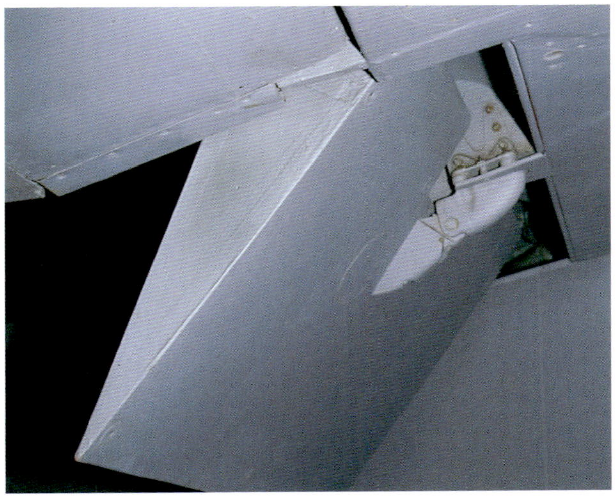
182

写真180-183: N.F.Mk.30の外側フラップ詳細。機体／主翼部分と同じく木製桁に合板張りの構造。エンジン・ナセルをはさんで内側フラップと連動して作動する。作動はエンジン・ナセル後部に置かれたフラップ操作ジャッキ／クランクによって行なわれる。外側、内側とも各１個のフラップヒンジによって主翼と結合している。フラップ作動角は最大45度であるが、通常は15度角で使用される。飛行中では175ノット(324km/h)以上では25度以下のフラップ下げ、130ノット(240km/h)以上でのフラップ全開は機体姿勢に悪影響をおよぼすため制限されている。

写真184: ラジエター下面に設けられたスリット。

写真185: 左右主翼下面に装備されている着陸灯。夜間着陸進入時に前方を照らすためにポップアップするが、普段は主翼下面に格納されている。

Photos 180-183: Close-ups of the outboard flap. Like the wings and fuselage, it's an entirely wooden construction. It moves in tandem with the inboard flap on the other side of the engine nacelle. It was operated by the flap jack/crank assembly behind the engine nacelle. Both the inboard and outboard flaps were attached to the wing via a single hinge. The flaps could pivot down to 45°, but were normally used at a 15° setting. More than 25° of flaps at over 175 knots, or full flaps above 130 knots would put the aircraft into a dangerously unstable configuration. For this reason, these settings were physically restricted.

Photos 184: A view of the slit assembly under the radiator.

Photos 185: A close-up of one of the two landing lights mounted under the wings. They would pop up for use during night approaches, but normally were stored as seen here.

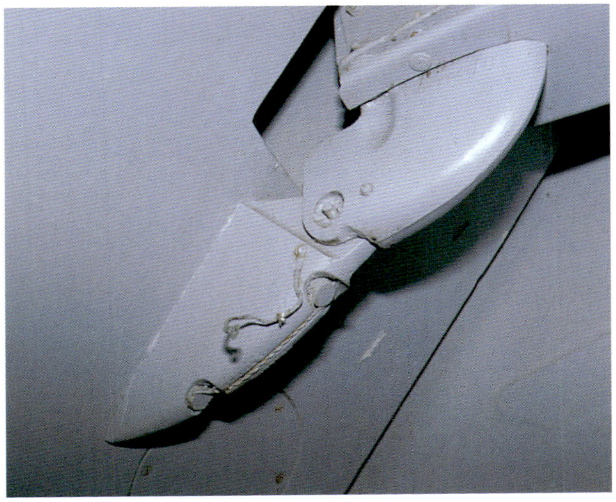
183

184

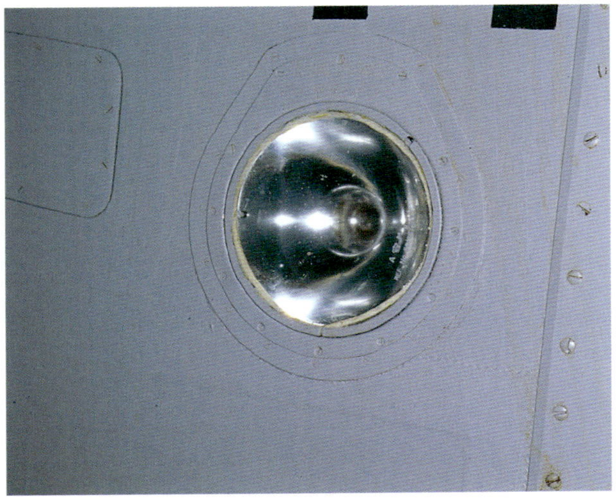
185

主翼 WINGS

186

187

写真186: 主翼ラジエター点検パネル。モスキートの機体のうちで数少ない金属製部分のひとつである。ただし、薄板なのでこの上に立つことは禁じられている。この見開きの写真はすべてB.Mk.35(S/N TA634)。

写真187: ラジエター後方の主翼内翼燃料タンク注入ハッチ。内翼燃料タンクは左右とも胴体よりに容量78英ガロン(355リッター)、ナセル寄りに65.5英ガロン(298リッター)の２個のタンクにわかれている。

写真188: 右主翼補助翼下面を見る。補助翼はアルミ合金製のリブに薄いアルミ合金皮を張って作られている。補助翼の寸法は、全長3,835ミリ、全幅559ミリで、作動角は＋26度30分、－11度30分である。この補助翼に金属外皮が採用されたのには、モスキートのような高速機では羽布張りの補助翼では操作に必要以上の労力を操縦士に強いることになり、かつ利きも遅くなるという事情があるからであろう。

写真189: 補助翼外側の翼端後縁に取り付けられたレジンランプ。ランプ自体は紛失。

写真190: 補助翼前方にあるパネルは補助翼関係のアクセスハッチのひとつ。その脇にあるリングは地上係留の際に機体を固定するためにロープをとおす係留金具。

Photo 186: The radiator access panel on the wings, one of the few metal parts on the Mosquito. Because of its thin construction, no stepping on this part was allowed.

Photo 187: The fuel fill hatch for the internal wing tanks located behind the radiator. The inboard tanks were located near the fuselage on both the right and left sides and could carry 78 imperial gallons (355 liters) each, while the outboard tanks near the engine nacelles held 65.5 imperial gallons (298 liters).

Photo 188: The underside of the right aileron. The ailerons were made entirely of aluminum, both the internal frame and the skin. The ailerons were 3,835mm long, 559mm wide and could move from +26°30" to –11°30". The aluminum construction was utilized instead of a more standard fabric design to lessen the amount of power required by the pilot to move the control surfaces as well as to provide quicker response, important issues in a high-speed aircraft like the Mosquito.

Photo 189: A close-up of the resin lamp mount located at the outboard side of the aileron on the trailing edge. The lamp itself is missing in this example.

Photo 190: The small panel ahead of the ailerons is an access hatch for their mechanism. The ring next to that is used to tie down the aircraft when it is stored out of doors.

188

189

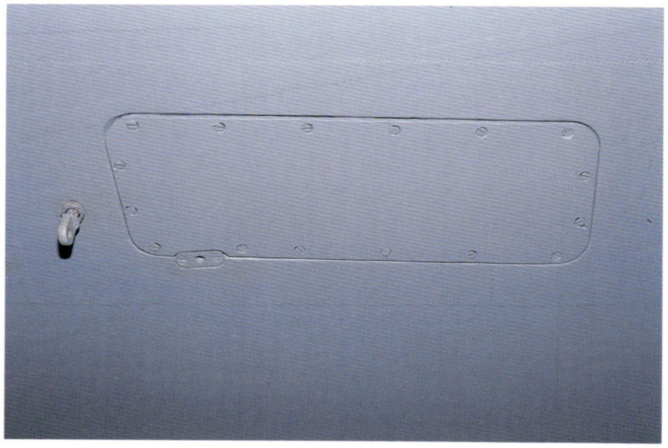

190

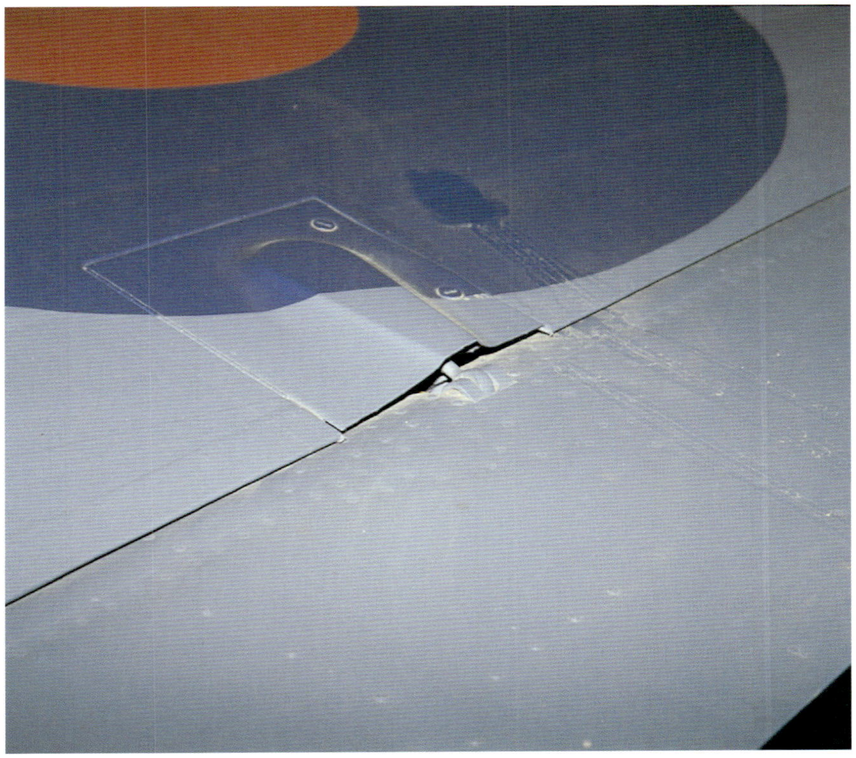

191

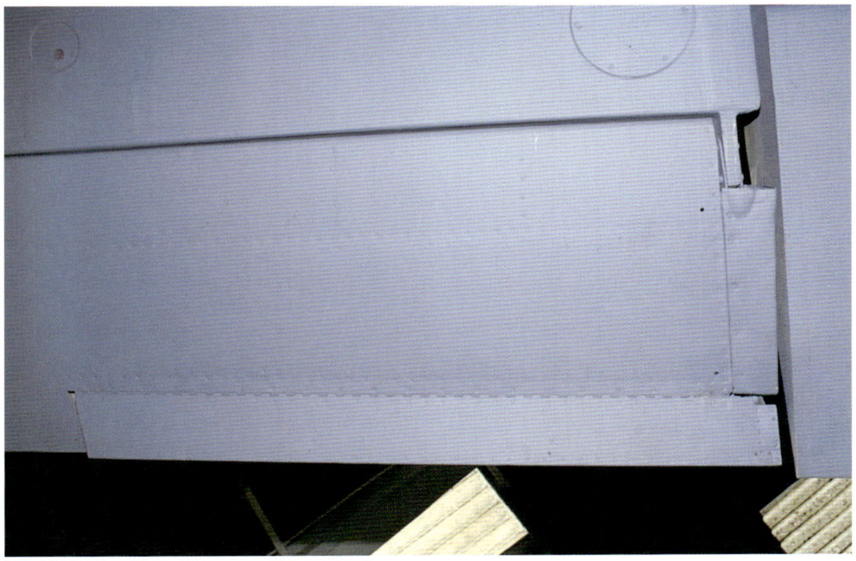

192

写真191：主翼上面の補助翼アクチュエーター・カバー。補助翼の作動は操縦桿の左右の動きがケーブルによってこのカバー内のプーリーに伝達される。プーリー上面にアクチュエーターが取り付けられており、プーリーの回転によりアクチュエーターが押し出されると補助翼下げ、逆に引き込まれると補助翼上げの状態になる。もちろん、この際には反対側の補助翼は作動が逆になる。

写真192：下面よりみた補助翼トリムタブ。作動角は＋8度30分、－9度15分の範囲で、コクピット内から調整できる。

写真193-194：主翼翼端方向からみた主翼下面。補助翼およびフラップとの位置関係がおわかりいただけよう。写真194はフラップヒンジのフェアリング。

Photo 191: The actuator cover for the ailerons on the upper side of the wing. Left and right motions of the control column were conveyed here by a cable and then to the pulley under the cover. The actuator is attached to the upper side of the pulley, so if pulley rotates one direction, the actuator is pushed out, lowering the aileron and if it goes the other way, it pulls the actuator, raising the aileron. Naturally, the opposite action takes place simultaneously on the other side of the plane.

Photo 192: The aileron trim tab, seen from the underside. It was controlled from the cockpit and could more from +8°30" to –9°15".

Photos 193-194: The underside of the wing, seen from the wingtip. The relative positions of the ailerons and flaps are discernable from these photos. Photo 194 shows the flap hinge fairing.

193

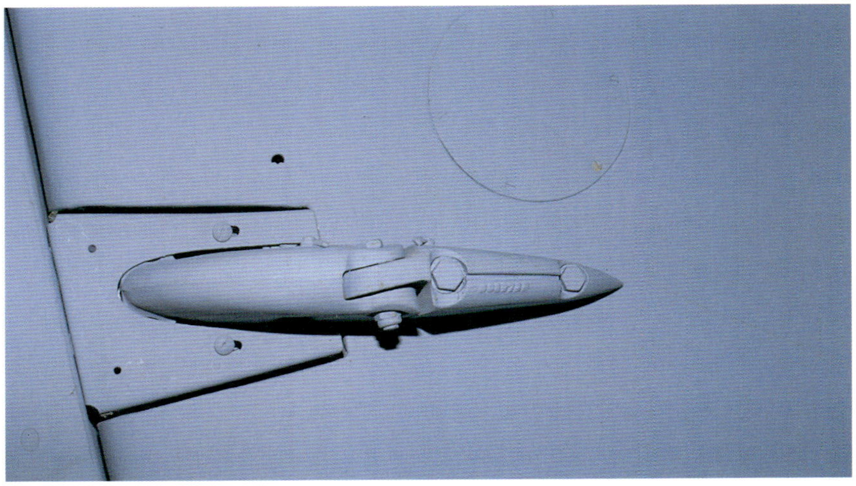

194

主翼 WINGS

195

196

写真195-196: エンジンナセル中央部にあるフラップ操作ジャッキ／クランク部点検パネル。写真はB.Mk.35(S/N TA634)。もちろん、このパネルもラジエター点検パネルと同じく金属製である。モスキートのエンジンナセルは試作機および極初期量産型では主翼後縁から突出してはいなかったが、テスト中にナセルによって生じる乱流が尾翼にバフェッティング(振動現象)を起こさせることがわかった。この対策としてエンジンナセルを後方に伸長させることによって解決をみたため、以後の量産機にも採用されてモスキート独特の平面形を作り出した。またそれまで１枚板タイプだったフラップは、写真のように内翼と外翼にわかれた２枚式になっている。当初はナセルの短い機体をシリーズ１、伸長された機体をシリーズ２と称した。本書に収録されている原型機も最初は短いナセルで、現在の姿は改修後を示している。

写真197-198: フラップ点検パネルの詳細と、開いた状態を示す。内側に見えるのがフラップ操作ジャッキ／クランクで、コクピット内のフラップ切換えレバーを操作することによりジャッキからのびた作動桿が押し出され、内翼フラップと外翼フラップを連結するクランク・ヒンジをダウン位置に作動させる。逆にレバーをアップにもどすとフラップはふたたび上げ位置に戻される。

Photos 195-196: The flap mechanism inspection hatch located in the mid-section of the engine nacelle. Like the radiator inspection hatch, this panel is also made of metal. The engine nacelles of the prototype and very early production aircraft did not stick out beyond the wing's trailing edge, but tests showed this design produced buffeting on the tail surfaces. The problem was solved by extending the engine nacelles beyond the back edge of the wing, creating the Mosquito's distinctive form. Early, short nacelles were described as Series 1, while the more standard nacelle was Series 2. The prototype in this book was originally equipped with short nacelles, but this was modified during testing.

Photos 197-198: The flap inspection panel. Photo 197 shows it removed. The flap control mechanism can be seen within. Operating the control in the cockpit would result in the connected inboard and outboard flaps being moved into the down position.

197

198

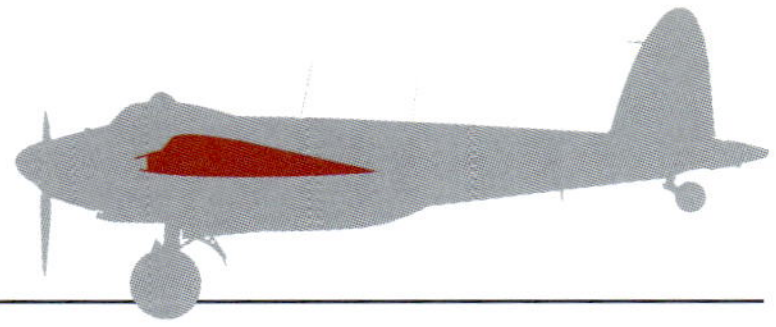

写真199: 主翼下面の補助翼アクチュエーター点検ハッチを開いたところ。アクチュエーターを作動させるプーリーの一部がみえる。

写真200: 左主翼内翼下面のラジエター部分にあるドレイン・ホール。

Photo 199: The aileron actuator inspection hatch on the underside of the wings in the open position. Part of the pulley can also be seen.

Photo 200: Some of the Mosquito's many drain holes, these located near the radiator on the lower, inboard side of the wing.

199

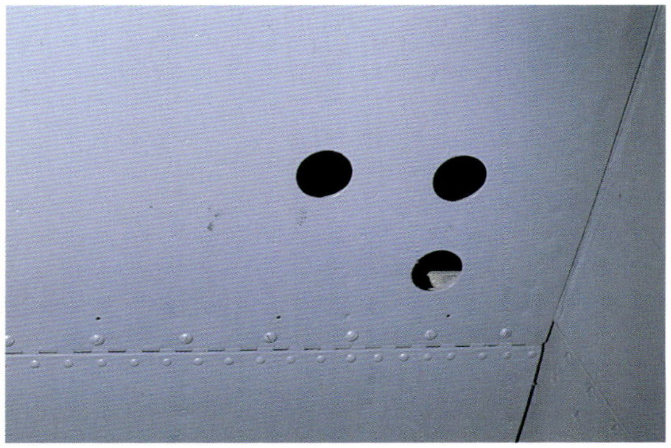

200

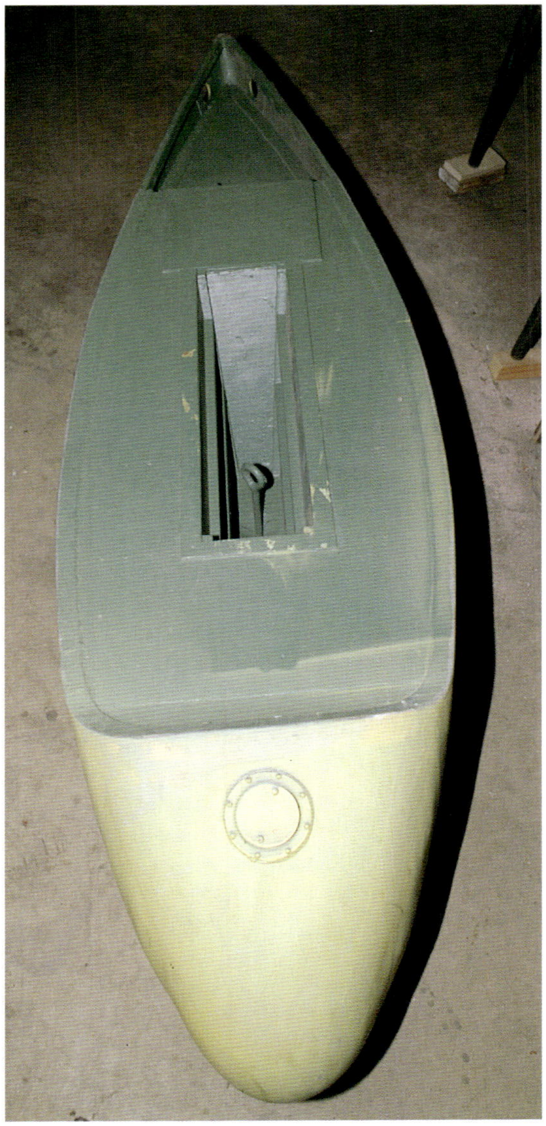

201

写真201-203: モスキートは、胴体の長距離用燃料タンク(容量69英ガロン=314リッター、75英ガロン=341リッター)を含め、全燃料タンクを満タンにすると合計547英ガロン(2,489リッター)もの燃料を搭載できた。しかし、さらに航続力を延伸させるために、主翼外翼下面にスリッパー式増加タンクを装備した。写真は容量50英ガロン(228リッター)タイプで、左右両翼下面に1個ずつ装備できた。もちろん緊急時には投棄することができた。これにより各型とも従来の航続距離を2,200kmから2,800kmに増大することができた。このほかに100英ガロン(455リッター)、200英ガロン(910リッター)容量の増加タンクも一部で使用された。余談ながら、この200英ガロンを2個、さらに爆弾倉内にも大型燃料タンクを装備した長距離写真偵察機P.R.Mk.34は実に最大5,600kmという大航続距離を誇った。

Photos 201-203: If a Mosquito topped off all of its fuel tanks, including the long-range tanks in the fuselage, it could carry 547 imperial gallons, or 2,489 liters of fuel. On top of that, "slipper" external tanks could be mounted on the wings as well. The photo shows the 50 imperial gallon (228 liter) type, one of which could be carried under each wing. These could be jettisoned in an emergency, and allowed the Mosquito to boost its range from 2,200km to some 2,800km. 100-gallon and 200-gallon type tanks were also in use. The photo-reconnaissance Mk. 34 model of the aircraft carrying two of these 200-gallon tanks and adding another fuel tank in the bomb bay could fly a remarkable 5,600km.

202

203

主翼　WINGS

写真204-206: レストア中のモスキートF.B.Mk.Ⅵの主翼部。当然ながら、大戦当時の主翼製作と同じ手順を踏んで復元作業が行われている。写真204は主翼上面に相当する部分にあたり、スプルース(エゾマツ)で作られたボックス型の前後主桁に、合板製のリブが整然と置かれている。さらに機体中心から主翼中央部までには細いストリンガーが主桁にそってビッシリと貼りつけられている。リブは内翼フラップ内側から始まって翼端までの間に左右とも16個ずつ配されている。ストリンガーも計16本で、この内中央部７本は1.6インチ(40.6ミリ)角、残りは１インチ(25.4ミリ)角のものが使用されている。材質は同じくスプルースである。この上にバルサをサンドイッチした薄い合板が表面に張られていくわけだが、接着剤の張り合わせだけでは強度的にも不安のある部分はネジ留め処理が行なわれている。ただし、その箇所にあたる合板の芯の部分には、バルサ材のかわりにスプルース材の積層がくるように、合板自体があらかじめ作られていた。写真205、206は下面部に相当する部分で、リブの断面が白く塗られているのも考証に忠実。その他の主翼内側部分は、プライマーを一度塗りしたあとに耐水塗料が一度塗りされているだけである。

204

205

206

Photos 204-206: The wing of the F.B. Mk.VI under restoration. The same construction techniques used during the war are being used to restore the aircraft. Photo 204 shows what will be the upper surface of the wing, revealing the box-type spars made of spruce with the ribs attached. Thin stringers are then attached to the main spars leading out from the center of the aircraft. There are sixteen wing ribs on each side of the aircraft, beginning just inside of the inboard flaps. There are also 16 stringers, the central seven of which were 1.6-inch (40.6mm) lumber, while the remaining nine were 1-inch (25.4mm) pieces, also of spruce. A light plywood skin with an inner layer of balsa was then attached over this. In areas where cement alone didn't provide enough strength, screws were used, but these portions lacked the balsa core, being all spruce instead. Photos 205 and 206 are views of the underside. The white painting of the ribs is also historically accurate. Other portions of the wing interior received just a single coat of primer and a coat of waterproofing paint before they were completed.

207

208

写真207-211：同じく復元中のモスキートF.B.Mk.Ⅵ。普段ではじっくり観察できない主翼内側の構造がよくわかり、その点からも貴重な写真であろう。モスキートに使われた木材の主体は軽いバルサ材のほか、スプルース(エゾマツ)が多用されているが、これはこの木材特有の耐水性、撥水性が何よりもまして選ばれた理由であろう。モスキートの成功の理由のひとつに木製機ということがあげられるが、これは他の軍用機生産の進捗を阻害することが少なく、しかも従来軍用機産業の外にあった木工業、家具産業に生産の拠点をおけるというメリットを生んだ。事実、モスキートの生産には上記の産業にまじって棺桶業者も参入していたといわれている。

Photos 207-211: More views of the under-restoration F.B. Mk.VI, providing usually unseen views of the wing structure. Besides balsa, the Mosquito was built primarily with spruce, chosen specifically for its resistance to water. This wooden construction was one of the primary reasons for the success of the Mosquito, as it made use of British manufacturing resources (wood workers) that were normally considered outside of the war production industries. It is often mentioned that even coffin manufacturers were put to work building Mosquitoes!

209

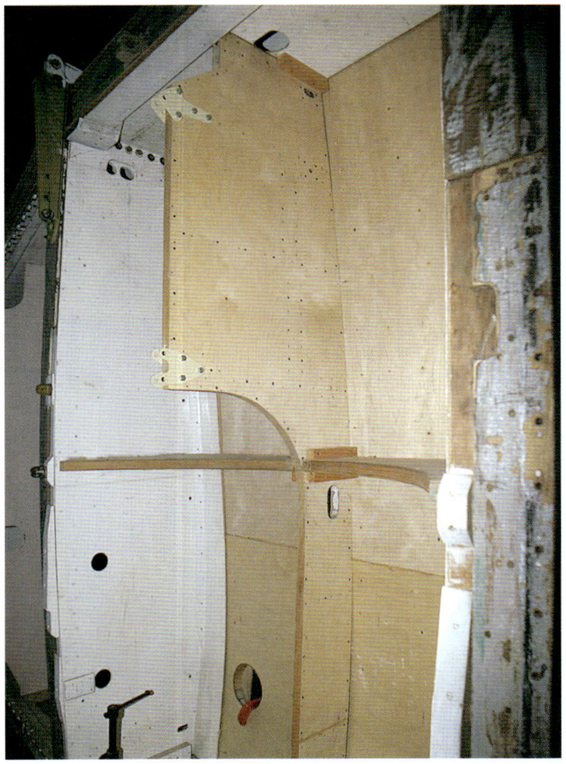
210

211

エンジン　ENGINES

写真212-215: モスキート原型機のエンジン・ナセル周囲。原型機とはいえ、その後にさまざまなテストや改修が施されしたため、初飛行当時の姿とはかなり異なる部分も少なくない。試作当時の機体ではエンジンはロールスロイスRM2SM(離昇1,280hp、のちのマーリン21)を搭載、排気管はナセル中央に小さく突き出した集合式排気管を装備していた。その後コクピット後方に7.7ミリ４連装銃座を搭載したり、マーリン61のテストベッドにと供され、その都度スタイルに変化をみせたが、現在は一段二速のマーリン搭載モスキートではオーソドックスなシュラウド付きのエンジンナセルを取り付けている。このシュラウド付半集合排気管はサキソフォンと呼ばれた独特な形をしており、シュラウドによるラムエア効果によって排気管冷却を図るというものであったが、これは成功したものとはいえず、のちにむきだしの単排気管がテストされ、そのロケット効果から若干の増速もみられたために、夜間戦闘機型をのぞいてほとんどの機体がそれにならった。なお、現在本機はマーリン25(離昇1,640hp)を搭載している。写真215は伸長されたナセル後端。初飛行当時のナセルは短く、フラップは内翼、外翼とも一体の一枚式であった。

212

213

214

Photos 212-215: The engine nacelle area of the prototype. Keep in mind that as this airframe underwent numerous modifications during its life, its present condition differs considerably from that when it made its maiden flight. The plane was originally powered by Rolls-Royce RM2SM engines (1,280hp takeoff power), later by Merlin 21s, and had the small, joined-type exhaust pipes in the center of the nacelle. Later, the plane had four 20mm cannons mounted behind the cockpit, Merlin 61 engines installed and other modifications, creating a slightly different appearance each time. At present, the aircraft has a nacelle standard for a one-stage two-speed supercharger Merlin engine. These half-collective exhaust pipes are known as "saxophone" type and are designed to cool the pipes via the ram air effect caused by the shrouds. This design did not prove effective however, and was changed to the single projecting-type exhaust pipes that produced a slight rocket effect and raised the top speed of the plane a bit. Consequently, this type was used in virtually every model of the plane except for night fighters. At present, this Mosquito mounts Merlin 25 (1,640hp takeoff power) engines. Photo 215 shows the trailing end of the extended engine nacelle. At the time of its first flight, the plane had the short nacelle and the single-piece flaps.

215

213

写真213-217：同じく原型機のエンジン・ナセル、および気化器空気取入口。写真213の水滴型のバルジは冷却液パイプの張り出しをカバーしている。写真215から217は気化器空気取入口。空気取入口前部に取り付けられているのは防氷／除雪ガード。北ヨーロッパという高緯度の戦域がモスキートの主戦場だったため、高高度、低空にかかわらず必須の装備である。写真215はアタッチメントのみ。

Photos 213-217: The engine nacelle and carburetor intake of the prototype. The teardrop-shaped bulge visible in photo 213 is to cover a protruding coolant pipe. Photos 215 to 217 are of the carburetor intake. The guard attached to the front is an anti-water and snow protector. Operating in northern Europe made this required equipment for the Mosquito. Photo 215 shows the attachment only.

214

215

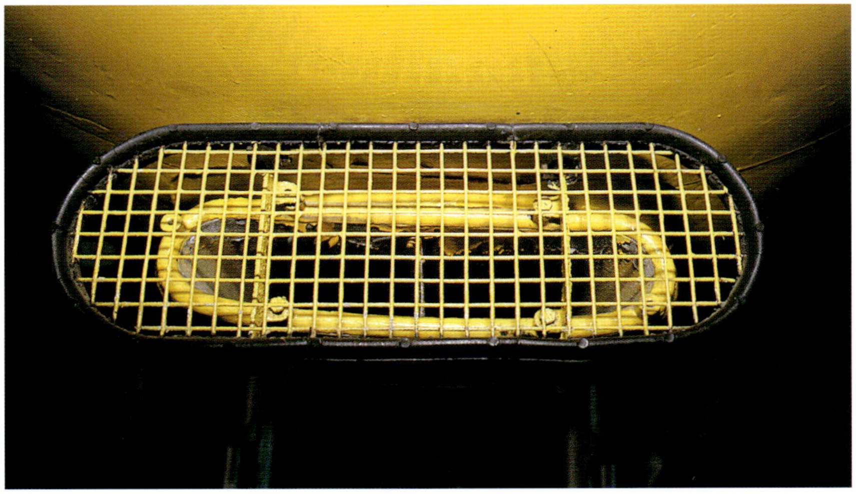
216

217

エンジン　ENGINES

218

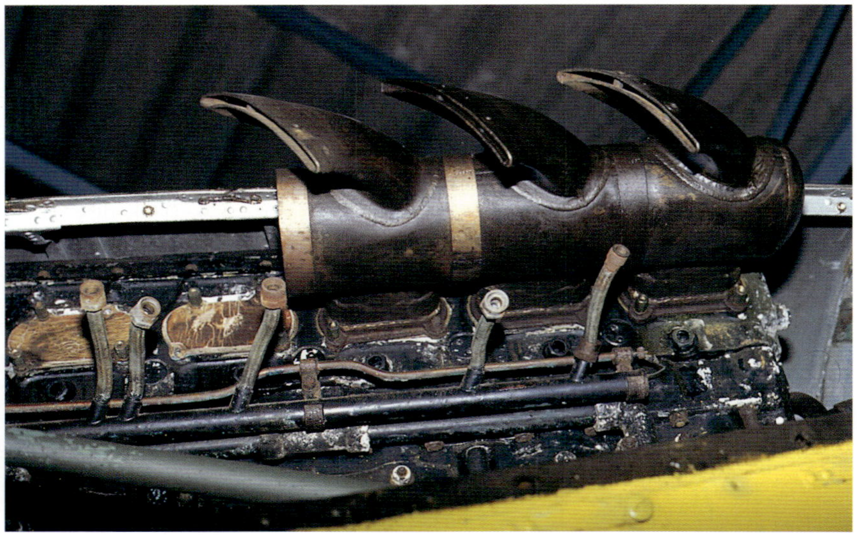

219

220

写真218-222: 原型機に搭載されているマーリン25エンジン。同エンジンは戦闘爆撃機型F.B.Mk.Ⅵ、ⅩⅧ、練習機型T.Mk.Ⅲ、夜間戦闘機型N.F.Mk.ⅩⅢ、ⅩⅨなどにも採用されており、一段二速マーリンの代表型。写真は前ページのシュラウドを外した状態を示すが、排気管はいわゆるサキソフォン形とは異なる。サキソフォン形は片側６個の排気管マニホールドを１本に集合、その前部に２本のフィッシュテイル形の排気口がついているというユニークな形状からその名の由来になっているが、写真の排気管はそれとも異なるし、後期型の推力式単排気管とも異なる。ただし、排気管以外はオリジナルなマーリンなので、各種補器やパイピング、コード類などディテールアップの恰好の資料となるだろう。写真221、222は後方よりみたシュラウド付き状態の排気管。手前の矩形のパネルは排気熱からナセル外板を保護するための耐熱板である。

Photos 218-222: The Merlin 25 engine of the prototype. The same engine was used in many of the fighter-bomber, trainer and night fighter versions of the plane. The photo shows the shroud described above removed, but the exhaust pipes differ from the standard saxophone type. The saxophone type has the six individual manifolds gathered into one pipe that has two fishtail shaped exhaust ports at the front. The pipes here also differ from the jet-type seen on later Mosquitoes. However, except for the exhaust pipes, everything else seen here is a classic Merlin engine and is an excellent reference for this standard WWII workhorse. Photos 221 and 222 are views from the rear with the shroud in place. The small plate in the foreground is a heat-resistant plate to protect the nacelle from the heat of the exhaust.

221

222

写真223-225: N.F.Mk.30のエンジンとプロペラ。モスキート後期型の代表的な搭載エンジンとなった二段二速マーリン・エンジン。写真はマーリン76で離昇1,710hpのパワーを発揮した。前にも述べたように、N.F.Mk.30は実戦に登場した最後のモスキートで、合計530機が量産された。この内はじめの79機はマーリン72(1,650hp)を搭載していたが、残りの451機がマーリン76を搭載していた。スピナー直下に開いているのはインタークーラー用の空気取入口で、その後ろにサメのエラのように開いているスリットはその排出口。気化器空気取入口も一段二速マーリン型にくらべ大型化されている。プロペラもパワーアップしたエンジンに対応してプロペラ効率の高い幅広の、いわゆるパドルブレード・タイプが装着されている。

Photos 223-225: The engine and propeller of the N.F. Mk.30. This is the standard two-stage two-speed supercharger powerplant of later model Mosquitoes. The photo shows a Merlin 76 (1,710hp takeoff power). As described above, the N.F. Mk.30 was the final model of the Mosquito to enter combat, with 530 examples being produced. Of these, the initial 79 were powered by Merlin 72 (1,650hp) engines, but the remaining 451 got the Merlin 76. The intake for the intercooler can be seen directly under the spinner, with the "shark's gill" slits behind that the exhaust for the same. The carburetor intake is noticeably larger than that on the single-stage, two-speed engines. The propeller was also changed to the paddle-blade type to take advantage of the additional power being provided by the engine.

223

224

225

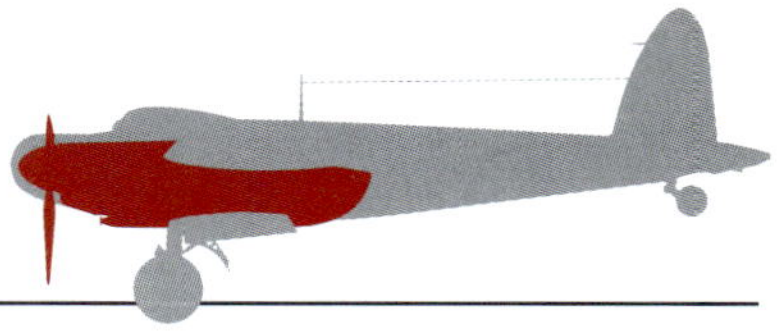

エンジン ENGINES

226

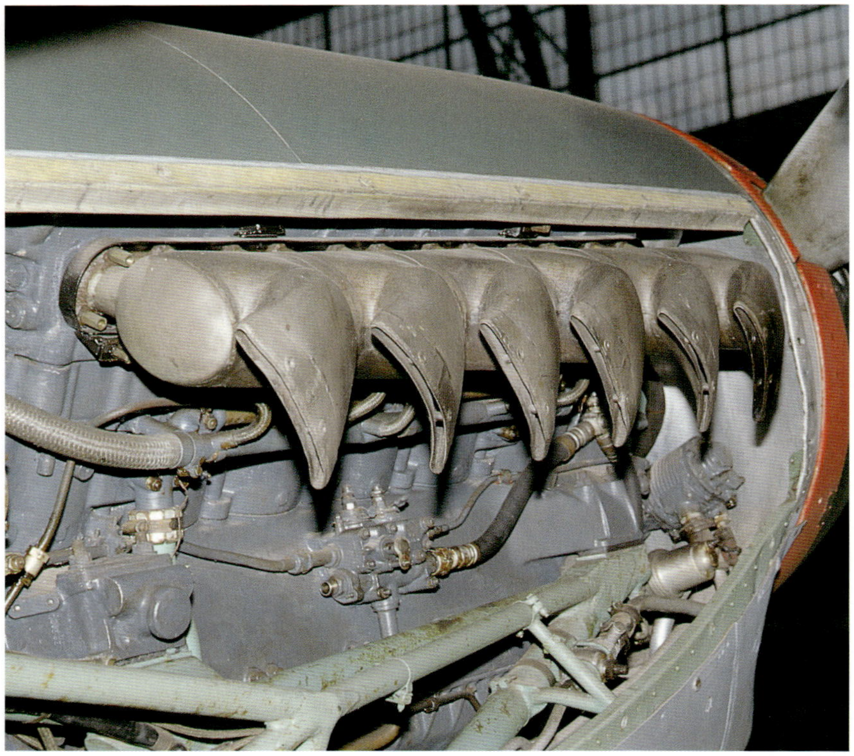

227

写真226-228: 同じくマーリン76エンジンのクローズアップ。第二次大戦で名を馳せた英国軍用機の例にもれず、モスキートもまた、その高性能と不朽の名声とは、ひとえに傑作エンジン、マーリンの賜物である。各国とも戦闘の激化にともない、第一線機の多くが高空性能の向上が求められていた。このうち、高高度でのエンジン出力向上については、米国では排気タービン駆動の過給器、ドイツではパワーブーストと、開発に余念がなかったが、英国では、従来の一段二速過給器から二段二速過給器を備えた新型エンジンの開発によって対処した。戦闘機型モスキートで最初にこの二段二速過給器付エンジンを搭載したのは高高度夜間戦闘機、N.F.Mk.ＸＶで、マーリン73ないし77(1,650hp)を搭載したが、試作型をふくめ、わずか５機生産されただけであった。続いて本格的な夜間戦闘機として登場したのがN.F.Mk.30で、最新鋭マイクロ波レーダー、A.I.Mk.Ｘを機首に装備、エンジンも出力の向上したマーリン76(1,710hp)を搭載、1944年中頃からは英夜間爆撃機部隊に随伴して、ドイツ夜間戦闘機群に対して果敢な戦闘を繰り広げた。なお、マーリン76はマーリン72から発達したエンジンで、機内与圧用スーパーチャージャーを追加したもの。同じくマーリン77はマーリン73にスーパーチャージャーを追加したものである。ただし、N.F.Mk.30は与圧キャビンではない。

Photos 226-228: Close-ups of the Merlin 76 engine, one of the major reasons for the Mosquito's brilliant success. As the war progressed, the major nations all worked at systems to boost the high-altitude performance of their aircraft engines. The U.S. approach employed exhaust-driven superchargers while the German approach involved power boost systems. The English, instead, focussed on upgrading the single-stage, two-speed supercharger-equipped engines of their planes to those that boasted two-stage, two-speed models. The ultra-high altitude night fighter Mk.XV was the first Mosquito model to employee such a powerplant in the Merlin 73 or 77 (1,650hp). However, only five examples of this model were produced. The N.F. Mk.30 which followed was powered by the Merlin 76 (1,710hp), and featured the latest microwave radar, the A.I. Mk.X in the nose. Employed by British night-fighting squadrons beginning in mid-1944, it proved extremely successful. The Merlin 76 was developed from the Merlin 72 and included a supercharger to provide cabin pressure. In the same way, the Merlin 77 was developed from the Merlin 73, adding a supercharger. However, the N.F. Mk.30 did not have a pressurized cabin.

228

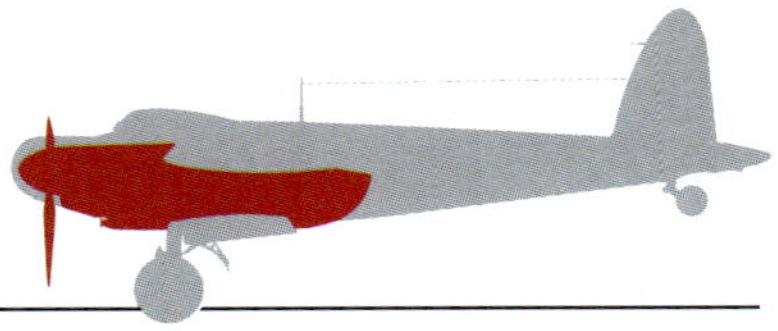

229

230

231

写真229-231： N.F.Mk.30独特の排気管シュラウド(覆い)をみる。夜間作戦機にとって排気管から吹き出す排気焔は操縦士の目を幻惑させ、さらに自機の存在を秘匿するうえでも好ましくない。このためさまざまな消焔式排気管が考案された。当初N.F. Mk.30でも従来の排気管／シュラウドが継承されたが、出力の向上したエンジンによるものか、排気管周囲を排気熱で傷めやすく、また排気焔も格段に激しかったため、より効率的な排気管／シュラウドの設計が要求された。そこで考案されたのがこのルーバー・タイプのシュラウドで、好結果をおさめたため、N.F.Mk.30にただちに採用され、さらに生産中の二段二速マーリン搭載の写真偵察機、爆撃機型にも広く採用されることになる。排気管もそれまでのサキソフォン型を廃し、丸型ないしフィッシュテイル(魚の尾)型に改められた。ただし、このルーバー・タイプを取り付けた機体でも主に昼間作戦に従事する機体では取り外して排気管をムキ出しにしたままのものも多かった。写真229でシュラウド内側下にみえるのは磁石発電機の冷却用ダクト。なお、排気管の形状は左右で微妙に異なる。

Photos 229-231: The unique exhaust shroud of the Mk.30. Exhaust flames were deadly for aircraft that operated at night, destroying the pilot's vision and giving the aircraft's presence away. The N.F. Mk.30 was originally to use existing designs, but the increased heat and flame of the new, more powerful engines forced a modification. This louver-type shroud proved successful and was quickly adapted for use in two-stage, two-speed supercharger-equipped photoreconnaissance and bomber models of the Mosquito as well. Simultaneously, the saxophone-type exhaust pipes were dropped in favor of a simple round, or a fishtail design. Aircraft involved primarily in day missions frequently removed the shroud and left the exhaust pipes exposed. The object visible inside the bottom portion of the shroud in photo 229 is the coolant duct for the magnetic generator. Note the slight differences in the shape of the exhaust pipes on the left and right.

232

写真232： 二段二速マーリン搭載により、気化器空気取入口もそれまでの機体にくらべ大型化されているのがわかる。

写真233： マーリン70系は中間冷却器を取り付けているため、スピナー直下に空気取入口が新たに設けられている。写真はその冷却器からの空気をにがすためのアウトレット。

Photo 232: The carburetor intake was increased noticeably in size on the two-stage, two-speed supercharger-equipped versions of the aircraft.

Photo 233: Engines of the Merlin 70 series employed an intercooler that resulted in another intake being added under the spinner. The photo shows the outlets that bleed off the air from the intake.

233

エンジン ENGINES

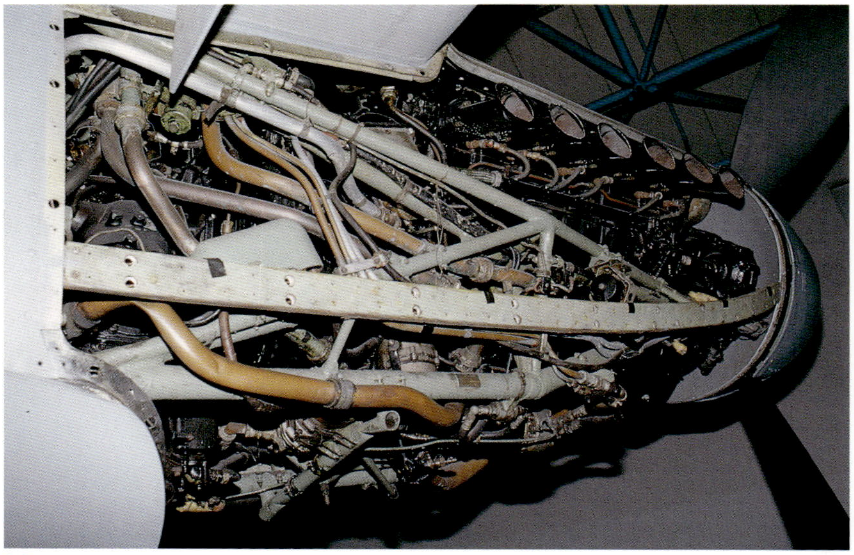

234

写真234-237: ナセルの各パネルを外したB.Mk.35(S/N TJ138)に搭載されているマーリン114(1,710hp)のディテール。一段目の過給器で圧縮され高温となった空気を二段目の過給器に送る前にいったん中間冷却器を通して冷やしてから送るのが二段式過給器で、飛行高度により過給器内のインペラ(羽根車)の回転速度を変更できるのが二速の意味である。これにより、シリンダー内でノッキング(異常爆発)を誘引する高温の空気を送ることなく、しかも濃度の高い空気を送り込むことができ、高空においても中高度域と変わらないエンジン出力を維持することができるわけである。写真では見えにくいが、中間冷却器および二段過給器はエンジン本体直後に取付けられている。

Photos 234-237: Detail views of a Merlin 114 (1,710hp) as installed in a B. Mk.35(S/N TJ138). In a two-stage supercharger, the air compressed and heated by the first supercharger is passed through the intercooler and cooled before moving on the to second supercharger. Depending on altitude, the impeller speed can also be altered, hence the "two-speed" reference. This system allows relative cool, dense air to be fed to the cylinders for combustion, preventing "knocking" (premature explosions in the cylinders due to super-heated air) and providing the same air pressure as middle altitudes. Though hard to see in the photos, the intercooler and second-stage supercharger are mounted directly behind the engine.

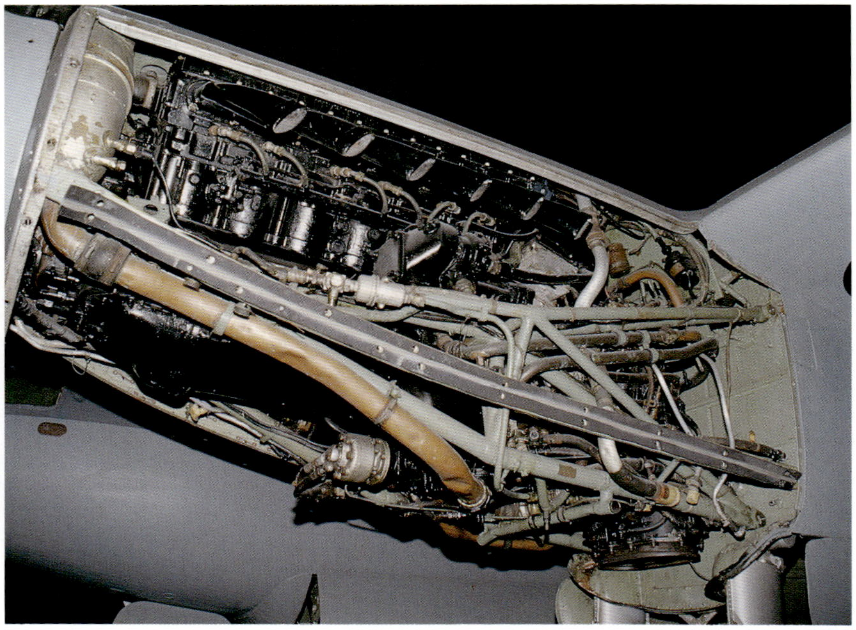

235

236

237

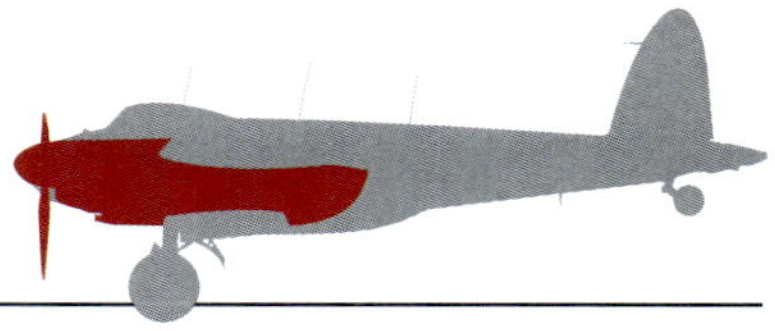

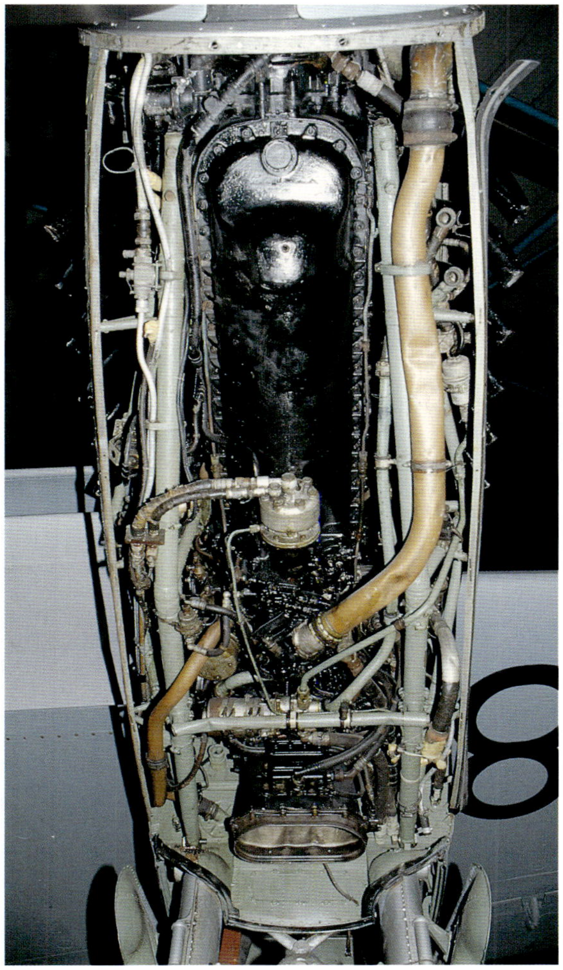

238

239

240

写真238-240: B.MK.35(S/N TJ138)のマーリン114エンジン(左側)を下方からみる。大きく開いているのが気化器および一段目過給器につづく空気取入口である。マーリンは気化器、過給器をエンジン後部に置いたため、新たに二段過給器を設置する際にもその分前後に寸法が伸びただけで、正面面積を変更することなくエンジンを搭載することができ、飛行性能に悪影響を及ぼすこともなかった。とくにモスキートではエンジン艤装後方に主脚収容口があるだけなので、大幅な改修を施すこともなかった。エンジン下面にはしるパイプ類は冷却液および潤滑油を、エンジンと各冷却器とに循環させるためのものである。

Photos 238-240: The left side of the Merlin 114 of B.Mk.35(S/N TJ138) seen from below. The large opening is the intake for the carburetor and first-stage supercharger. Merlin placed the carburetor and superchargers of its engines behind them, so the addition of another supercharger did not increase the frontal area of the engine, which would have hurt the aircraft's performance. As the only major structure behind the Mosquito's engines were the landing gear, the change did not require major modifications to the plane. The various pipes on the underside carry coolant and oil.

写真241-243: 取り外した状態のエンジン・ナセル・カバー。下面部分は三分割ながら一体式になっている。気化器空気取入口にあたるルーバー状のダクトは、フィルターを装備するためと思われる。パネル内側はコクピットと同じくグレイグリーンで塗られている。

Photos 241-243: The engine nacelle cover removed. The lower piece is one unit, but divided into three sections. The louver-shaped duct at the carburetor intake location is thought to be present to hold a filter. The inside of the panel is painted in the same gray-green as the cockpit.

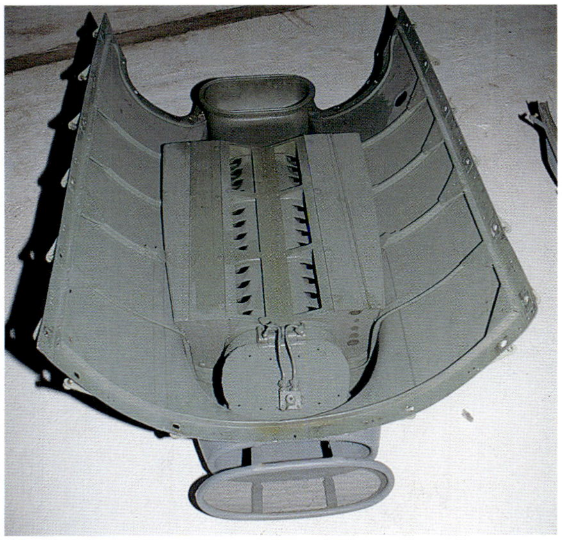

241

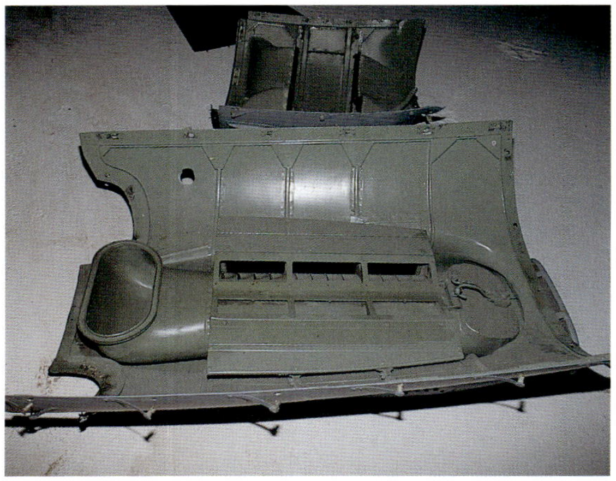

242

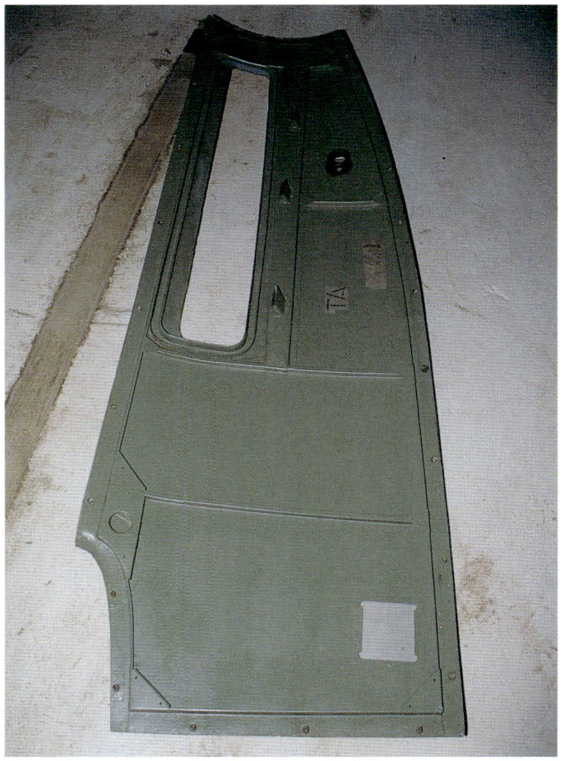

243

エンジン ENGINES

244

写真244-246: B.Mk.35(S/N TA634)のエンジン周囲およびスピナー。マーリン20シリーズ搭載型と比較すると外観からもパワーアップされた力強さがうかがえる。排気管下の細長いインテイクはインタークーラー用ポンプの発電機を冷却させるためのもの。細かいことだが、ナセル・カバーの取り外しは従来の機体では手間のかかるネジ止め式であったが、本機のころから簡便なファスナー式に改められており、整備効率のうえでも迅速な作業をすすめることができるようになった。スピナー形状は変わりないようだが、プロペラは、従来と同じくデ・ハヴィランド社製金属３翅プロペラながら、ブレード幅の広い、いわゆるパドル(水掻き)ブレードとよばれるタイプが装備されている。これはパワーアップされたエンジン出力が十分に発揮でき、かつ高空でのプロペラ効率を高めるためでもある。事実、このパドルブレードを装備した機体では実用作戦高度が100メートルもアップしたといわれている。

Photos 244-246: More details of the area around the B. Mk.35's engine and spinner of B.Mk.35(S/N TA634). The long, narrow intake under the exhaust pipes is for cooling the generator for the intercooler's pump. Note that the old screw-fastened panel attachment system has been replaced by the more efficient fasteners on this aircraft. The spinner is the same as previous models, but the propeller blades, though being de Havilland-built metal three-blade models, are of the wider "paddle-blade" type. This propeller was designed to make the most of the additional power that the new engine was providing. There are reports that it increased the aircraft's operational ceiling by some 100 meters.

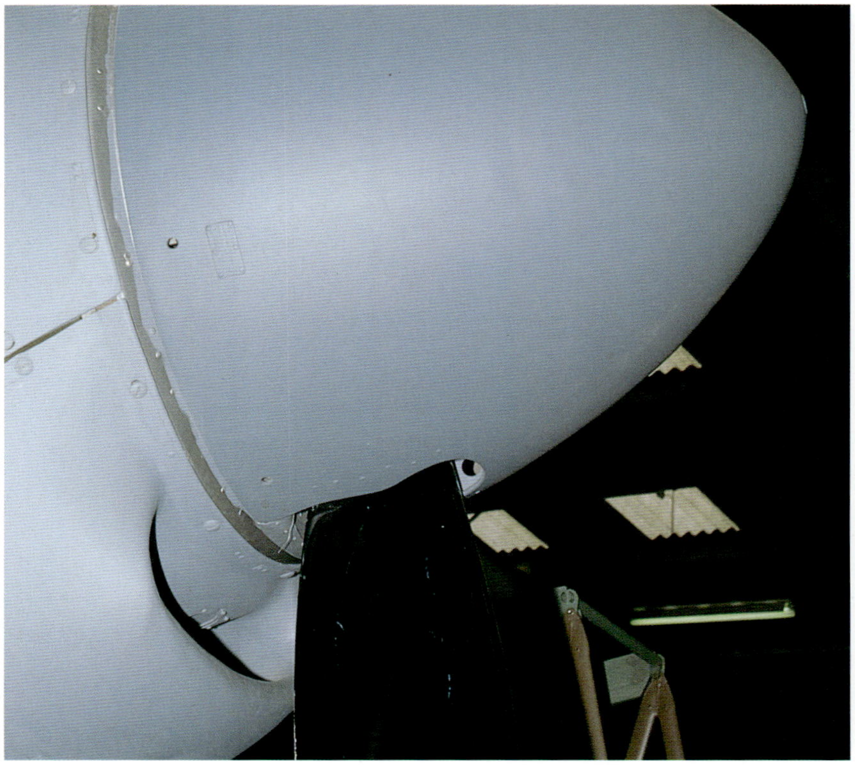
245

246

247

写真247: B.Mk.35(S/N TA364)右側エンジン・ナセル内側の状況。

写真248: エンジンナセル上部左側にあるハッチは冷却液注入口で、ハッチを開くと内側に冷却液タンクが収容されているのは、マーリン・エンジンでは共通のレイアウト。

写真249: 出力はマーリン113と同じであるが、左側エンジンにはマーリン114が搭載されている。これはマーリン114には機内与圧用スーパーチャージャーが付加されているためにエンジン番号が変わっているのである。写真の排気管下にあるインテイクはその与圧空気用のものかもしれない。

写真450: スピナー直下にある中間冷却器の空気取入口。二段二速マーリン装備機の一番目につく特徴である。

248

249

250

Photo 247: The inside of the right engine nacelle of the B.Mk.35 (S/N TA634).

Photo 248: On the upper left of the engine nacelle can be found a coolant fill. The coolant tank is mounted directly below this hatch. This is a standard configuration for all Merlin engines.

Photo 249: Though the same power as the Merlin 113, the left engine is the Merlin 114, which includes a supercharger for cockpit pressure. The intake visible below the exhaust pipes in the photo is thought to be for that purpose.

Photo 250: The intercooler intake directly below the spinner. This is the quickest way to identify which Mosquitoes have the two-stage, two-speed turbocharger-equipped engines.

エンジン ENGINES

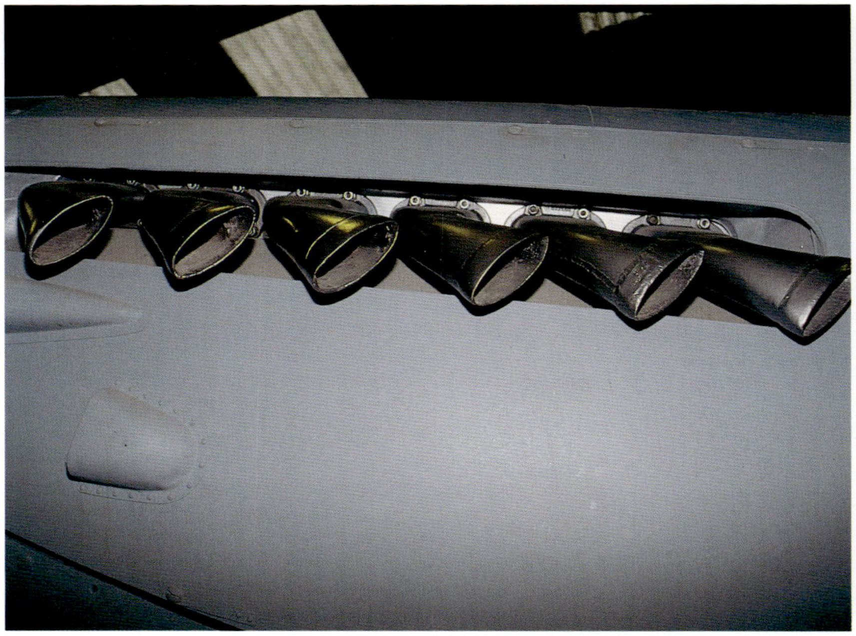

251

252

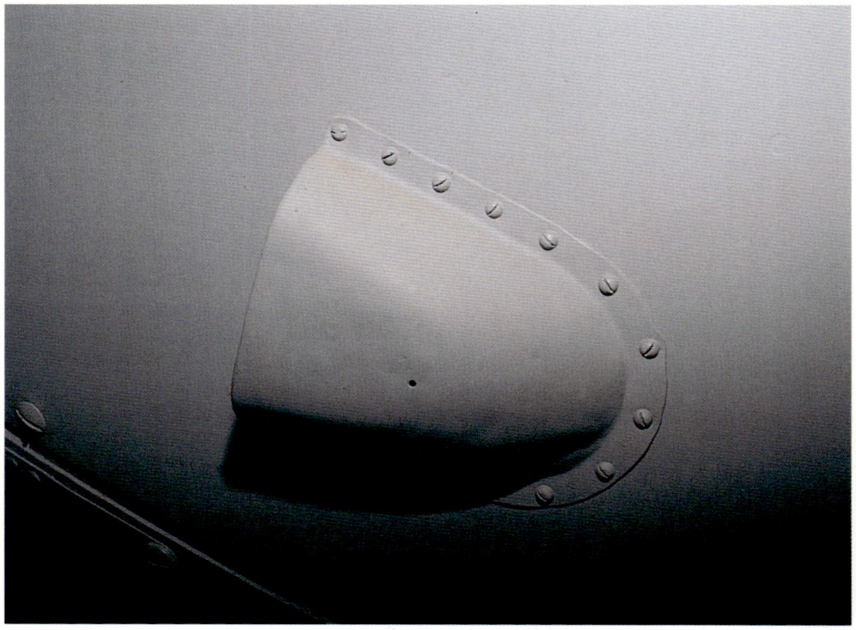

253

写真251-252: B.Nk.35(S/N TA634)のエンジン排気管詳細。当初あまり芳しくなかったサキソフォン型排気管を一新し、オーソドックスな単排気管型に改められたが、これは出力のアップしたエンジンでは従来の排気管、シュラウドが実用に耐えられないという事情もあったようだ。加えて排気ガスの推力による若干の速度アップも意図されていた。

写真253: 排気管後方下にある小さな空気取入口は燃料供給ポンプ冷却用で、これは必ずしも二段二速マーリン・エンジンだけの装備とは限らない。

写真254: エンジンナセル下部左右にある中間冷却器用のアウトレット。

Photos 251-252: Close-ups of the exhaust pipes of B.Mk.35(S/N TA634). The relatively complex saxophone-style pipes were dropped in favor of something simpler, but this couldn't accommodate the additional power of the engine. Nevertheless, their jet effect is said to have added a few knots to the aircraft's speed.

Photo 253: The small air intake located below and to the rear of the exhaust pipes is the coolant for the fuel pump. Its presence does not necessarily identify a plane with the two-stage, two-speed supercharger installed.

Photo 254: The outlets for the intercooler, located on the underside of the engine nacelle .

254

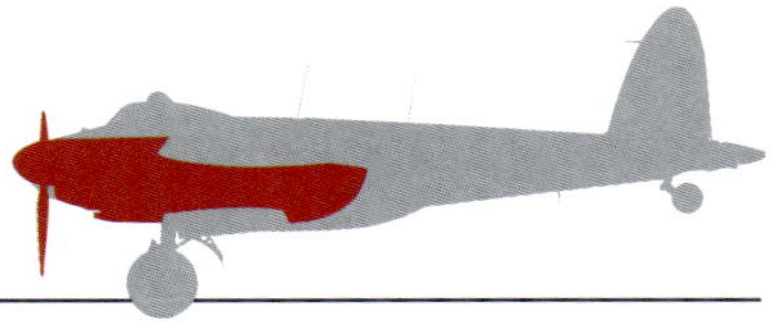

255

257

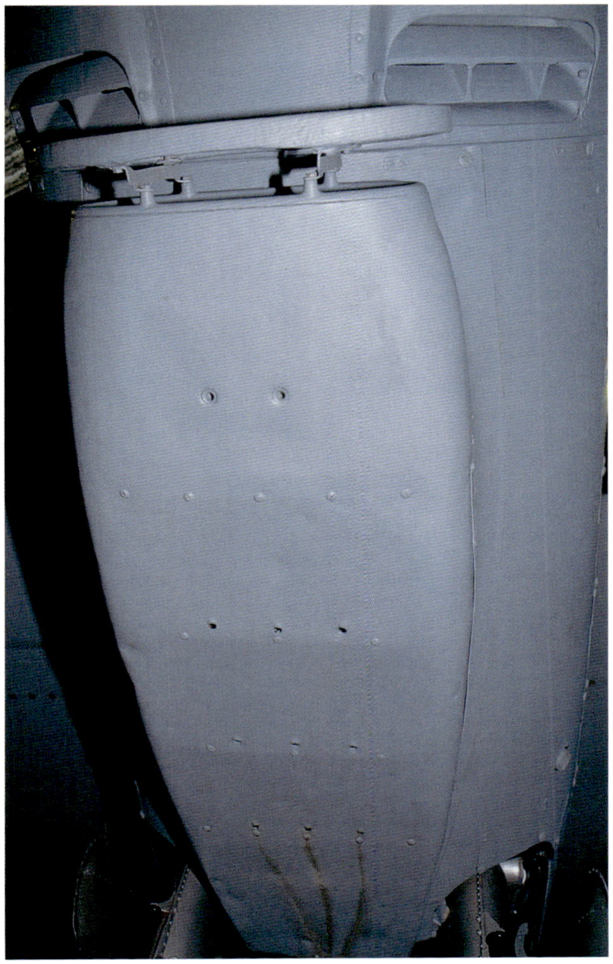

256

写真255-256: 下面よりみたB.Mk.35(S/N TA634)のエンジンナセル下部。気化器空気取入口、左右の中間冷却器アウトレットの位置関係が把握できる。気化器空気取入口の前には防氷／除雪ガードが取り付けられている。

写真257: 気化器空気取入口の側面はルーバー状にスリットが開けられている。これは防塵フィルターを取り付けた場合の空気取入口と思われる。

写真258: エンジンナセル後方主脚ドア上の赤く色分けされているハッチはエンジン始動用のカートリッジスターターの差し込み穴。ここに火薬カートリッジをセットしてその爆発でエンジンを始動させるための装備。通常は電源を用いるが、供給できない場合にはカートリッジによって行なわれる。

Photos 255-256: The underside of the engine nacelle of B.Mk.35(S/N TA634), seen from below. The various intakes and outlets are well shown. A snow and water guard was installed in front of the carburetor intake.

Photo 257: Louver-slits are present near the carburetor's air intake. It's believed that these were used as the as the intake when an anti-dust filter was attached.

Photo 258: Toward the rear of the engine nacelle, above the main gear doors is a red-painted hatch covering the cartridge starter hole for non-powered engine starts. A gun-type cartridge was inserted here to turn the engine with the power of its explosion in place of electric power when it wasn't available.

258

エンジン　ENGINES

259

260

写真259-261: モスキートの心臓ともいうべきロールスロイス社製マーリン・エンジン。くわしい型式は不詳ながら、一段二速過給器装備のマーリン20系と思われる。モスキートに搭載されたマーリン20系には、21、22、23、25の４型式があるが、この内マーリン25について簡単に述べてみると、乾燥重量658kg、離昇出力1,620hp(3,000回転／分)、最大出力1,640hp(3,000回転／分、高度610m)とされている。このエンジンを搭載したモスキートの代表型、F.B.Mk.VIを例にとると、最大速度612km/h、実用上昇限度10,970m、航続距離1,940km、2,745km(増加タンク装備時)という、実戦配備が始まった1943年３月当時のあらゆる機体を凌駕する高性能を誇った。実際、ドイツ軍占領下のヨーロッパ諸国、そしてドイツ本土に侵入するモスキートを捕捉、撃墜するのは至難とされ、撃墜すれば公報に載るとさえいわれたくらいである。この当時にモスキートが多用した戦術は超低空で侵入し、目標直前で高度を上げて緩降下して目標をたたいたあと、素早く退避するというもので、しかも単機ないしは少数機で作戦を実施しており、高空での性能はさほど要求されなかった。のちに高高度での性能を要求された際にも既存のエンジンを二段二速過給器付きに改修することによって対処するあたり、あざやかというほかない。

Photos 259-261: The Rolls-Royce Merlin engine, the heart of the Mosquito. Though the exact type is hard to identify, it appears to be a single-stage, two-speed supercharger-equipped Merlin 20 series powerplant. In the Merlin 20 series, the 21, 22, 23 and 25 saw use in the Mosquito. These engines, using the 25 as a representative example, weighed 658kg, and produced 1,620hp of takeoff power at 3,000 rpm. With two of these, a typical Mosquito (F.B. Mk.VI) could reach 612km/h, climb to 10,970m and fly over 1,940km (or 2,745km if external tanks were used). This performance gave the Mosquito the ability to strike virtually any target in occupied Europe while at the same time making interception virtually impossible. German pilots who managed to destroy a Mosquito are said to have sometimes received special decorations for their deed. At the time, the most common Mosquito attack strategy was to sneak in a low level, pop up just before reaching their targets, drop their bombs and then hurry away. Single aircraft, or relatively small flights generally carried out these raids. As such, high-altitude performance was not that crucial for the plane, but it's interesting to note that when then did decide to strengthen it, the simple addition of another supercharger stage was enough to meet the need.

261

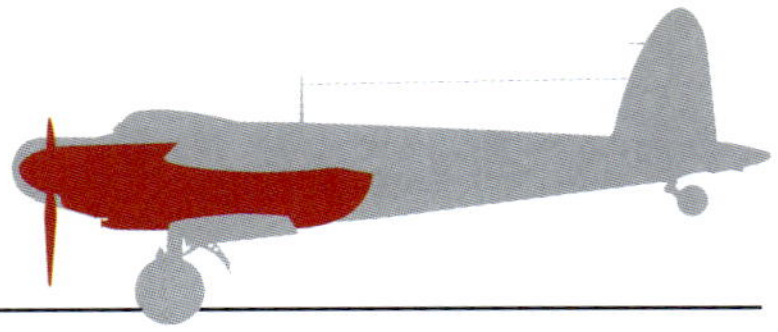

262

263

写真262-265: 同じくマーリン・エンジン。写真263、264が気化器、265がシャフトである。V型12気筒、総排気量27,000cc、昇流式気化器という基本的デザインは変更されることはなかった。エンジン自体の性能向上については乱暴な表現を使えば、気化器周辺部の改修だけで達成したといっても過言ではない。その改修もエンジン後部に集中しており、飛行性能に影響を及ぼす正面面積にはほとんど変更をあたえずに解決している。モスキートの基本設計はたしかにすぐれたものではあったが、その要求する性能を保証するのは、やはりすぐれたエンジンがあればこそである。同時にそれは不断の技術の研磨と蓄積、そして素早く対応できる広大なバックグラウンドがあって初めて始まる話ではあるが。

Photos 262-265: More shots of the same Merlin engine. Photos 263 and 264 show the carburetor. The basic V-12, 27-liter displacement design was unchanged throughout all models. The only major modifications were made to the area around the carburetor that, as mentioned before, allowed them to be carried out without changing the engine's front profile and harming performance. Certainly the Mosquito was an excellent aircraft design from the start, but without a superb engine such as this to be coupled with it, the type could have never achieved what it did.

264

265

主脚 MAIN GEAR & GEAR BAY

266

267

268

写真266-268: モスキート原型機の主脚部。この主脚もユニークなメカニズムを秘めたモスキートの特長のひとつである。航空機の主脚というのはそのみかけによらず設計者を悩ます箇所である。まず、地上にあって機体を支える必須の装備であること、しかも離着陸での激しい衝撃にも十分耐えられる頑丈さも要求される部分である。当然そのためには材料である金属を多量に必要とする。「主脚は金属の塊」といわれる所以である。ところが、飛行にはなるべく軽量であることが要求される。この相反する矛盾が設計者を四苦八苦させるのである。ましてやモスキートでは戦略物資である金属は可能なかぎり使わない、という前提で設計が始まっているのである。本機では従来の油圧オレオ式の主脚メカニズムの採用を見送り、かわりにゴムの弾性を応用した独自の主脚メカニズムを開発、採用したのである。まず、主脚柱は断面が楕円形になるように左右張り合わせた一対の金属板から作り、内側は断面に合わせた11枚の金属板と12個のゴムのブロックを交互に重ねて充填したあと、オレオに相当する部分はベークライト製のピストンとロッド式に伸縮する車軸との組合せで仕上げられている。このようにして金属使用を極力抑え、かつ頑丈で軽量な主脚の完成を見たのである。なお、主脚柱横の主脚ドア・ガイドが後掲の後期型のそれと異なっているのに注意。

写真269-270 タイヤが跳ね上げる泥や水などが脚収容口に入るのをふせぐため、主脚タイヤのうしろに取り付けられているマッドガード。

269

270

Photos 176-177: The main gear of the Mosquito prototype. These gear are another of the Mosquito's unique traits. Aircraft landing gear generally use metal quite liberally to meet the needs of being strong enough to handle heavy landings, while not adding too much to the plane's weight. However, one of the main thrusts of the entire Mosquito project was to use as little metal as possible. For this reason, the standard pneumatic oleo design was dropped, and a unique system using rubber blocks as shock absorbers was developed and employed. Note that the door guides seen here differ from those on the later models of the plane.

Photos 269-270: Views of the mudguard installed behind the main tire, designed to help prevent landings on wet, unpaved airstrips from flinging mud and other foreign matter into the gear bays.

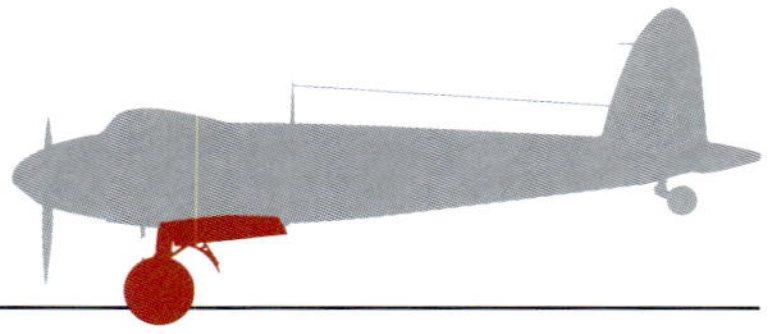

271

写真271、274: 原型機の主脚収容口をみる。大戦中の英軍機の脚収容口の内部はアルミニウム・ドープ、すなわち銀塗装が基準。機内の塗装色であるグレイグリーンやジンクロっぽい色で塗るとおかしなものになってしまう。写真271はエンジン方向をとらえたもので、防火壁直後に設置されているのは潤滑油タンクである。容量は72リッター。なお潤滑油冷却器は主翼前縁のラジエター・ユニット内のエンジンナセル寄りに置かれている。

写真272-273: 主脚収容口後部。ナセル・フェアリングに続く部分で、バルクヘッドでふさがれている。

Photos 271, 274: Views of the gear bays on the prototype. Landing gear bays on British warplanes were generally painted in aluminum dope. In other words, they should be silver. Photo 271 is taken in the direction of the engine, showing the oil tank directly behind the firewall. The tank held 72 liters. The oil cooler is part of the leading edge radiator assembly (the portion closest to the nacelle).

Photos 272-273: The rear portion of the landing gear bays. A bulkhead cuts it at the point where it blends into the engine nacelle behind the trailing edge.

272

273

274

主脚 MAIN GEAR & GEAR BAY

275

写真275-279: B.Mk.35(S/N TA634)の主脚柱およびタイヤ。写真の機体にはグッドイヤー社製タイヤが装着されているが、実際にはダンロップ社製タイプI.G.11が使われている。

Photos 275-279: The main struts and tire of the B. Mk.35(S/N TA634). This aircraft is currently sporting Goodyear tires, but during the war, it would have mounted Dunlop Type I.G.11 tires.

276

277

278

279

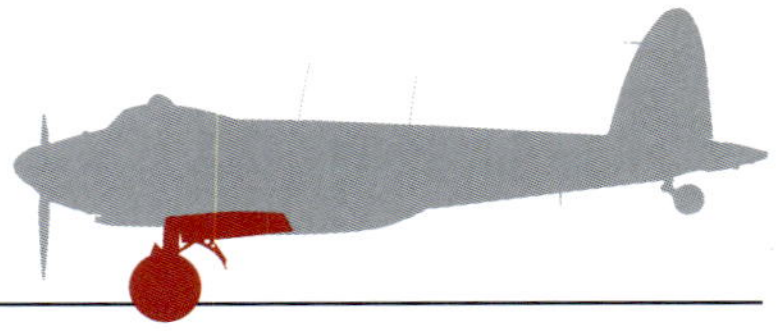

280

281

写真280-281：同じくB.Mk.35の主脚収容口をみる。写真280は真上、写真281は後部バルクヘッド。エンジン防火壁には潤滑油タンクがみえる。

写真282-283：後方よりみた主脚ドアの詳細。

Photos 280-281: Views of the landing gear bays of B.Mk.35 (S/N TA634). Photo 280 is the upper, and photo 281 is the rear bulkhead. The oil tank is visible on the engine firewall.

Photos 282-283: Close-ups of the gear bay doors seen from the rear.

282

283

284

285

写真284-285: デ・ハヴィランド社の機体に共通する独特のアウトラインを継承する垂直安定板および方向舵。構造は主翼と同じく垂直安定板は木製の桁に合板張り、方向舵はアルミ合金製のリブに羽布張りである。

写真287: 原型機の胴体後部に描かれた機体番号。E0234はデ・ハヴィランド社のクラスBの登録番号で、英空軍に受領されたあと、W4050のシリアルをつける1941年1月はじめまでの間、本機に描かれていた。

Photos 284-285: The vertical stabilizer and rudder, in the same general shape as nearly all de Havilland aircraft. The structure is identical to that of the wings. The stabilizer itself is wooden spars and plywood skin, while the rudder is fabric-covered aluminum frame.

Photo 287: The serial number on the rear of the prototype's fuselage. "E0234" is a class-B de Havilland serial number. It was present on the aircraft until it was delivered to the RAF in January of 1941 and re-serialed as "W4050."

写真286: 垂直尾翼前縁に取り付けられているピトー管。ピトー管の原理は、正面に開いた穴からの動圧と、側面に開いた穴からの静圧とによって生じる水銀柱の動きを速度計の針の移動に移して機体速度を表示するための装備で、なるべく気流が乱れない機体位置に取付けられることが多い。

Photo 286: The pitot tube on the leading edge of the vertical stabilizer. In order to provide the most accurate airspeed reading, pitot tubes were located on areas of the aircraft with relatively stable airflow.

286

287

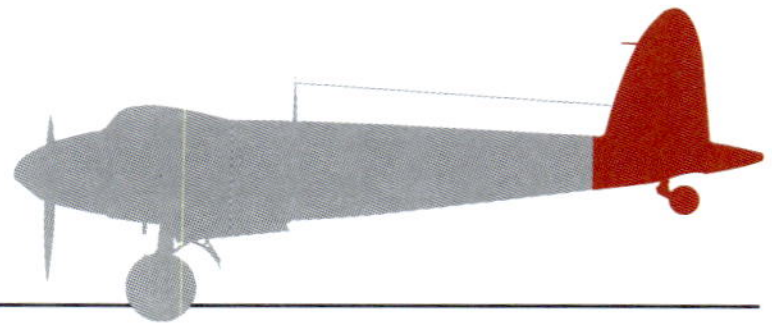

288

290

289

写真288-289：後方よりみたモスキート原型機の尾部。なだらかなカーブを描いて収束するこの尾部コーンによって、美しい紡錘形の胴体アウトラインが完成する。コーンの先端には航法灯が取り付けられている。

写真290：垂直尾翼付根の後方左側にあるふくらみは昇降舵作動リンケージのフェアリング。中央の小さなハッチはトリムジャック点検用。

写真291：同じく右側にあるふくらみは昇降舵内側マスバランスのためのフェアリング。

Photos 288-289: The Mosquito prototype seen from the rear. The aircraft's clean lines are completed nicely by the tail cone at the rear. A navigation light is mounted at the tip of the cone.

Photo 290: The bulge on the left side of the vertical stabilizer at the root is a fairing over the control linkage for the plane's elevator. The small hatch in the center provides access to the trim jack.

Photo 291: The bulge on the right side of the stabilizer is the fairing to clear the elevator's-mass balance.

291

尾部 TAIL

292

293

294

写真292: 原型機の昇降舵／修正タブと尾部コーン部周囲。昇降舵の作動範囲は＋21度30分、－12度30分、修正タブは＋７度30分、－７度30分である。

写真293: 昇降舵はアルミ合金製のリブにアルミ合金皮製である。原型機の水平尾翼(昇降舵をふくむ)の全幅は19フィート６インチ(5.948m)であるが、量産機では20フィート９インチ(6.325m)に拡げられている。

写真294: 尾部コーンは整備などの際には外すことができる。胴体第７バルクヘッドとは6個のネジで結合されている。

写真295: 昇降舵上の修正タブの作動ロッドのフェアリング。

Photo 292: The elevator and trim tab. The elevator itself could move from +21˚30" to –12˚03" while the tab moved 7˚30" in either direction.

Photo 293: The elevator was an all-aluminum construction, both the inner ribs and the skin covering them. The prototype's horizontal stabilizer measured 19 feet 6 inches (5.948m) across, but this was expanded to 20 feet 9 inches (6.325m) in the production version of the plane.

Photo 294: The tail cone could be removed for maintenance purposes. It was latched to the seventh bulkhead.

Photo 295: The fairing for the elevator trim tab control rod on the top of the elevator.

295

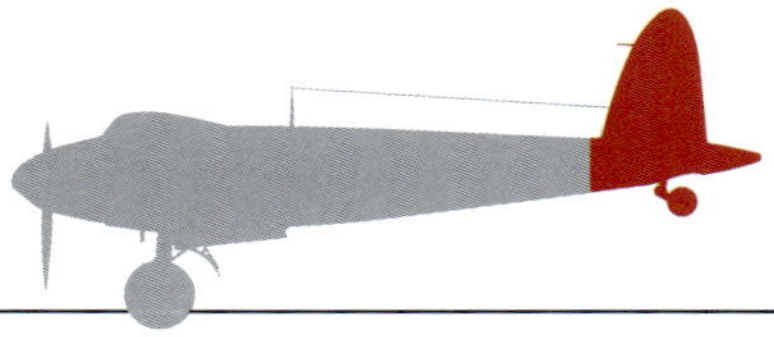

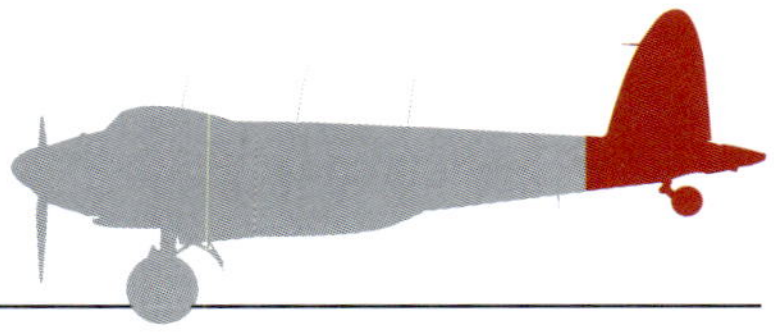

写真296：N.F.Mk.30の尾部コーン。方向舵の作動範囲は左右各26度。

写真297：方向舵修正タブ。作動範囲は左右16度。

写真298-299:取り外し式テイルコーンを外したモスキート尾部のディテール。

Photo 296: The tail cone of the N.F. Mk.30. The rudder could pivot 26° to either side of center.

Photo 297: The rudder trim tab. Its range of motion was 16° to either side.

Photos 298-299: The rear of the tail seen with the tail cone removed.

296

297

298

299

尾部 TAIL

写真300、302-303: モスキート原型機の尾輪のディテール。主脚と同じく、ショックアブソーバーはゴムのブロックを利用した独自のメカニズムが採用されている。フォーク型のタイヤ支柱は地上ではステアリングでき、飛行時には半引込式に胴体尾部に格納される。なお、尾輪の引込みには油圧が使われている。なお、この尾輪はスイベルはしてもロック機構はないようで、直進位置にする場合には数m前進させる必要があった。

Photos 300, 302-303: Close-ups of the tail gear on the Mosquito prototype. Like the main gear, it also uses rubber blocks as shock absorbers. The tail wheel could be steered when the plane was on the ground, and half-retracted into the fuselage in flight. The retraction mechanism was hydraulically powered. Although it could be steered, there was no lock mechanism to hold it straight.

300

301

302

303

写真301: 尾輪タイヤは断面が中央部が凹んだ、いわゆるシミー・ダンパーが付けられている。このシミー・ダンパーとは、地上タキシング、あるいは滑走中に尾輪(前輪式機体の場合は前輪)に生じる横揺れ現象(シミー)をおさえるためのもの。

Photos 301: Note the indented cross-section of the tail wheel. This is a "shimmy damper" arrangement to prevent excessive vibration (shimmy) of the aircraft while taxiing.

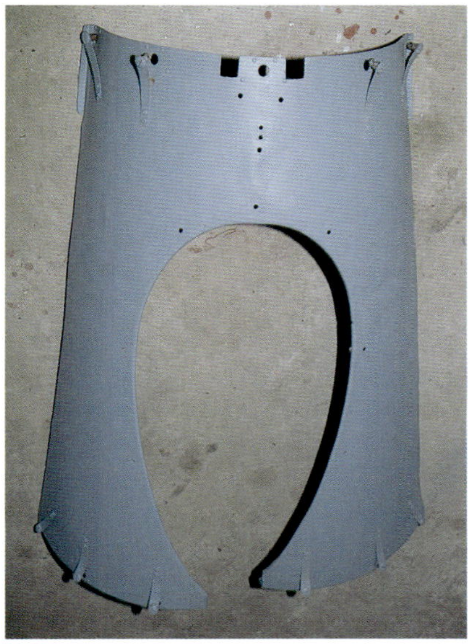

304

305

写真304-305: B.Mk35(S/N TA634)の引込式尾輪点検用着脱パネル。

Photos 304-305: Retracting tail gear access panel of B.Mk.35(S/N TA634).

306

307

308

写真306: F.B.Mk.Ⅵの尾輪タイヤに付けられたシミー・ダンパー。このシミー・ダンパーは米陸軍機のムスタングやサンダーボルトといった機体にもみられるが、このモスキートのようにタイヤ中央部と左右端とがはっきりとしたものではなく、なだらかに凹んだような断面形状になっている。

写真307: F.B.Mk.Ⅵの尾輪メカニズム。レストア中のために通常では見られない機構の詳細が理解できる。また、このメカニズム自体は第7バルクヘッドに取り付けられている。

写真308: 尾輪タイヤの上には主脚と同じくマッドガードが取り付けられている。ただし、胴体尾部内に隠れているので地上姿勢時には外側からは視認できない。写真はB.Mk.35(S/N TA634)。

Photo 306: The shimmy damper on the tail wheel. This type of system was also used on U.S. aircraft such as the Mustang and Thunderbolt, but those aircraft don't have such a pronounced depression in the tire, rather a more smooth cross-section.

Photo 307: The tail gear mechanism of F.B. Mk.VI. Usually hidden structures are well displayed in this photo. This structure was also fastened to the seventh bulkhead.

Photo 308: Like the main gear, a mudguard is installed above the tail wheel, too. However, it is located inside the tail gear bay and cannot be seen when the aircraft is on the ground.

武装 ARMAMENT

309

310

写真309-310: 爆撃機型モスキートが多用した4,000ポンド爆弾。ブロックバスター爆弾、あるいはその形状からクッキーとも呼ばれた大型爆弾で、ランカスター、ハリファックス、ウェリントン、そしてこのモスキートだけが搭載することができた。正面にみえる３個の突起は信管。大戦末期の1945年の５カ月間だけでも約３千発もの4,000ポンド爆弾がモスキートによって投下されている。

写真311-313: 爆撃機型、戦闘爆撃機型で汎用された各種爆弾。当初、通常の500ポンド爆弾は２発しか爆弾倉に収容できないと思われていたが、爆弾後部の安定フィンを縮めれば４発収容できることが判明。伸縮するフィンなども計画されたが、フィンを切り詰めただけでも爆弾の安定性に問題がないと分かったため、写真のように安定フィンを短くした爆弾がモスキートでは使用された。

Photos 309-310: Views of the 4,000-lb. "cookie" bomb, used with great frequency by Mosquitoes. The "blockbuster" and "cookie" bombs could only be carried by Lancaster, Halifax, Wellington or Mosquito bombers. The three projections on the nose of the bomb are the fuses. In the final five months of the war, Mosquitoes dropped more than 3,000 of these massive weapons.

Photos 311-313: The multi-purpose 500-pound bomb carried frequently by bomber and fighter-bomber versions of the Mosquito. It was originally thought that only two of these bombs would fit in the bomb bay of the Mosquito, but tests should that if the fins were folded or reduced in size, four could be carried. As such, designs for bombs with collapsible fins were drawn up, but further tests revealed that reduced-size fins, such as those seen here, didn't harm the bomb's stability while falling, so such weapons became the norm.

311

312

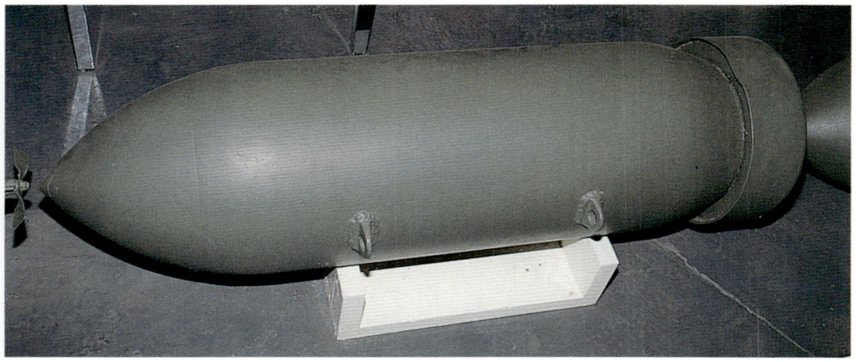

313

314

写真314: F.Mk.Ⅱ、F.B.Mk.Ⅵなどの機首に装備されたブローニングMk.Ⅱ7.7mm機銃。小口径で炸薬を封入できないため威力不足と思われがちだが、反面、発射速度が高く、一度に多量の弾丸を標的に集中でき、近接戦闘ではなかなか有効な武装であった。モスキートでは各銃に500発が装弾された。

写真315: 戦闘機型の標準火器となったイスパノMk.Ⅱ20mm機関砲。本機に限らず英戦闘機の多くで採用された火器である。もちろんベルト式給弾で装弾数は150発。

写真316-318: 対艦船攻撃機型のモスキートF.B.Mk.XⅧが胴体下面に装備した57mm砲。もともとは英陸軍の6ポンド対戦車砲で、装弾数は25発、自動装填式で20秒間で全弾発射できた。この自動装填機構を開発したのが、煙草製造機械を作っていたモリンズという人だったため、別名モリンズ砲とも呼ばれている。

写真319-320: 大戦後半のモスキートの主要武器のひとつとなった60ポンドロケット弾。弾頭部は作戦任務に応じて、通常、高性能炸薬、徹甲などの弾頭が選ばれた。モスキートでは主翼下面のランチャーに計8発を吊下できた。この8発を一斉発射するとその威力は巡洋艦クラスの全主砲の斉射に匹敵するといわれている。

315

316

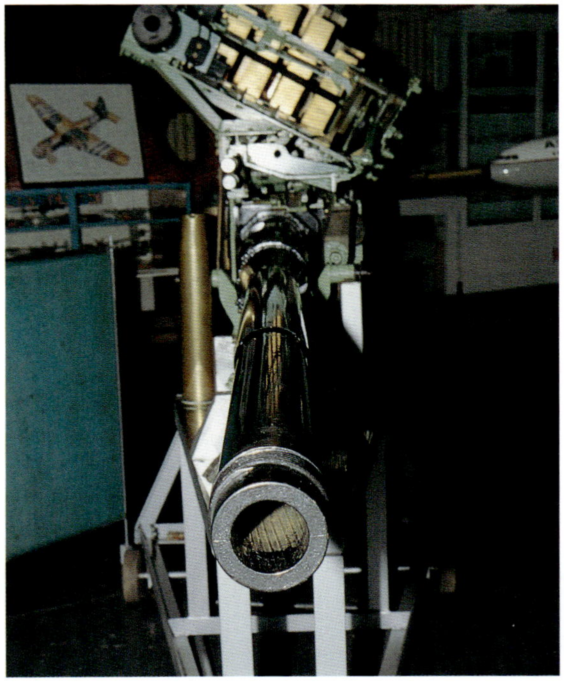
317

318

Photo 314: The Browning Mk.II 7.7mm MG that was standard nose armament in the F. Mk.II, F.B. Mk.VI and other variants. Though often considered an under-powered weapon, the high muzzle velocity and rapid rate of fire made it possible to pour large numbers of rounds into an enemy accurately, particularly at close range. In the Mosquito, each gun could carry 500 rounds.

Photo 315: The Hispano Mk.II 20mm cannon, which was standard armament on the fighter versions of the Mosquito, as well as numerous other British warplanes of WWII. It was belt fed and could carry 150 rounds.

Photos 316-318: The 57mm cannon that the anti-shipping F.B. Mk.XVIII mounted under its fuselage. This weapon was originally the 6-pound anti-tank gun. It carried 25 rounds and could fire them all in 20 seconds by means of an auto-loading system. The gun was also sometimes referred to as the Molins gun after its creator.

Photos 319-320: One of the Mosquito's main weapons in the latter half of the war was these 60-lb. rockets. Standard, high-explosive and armor-piercing rounds were available depending on the mission at hand. The Mosquito carried up to eight rockets in launchers under the wings. If all eight rockets were fired simultaneously, the effect on the target was said to be similar to that of taking a direct hit from a gun salvo of a cruiser-class naval vessel.

319

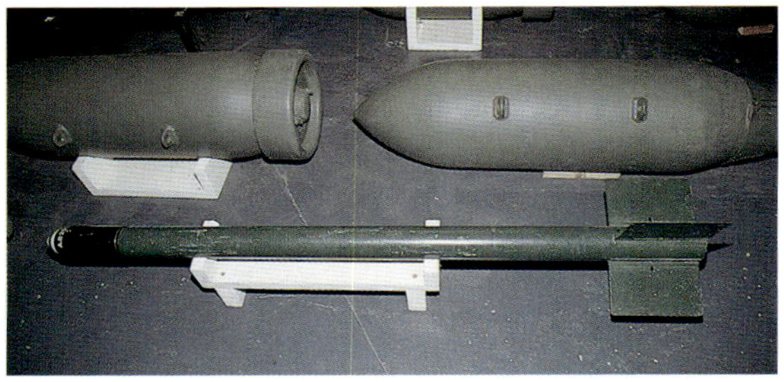
320

モスキートの基本塗装&マーキング

作図／カルミネ・ディ・ナポリ、ラファエレ・マンチーニ　解説／ラファエレ・マンチーニ
Illustrations by Carmine Di Napoli & Raffaele Mancini, Commentary by Raffaele Mancini

Camouflage & Markings of the MOSQUITO

昼間爆撃型モスキートの基本迷彩（側面図）
Day Bomber Basic Camouflage (Side View)

図はシリアルナンバーのオリジナルサイズおよび記入位置を示す。しかし、第二次大戦中の機体では、ごくまれに大きさ、記入位置ともに異なる例がみられる。
Note that these were the original dimensions and specifications, but they were rarely followed. One can easily find WWII photos documenting both different dimensions and marking locations.

Fuselage left side
胴体左側面

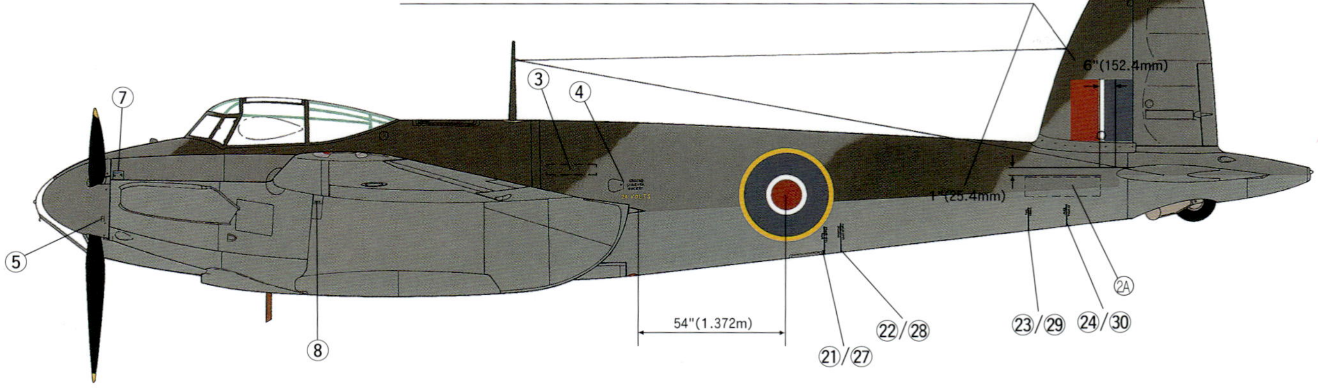

Fuselage right side
胴体右側面

③（ごくまれだがこの位置に記入されている場合もあり）
This stenciling was seldom marked.

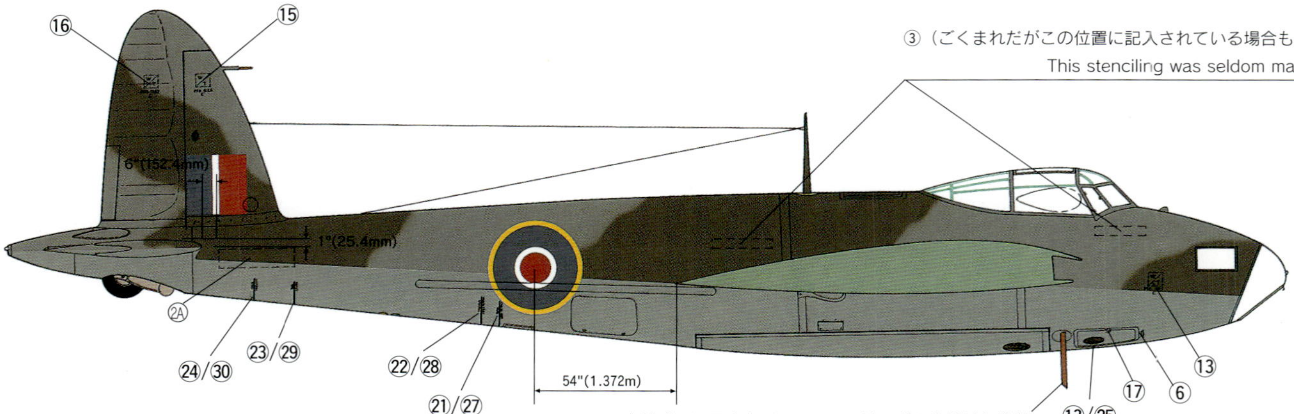

曳航式アンテナカバー・フェアリングの先端は無塗装
No paint over the trailing antenna fairing cuff.

主翼ラウンデル　Wing Roundels

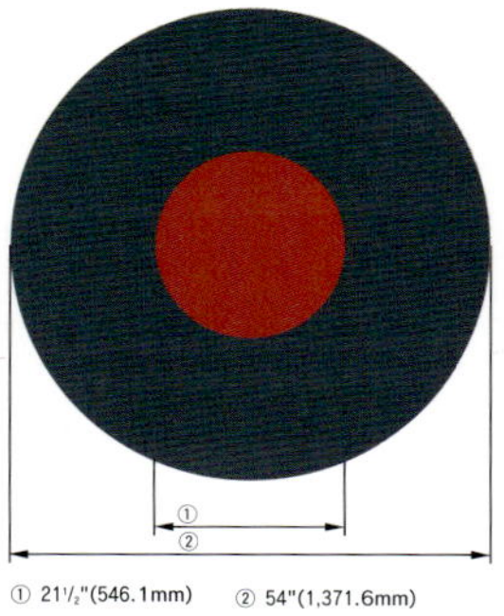

タイプB　Type B

タイプBは1945年1月にタイプCに置き換えられるまで、プロトタイプとすべての生産型モスキートで使用された。直径は54インチ、色は青（ダルブルー、外径54インチ／1,371.6mm）と赤（ダルレッド、直径21.5インチ／546.1mm）の組みあわせ。なお、生産型モスキートの主翼下面には国籍標識は描かれていない。

Upper wing Type B roundels were standard on prototypes and all production Mosquito variants until they were replaced by Type C roundels in early January 1945. 54in diameter, Dull Blue (54in - 1,371.6mm) and Dull Red (21.5in - 546.1mm). Production Mosquitos carried no insignia on the underside of the wings.

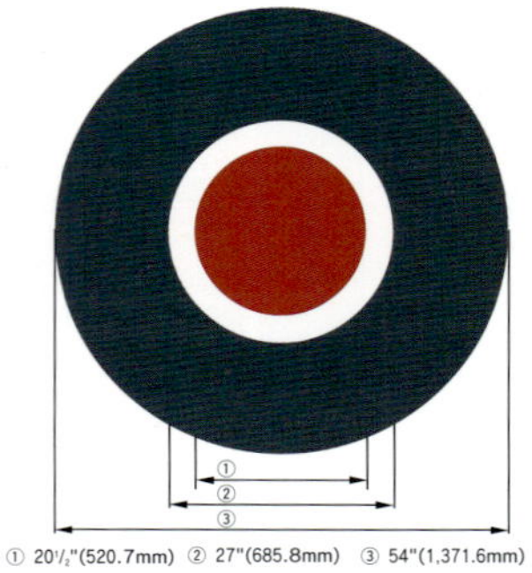

タイプC　Type C

タイプCはタイプBにかわって、1945年1月5日付でまず第2戦術空軍のモスキートに使用され、同月7日からすべての部隊のモスキートがこれにならった。このタイプのラウンデルは終戦後も1947年まで使用されている。直径は54インチ、色は青（外径54インチ／1,371.6mm）、白（外径27インチ／685.8mm）と赤（直径20.5インチ／520.7mm）の組みあわせ。

Upper wing Type B roundels were replaced by Type C from January 5, 1945, on Mosquitos attached to 2nd Tactical Air Force, and from January 7, 1945 on Mosquitos of all other Commands. This type of roundel was in use until 1947. 54in diameter, Dull Blue (54in - 1,371.6mm), White (27in - 685.8mm) and Dull Red (20.5in - 520.7mm).

昼間爆撃型モスキートの基本迷彩(平面図)
Day Bomber Basic Camouflage (Top & Bottom View)

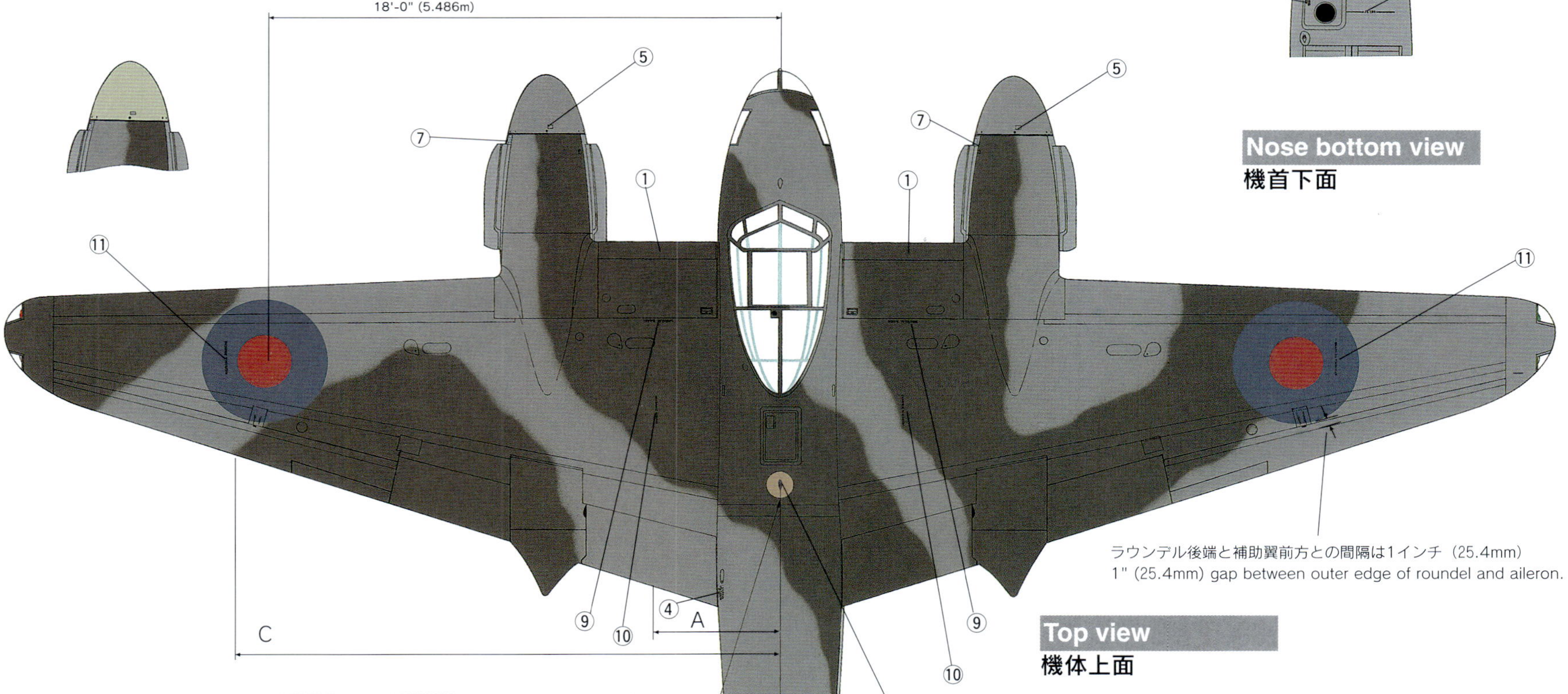

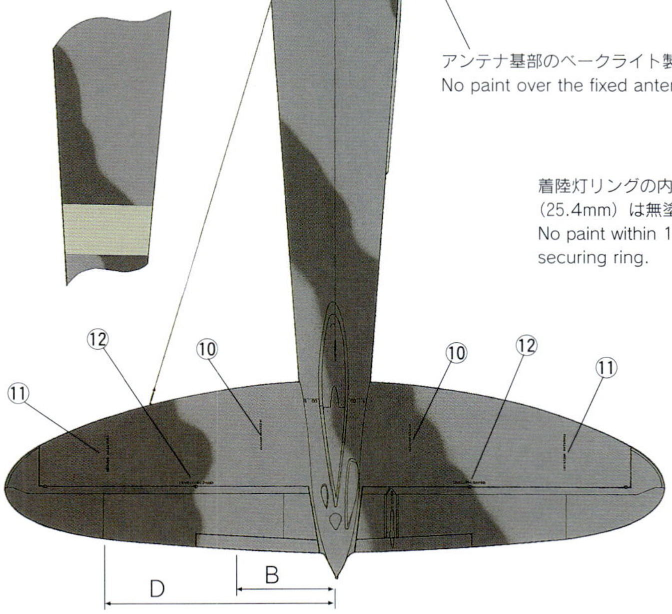

Top view
機体上面

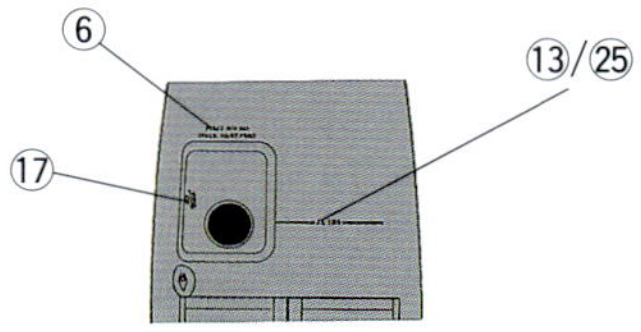

Nose bottom view
機首下面

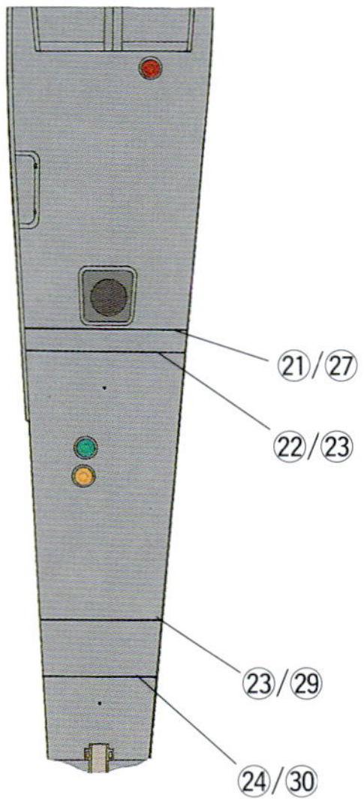

Rear fuselage Top & bottom view
胴体後部上、下面

着陸灯リングの内側1インチ(25.4mm)は無塗装
No paint within 1" of landinglight securing ring.

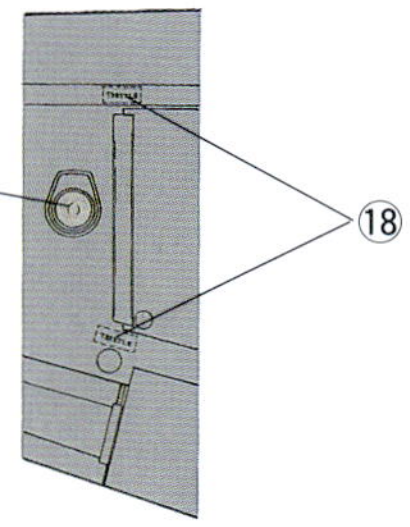

Right wing bottom view
右主翼外翼下面

第二次大戦当時の写真でモスキートに一般的にみられるマーキングとステンシルの例。機体表面が黒で塗装されている場合は赤（ダルレッド）で記入された。

Most Mosquito airframes observable in WWII photographs show the common use of the following main stenciling and markings. Note that stenciling and markings located on black-painted surfaces were usually in Dull Red.

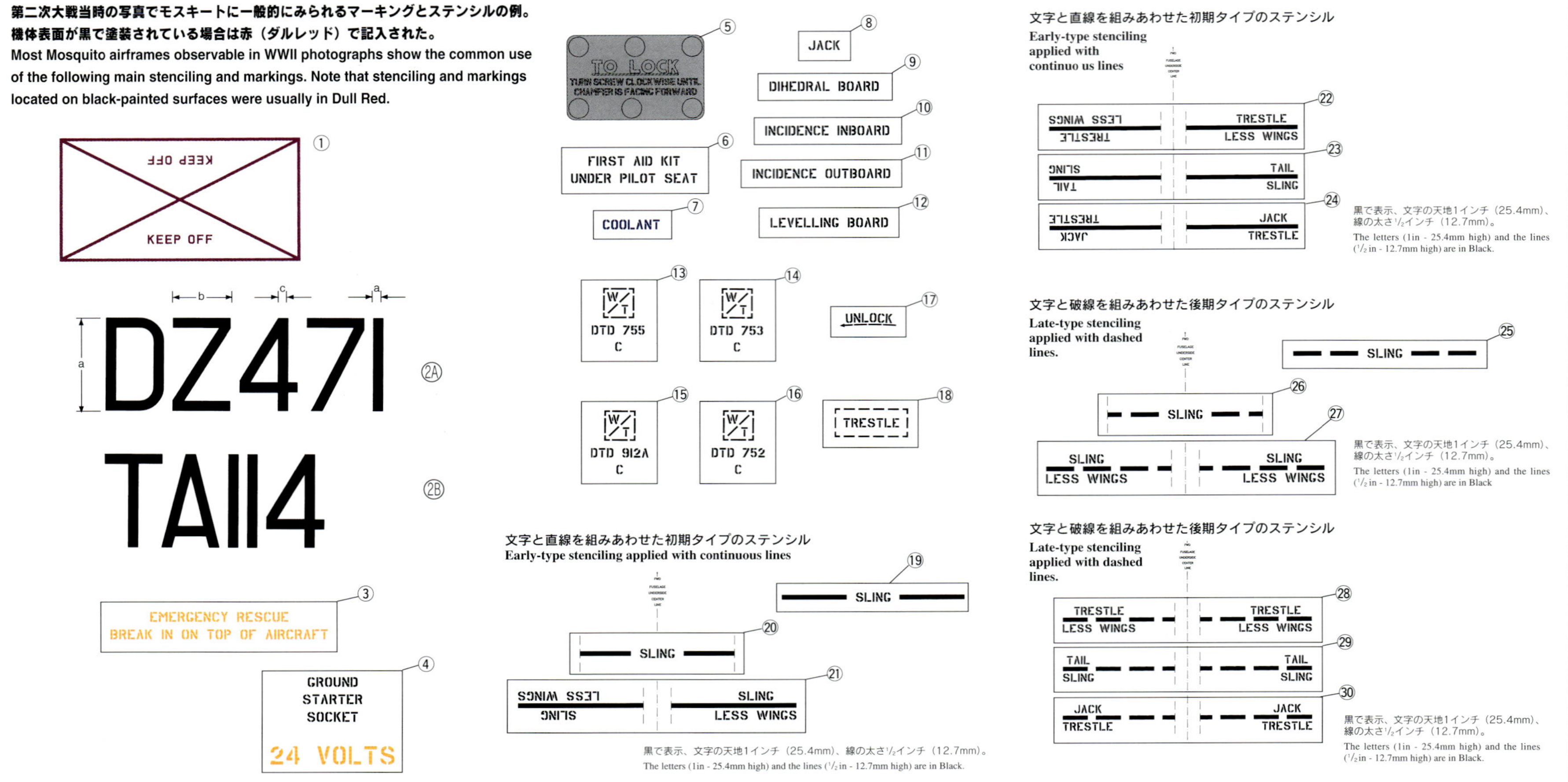

① ラジエター点検ハッチの注意表示
実戦部隊では消されている場合が多い。

② A／B　シリアル・ナンバー
a=8インチ／203.2mm
b=5インチ／127mm
c=1インチ／25.4mm
d=1インチ／25.4mm
機体の迷彩パターンにより、黒または赤で記入。

③ 非常用救命具に関する注意表示

④ 地上スターター電源ソケット接続部の表示
黒の文字は天地1インチ（25.4mm）、黄色の文字は天地2インチ（50.8mm）。

⑤ ネジ式ロックに関する注意表示
アルミニウム製のプレートにロックの方法を表示。

⑥ 救急用応急処置セット収容部の表示
文字は黒、天地1インチ（25.4mm）。

⑦ 冷却液注入ハッチのマーキング
文字は青、天地1インチ（25.4mm）。

⑧ ジャッキ・アップ・ポイントの表示
着脱式アクセスハッチを外したあとのジャッキ・アップ・ポイントをあらわす。文字は黒、天地1インチ（25.4mm）。

⑨ 主翼上反角確認位置の表示
前方主桁の上反角確認位置をあらわす。上反角は前方主桁上面で1度24分±0度10分。文字は黒、天地1インチ（25.4mm）。

⑩ 主翼および水平尾翼迎え角確認位置に関する表示
主翼および水平尾翼付け根側の迎え角確認位置をあらわす。主翼は第2リブ位置に記入。主翼の迎え角は1度30分±0度10分。尾翼は第3リブ位置に記入。尾翼には迎え角がつかず、中立。A=4ft-4.56in／1.335m、B=2ft-10.20in／0.869m。文字は黒、天地1インチ（25.4mm）。

⑪ 主翼および水平尾翼迎え角確認位置に関する表示
主翼および水平尾翼翼端側の迎え角確認位置をあらわす。主翼は第10リブ位置に表示。C=19ft-0.22in／5.797m、D=7ft-6.75in／2.305m。文字は黒、天地1インチ（25.4mm）。

⑫ 胴体水平位置確認に関する表示
水平尾翼後桁で胴体の水平を確認する位置をあらわす。文字は黒、天地1インチ（25.4mm）。

⑬ 爆撃機型と写真偵察機型機首右側のステンシル
黒で表示、文字の天地1インチ（25.4mm）。

⑭ 戦闘機型と戦闘爆撃機型機首右側のステンシル
黒で表示、文字の天地1インチ（25.4mm）。

⑮ 垂直安定板右上方のステンシル
黒で表示、文字の天地1インチ（25.4mm）。

⑯ ラダー右上方のステンシル
黒で表示、文字の天地1インチ（25.4mm）。

⑰ 爆撃機型と写真偵察機型搭乗員乗降ハッチノブの表示
文字は黒、まれに黄色。文字の天地1インチ（25.4mm）。

⑱ 主翼架台（トレッスル）使用位置表示
主桁前桁／後桁の、主翼を架台にのせる部分をあらわす。文字の天地1インチ（25.4mm）。

⑲ 爆撃機型と写真偵察機型の胴体吊上位置表示
胴体第2隔壁部分に記入。

⑳ 戦闘機型と戦闘爆撃機型の胴体吊上位置表示
胴体第2隔壁部分に記入。

㉑ 胴体吊上位置表示
胴体第5隔壁部分に記入。

㉒ 胴体架台（トレッスル）使用位置表示
胴体第5隔壁に記入。

㉓ 胴体ジャッキ・アップ位置表示
胴体第6隔壁に記入。

㉔ 胴体ジャッキ・アップおよび架台（トレッスル）使用位置表示
胴体第6隔壁に記入。

㉕ 爆撃機型と写真偵察機型の胴体吊上位置表示
胴体第2隔壁部分に記入。

㉖ 戦闘機型と戦闘爆撃機型の胴体吊上位置表示
胴体第2隔壁部分に記入。

㉗ 胴体吊上位置表示
胴体第5隔壁部分に記入。

㉘ 胴体架台（トレッスル）使用位置表示
胴体第5隔壁に記入。

㉙ 胴体ジャッキ・アップ位置表示
胴体第6隔壁に記入。

㉚ 胴体ジャッキ・アップおよび架台（トレッスル）使用位置表示
胴体第6隔壁に記入。

① **Radiator Inspection Panel Caution Marking**
Usually Squadrons removed these red markings for war operations.

② **AB.Serial Number**
Examples of layout and dimensions.
a = 8 in - 203.2mm
b = 5 in - 127mm
c = 1 in - 25.4mm
d = 1 in - 25.4mm
The lettering is usually either in Black or in Dull Red, depending on the camouflage scheme.

③ **Emergency Rescue Marking**
The letters are in Yellow (1 in - 25.4mm high).

④ **Ground Starter Socket Marking**
The letters are in Black (1 in - 25.4mm high). The lettering "24 Volts" is in Yellow (2 in - 50.8mm high)

⑤ **Screw Lock Procedure Marking**
Aluminum plate on which the procedure for locking the screws is explained.

⑥ **First Aid Kit Stowage Marking**
The letters are in Black (1 in - 25.4mm high).

⑦ **Coolant Fill Marking**
The letters are in Blue (1 in - 25.4mm high).

⑧ **Lifting Jack Marking**
Location point for jacking after removing access hatch. The letters are in Black (1 in - 25.4mm high).

⑨ **Wing Dihedral Board Marking**
Location point for rigging board on front main spar. Dihedral on top surface of front main spar 1°24" ± 0°10".
The letters are in Black (1in - 25.4mm high).

⑩ **Wing and Tail Plane Incidence Inboard Marking.**
Wing location point for rigging board on rib no.2. Main plane incidence 1°30" ± 0°10".
Tail plane location point for rigging board on rib no.3. Tail plane incidence neutral.
A = 4ft-4.56in - 1.335m
B = 2ft-10.20in - 0.869m
The letters are in Black (1in - 25.4mm high).

⑪ **Wing and Tail Plane Incidence Outboard Marking**
Wing plane location point for rigging board on rib no.10.
Tail plane location point for rigging board on rib no.9.
C = 19ft-0.22in - 5.797m
D = 7ft-6.75in - 2.305m
The letters are in Black (1in - 25.4mm high).

⑫ **Fuselage Levelling Board Marking**
Tail plane location point for fuselage rigging board on rear spar.
The letters are in Black (1in - 25.4mm high).

⑬ **Location point on right side of the nose (bombers and photo-reconnaissance aircraft)**
The lettering is in Black (1in - 25.4mm high).

⑭ **Location point on right side of the nose (fighters and fighter-bombers)**
The lettering is in Black (1in - 25.4mm high).

⑮ **Location point on right side of the vertical stabilizer**
The lettering is in Black (1in - 25.4mm high).

⑯ **Location point on right side of the rudder**
The lettering is in Black (1in - 25.4mm high).

⑰ **Door Release Marking (bombers and photo-reconnaissance aircraft)**
The letters are in Black, sometimes in Yellow (1in - 25.4mm high).

⑱ **Trestling Point Marking**
Location points on front and rear main spars. The letters are in Black (1in - 25.4mm high).

⑲ **Fuselage Slinging Marking**
Location point at bulkhead no.2 (bombers and photo-reconnaissance aircraft).

⑳ **Fuselage Slinging Marking**
Location point at bulkhead no.2 (fighters and fighter-bombers).

㉑ **Fuselage Slinging Marking**
Location point around bulkhead no.5.

㉒ **Fuselage Trestling Marking**
Location point around bulkhead no.5.

㉓ **Fuselage Jacking Marking**
Location point around bulkhead no.6.

㉔ **Fuselage Jacking and Trestling Marking**
Location point around bulkhead no.6.

㉕ **Fuselage Slinging Marking**
Location point at bulkhead no.2 (bombers and photo-reconnaissance aircraft).

㉖ **Fuselage Slinging Marking**
Location point at bulkhead no.2 (fighters and fighter-bombers).

㉗ **Fuselage Slinging Marking**
Location point around bulkhead no.5.

㉘ **Fuselage Trestling Marking**
Location point around bulkhead no.5.

㉙ **Fuselage Jacking Marking**
Location point around bulkhead no.6.

㉚ **Fuselage Jacking and Trestling Marking**
Location point around bulkhead no.6.

夜間戦闘機型モスキートの基本迷彩（側面図）
Night Fighter Basic Camouflage (Side View)

図はシリアルナンバーのオリジナルサイズおよび記入位置を示す。しかし、第二次大戦中の機体をみると、ごくまれに大きさ、記入位置ともにことなる例がみられる。

Note that these were the original dimensions and specifications, but they were rarely followed. One can easily find WWII photos documenting both different dimensions and marking locations.

③（ごくまれだがこの位置に記入されている場合もあり）
This stenciling was very rarely painted.

Fuselage right side
胴体右側面

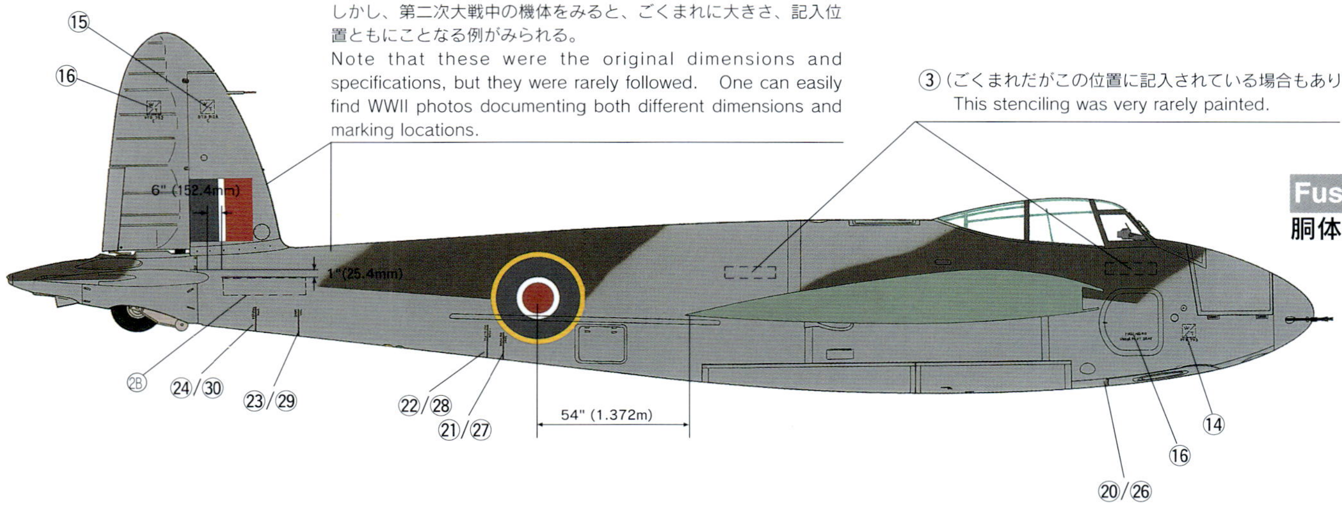

胴体ラウンデル　Fuselage Roundels

タイプA1　Type A1

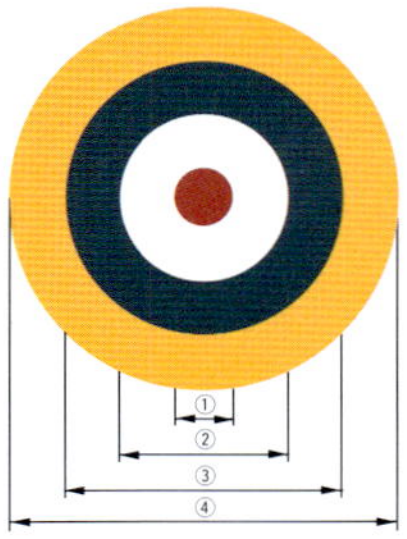

① 6"(152.4mm)
② 18"(457.2mm)
③ 30"(762mm)
④ 42"(1,066.8mm)

モスキート原型機とF./N.F.Mk.II初期生産分が1942年4月30日まで使用した胴体ラウンデル。直径は42インチ。色は黄（トレーナーイエロー、外径42インチ／1,066.8mm）、青（ダルブルー、外径30インチ／762mm）、白（外径18インチ／457.2mm）と赤（ダルレッド、直径6インチ／152.4mm）の組みあわせ。

First prototypes and early production F./N.F.Mk.IIs carried Type A1 fuselage roundels until April 30, 1942. 42in diameter, Trainer Yellow (42in - 1,066.8mm), Dull Blue (30in - 762mm), White (18in - 457.2mm) and Dull Red (6in - 152.4mm).

タイプC1　Type C1

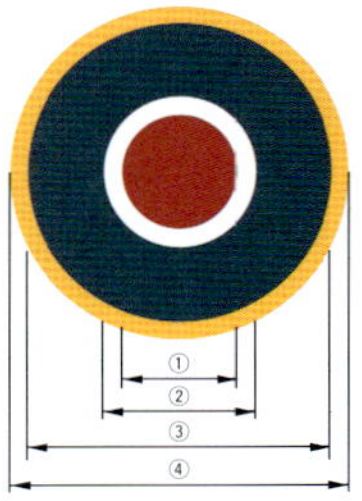

① 12"(304.8mm)　③ 32"(812.8mm)
② 16"(406.4mm)　④ 36"(914.4mm)

F./N.F.Mk.II後期生産分とすべてのモスキート（写真偵察型は除く）が1942年4月30日から第二次大戦中に使用。直径は36インチ、色は黄（外径36インチ／914.4mm）、青（外径32インチ／812.8mm）、白（外径16インチ／406.4mm）と赤（直径12インチ／304.8mm）の組みあわせ。

Late production F./N.F.Mk.IIs and all other Mosquito variants, except P.R. variants, carried Type C1 fuselage roundels during WWII, beginning April 30, 1942. 36in diameter, Trainer Yellow (36in - 914.4mm) Dull Blue (32in - 812.8mm), White (16in - 406.4mm) and Dull Red (12in - 304.8mm).

タイプB　Type B

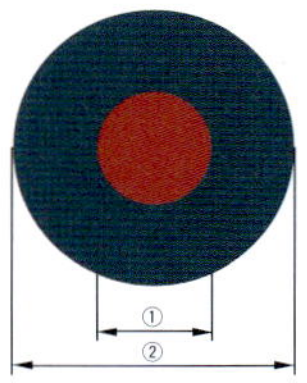

① 12"(304.8mm)
② 30"(762mm)

第2次大戦中のモスキートのうち、写真偵察型のみが使用した胴体ラウンデル。直径は30インチ、色は青（外径30インチ／762mm）と赤（直径12インチ／304.8mm）の組み合わせ。

Only Mosquito P.R. variants had this type of roundel marked on the fuselage sides during WWII. 30in diameter, Dull Blue (30in - 762mm) and Dull Red (12in - 304.8mm).

フィン・フラッシュ　Fin Flashes

Ⓐ

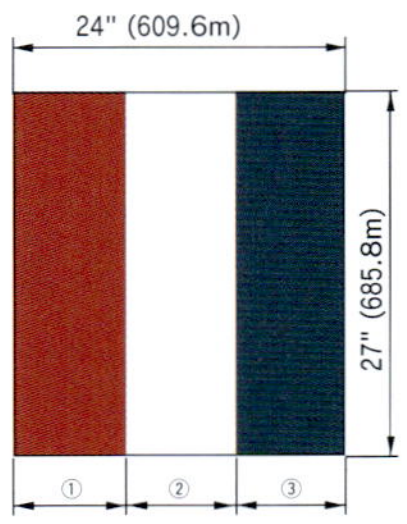

①～③ 8"(203.2mm)

Ⓑ

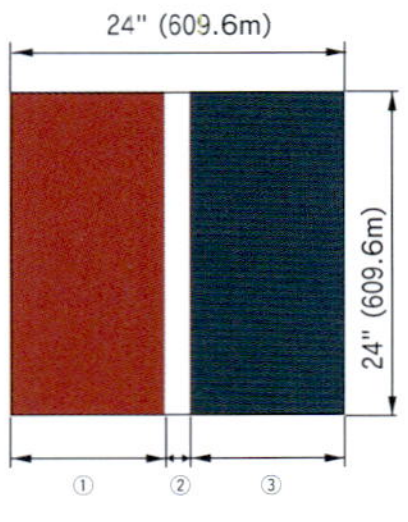

①③ 11"(279.4mm)
② 2"(50.3mm)

Ⓒ

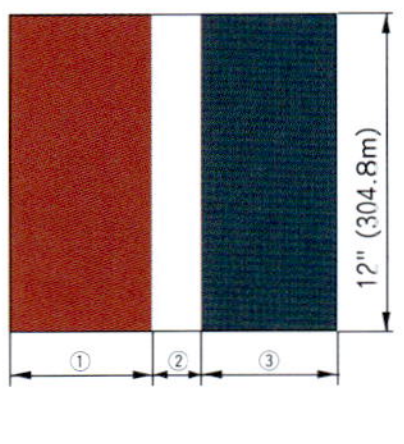

①③ 5"(127.4mm)
② 2"(50.8mm)

Ⓐ **24インチ x 27インチ(609.6 x 685.8mm)タイプ**
高さ27インチ（685.8mm）、幅8インチ（203.2mm）の赤、白、青のストライプを組みあわせたフィン・フラッシュ。モスキート原型機とF./N.F.Mk.II初期生産分が1942年4月30日まで使用。

Ⓑ **24インチ x 24インチ(609.6 x 609.9mm)タイプ**
F./N.F.Mk.II後期生産分と、以降すべてのモスキートが1942年4月30日から使用したフィン・フラッシュ。赤と青のストライプが高さ24インチ（609.6mm）、幅11インチ（279.4mm）、白のストライプのみ幅が2インチ（50.8mm）。

Ⓒ **12インチx12インチ(304.8 x 304.8mm)タイプ**
写真偵察機型モスキートのみが使用した小型のフィン・フラッシュ。1944年初頭からP.R.Mk.XVIとP.R.Mk.34に用いられた。垂直安定板に記入。赤と青のストライプが高さ12インチ（304.8mm）、幅5インチ（127.4mm）、白のストライプのみ幅が2インチ（50.8mm）。

Ⓐ **24in x 27in (609.6 x 685.8mm)**
First prototypes and early production F./N.F.Mk.IIs carried 24in x 27in high flashes until April 30, 1942. Dull Red, White and Dull Blue equal width 8in (203.2mm) stripes, each 27in (685.8mm) high.

Ⓑ **24in x 24in (609.6 x 609.6mm)**
Late production F./N.F.Mk.IIs and all other Mosquito variants carried 24in x 24in high flashes during the war, beginning April 30, 1942. Dull Red and Dull Blue 11in (279.4mm) wide, White 2in (50.8mm) wide, each 24in (609.6mm) high.

Ⓒ **12in x 12in (304.8 x 304.8mm)**
Only Mosquito P.R. variants had this smaller-type Fin Flash marked on the vertical stabilizer, in particular P.R.Mk.XVIs and P.R.Mk.34s beginning early 1944. Dull Red and Dull Blue 5in (127.4mm) wide, White 2in (50.8mm) wide, each 12in (304.8mm) high.

FS595A 33538
トレーナーイエロー
Trainer Yellow

FS595A 30109
ダルレッド
Dull Red

FS595A 35044
ダルブルー
Dull Blue

FS595A 36118
ダークシーグレイ
Dark Sea Grey

FS595A 36187
オーシャングレイ
Ocean Grey

夜間戦闘機型モスキートの基本迷彩（平面図）
Night Fighter Basic Camouflage (Top & Bottom View)

Top view
機体上面

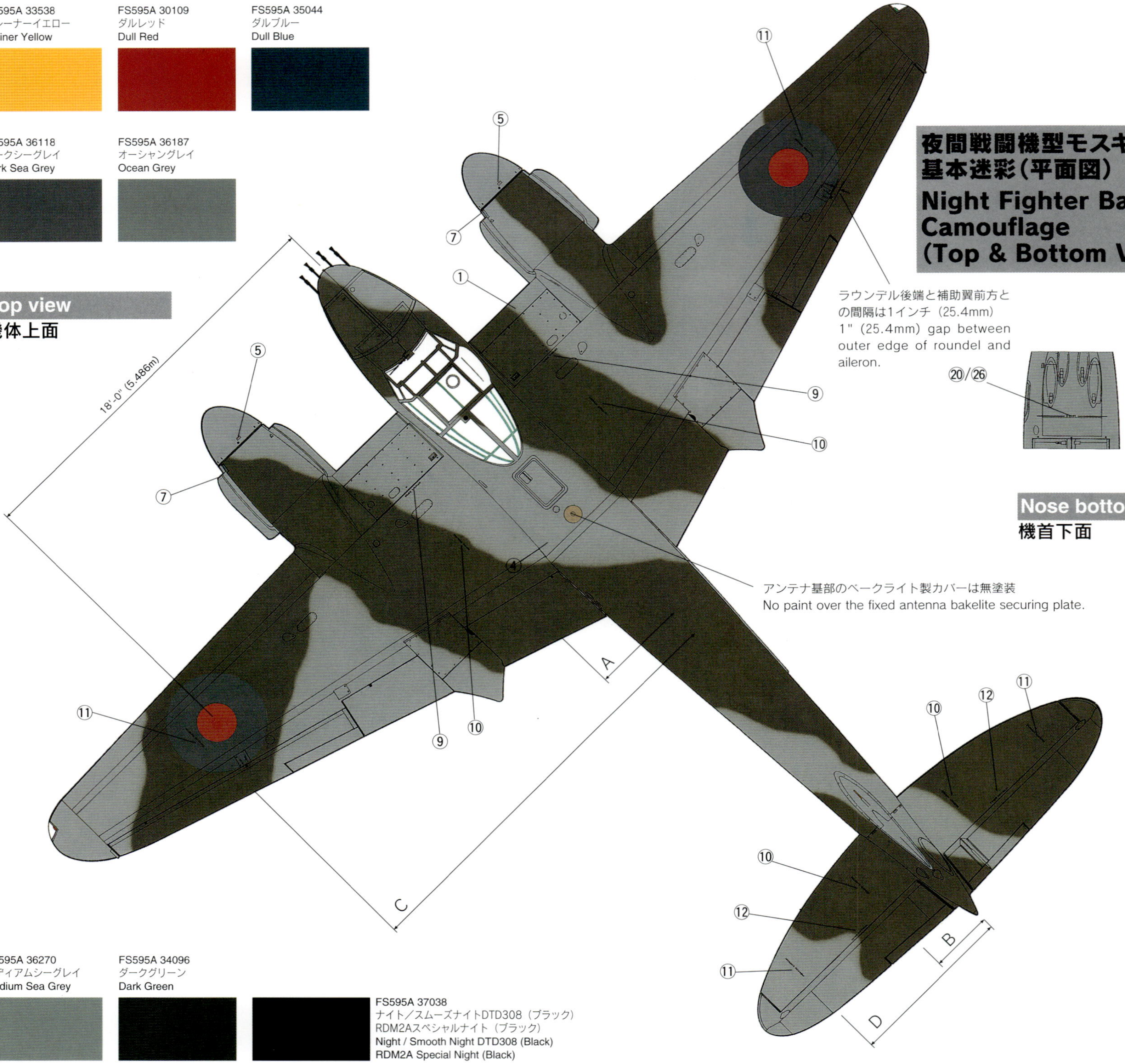

Nose bottom view
機首下面

FS595A 36270
ミディアムシーグレイ
Medium Sea Grey

FS595A 34096
ダークグリーン
Dark Green

FS595A 37038
ナイト／スムーズナイトDTD308（ブラック）
RDM2Aスペシャルナイト（ブラック）
Night / Smooth Night DTD308 (Black)
RDM2A Special Night (Black)

FS595A 30118
ダークアース
Dark Earth

FS595A 36463
スカイグレー
Sky Grey

FS595A 34159
ダークスレートグレー
Dark Slate Grey

FS595A 35164
PRUブルー
PRU Blue

FS595A 35183
アズールブルー
Azure Blue

FS595A 34424
スカイ
Sky

FS595A 34226
グレイグリーン（プライマー）
Grey Green (Primer)

モスキートの塗装およびマーキング例

作図／カルミネ・ディ・ナポリ　解説／ラファエレ・マンチーニ、ステファニア・グイドリッツィ
Illustrations by Carmine Di Napoli, Commentary by Raffaele Mancini & Stefania Guidorizzi

Actual Mosquito Marking Examples

1/72 scale

昼間爆撃機の迷彩塗装
Day Bomber Camouflage Scheme

昼間戦闘任務、沿岸航空軍団および爆撃機型のモスキートは、1942年3月6日付の指令により、上面ダークグリーン／オーシャングレイ、下面ミディアムシーグレイの基本塗装が採用されることになった。シリアル・ナンバーは黒、プロペラも黒で塗られ、その先端4インチがトレーナーイエローに塗られた。爆撃機型の上面グレイ／グリーン迷彩のパターンは戦闘機型のパターンを左右反転した鏡像となっている。また、初期に配備されたB.Mk.IVには、非武装の機体を戦闘機型に欺瞞するため、スピナーと機体コードをスカイに、また後部胴体にも戦闘機を示す幅18インチ（457.2mm）の帯が塗装されていた。1942年3月6日以前のB.MK.IV シリーズ I は、はじめは工場のスタンダード迷彩である上面ダークグリーン／ダークアースに下面スカイ。機体コードはスカイグレーで記され、のちに下面色はミディアムシーグレーに変更された。なお、1942年12月1日付でスピナーの色はオーシャングレイに、機体コードの色は赤（ダルレッド）に変更されている。**（イラスト1-4）**

Following the instructions issued on March 6, 1942, which also involved aircraft with a day role in Fighter and Coastal Command, bombers were to be painted in the standard finish of Dark Green and Ocean Grey upper surfaces and Medium Sea Grey undersurfaces. The serials were Black. The propellers were Black with 4in Trainer Yellow tips. The bombers had the grey/green camouflage scheme of the upper surfaces reversed from that of the fighters. In addition, with the intention of fooling the Luftwaffe into thinking that unarmed variants were armed day fighters, the early-delivered B.Mk.IVs were to have spinners, codes and 18in (457.2mm) wide rear fuselage bands painted in Sky. Before March 6, 1942, B.Mk.IV Series Is had been painted in the standard factory camouflage of Dark Green and Dark Earth on the upper surfaces with undersurfaces in Sky. The codes were Sky Grey. According to the new directive, the undersurfaces of this variant were restored to Medium Sea Grey. From December 1, 1942, the spinners were to be painted in Ocean Grey and the codes in Dull Red. **(Illustrations 1-4)**

イラスト 1

B.Mk.IVシリーズI　W4069 "GB◎M"
No.2グループ　No.150スコードロン
1942年5月 ホーシャム・セントフェイス基地

上面ダークグリーン／ダークアース、下面スカイ。スピナーとシリアル・ナンバーは黒。機体コードはスカイグレー。

Illustration 1

B.Mk.IV Series I, W4069, "GB◎M", No.105 Squadron, No.2 Group, May 1942, Horsham St.Faith.

The upper surfaces are Dark Green and Dark Earth. The undersurfaces are Sky. The spinners and the serials are Black. The codes are Sky Grey.

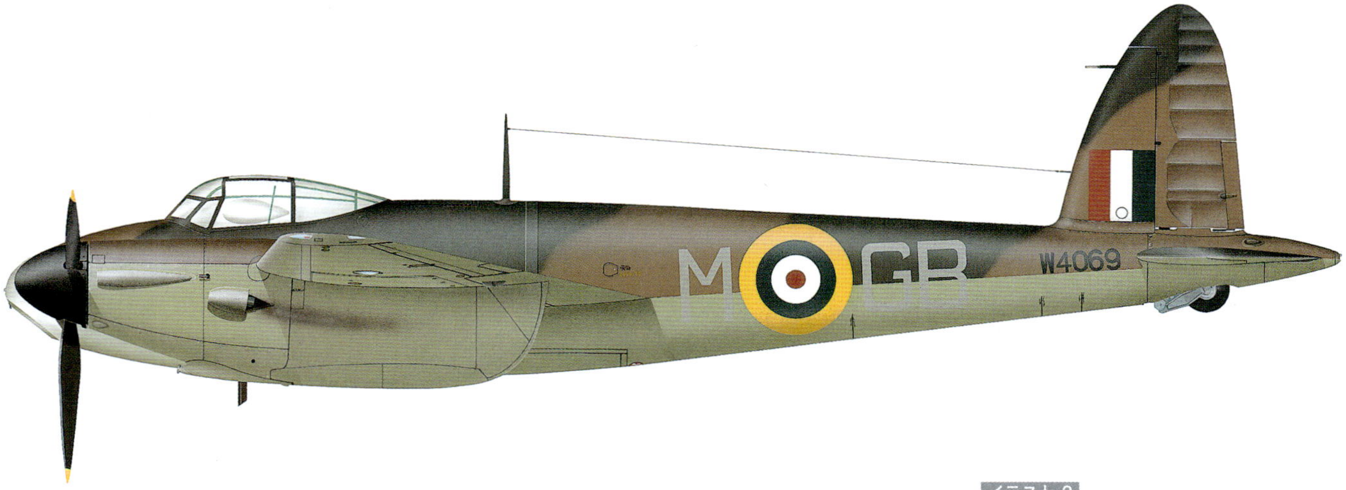

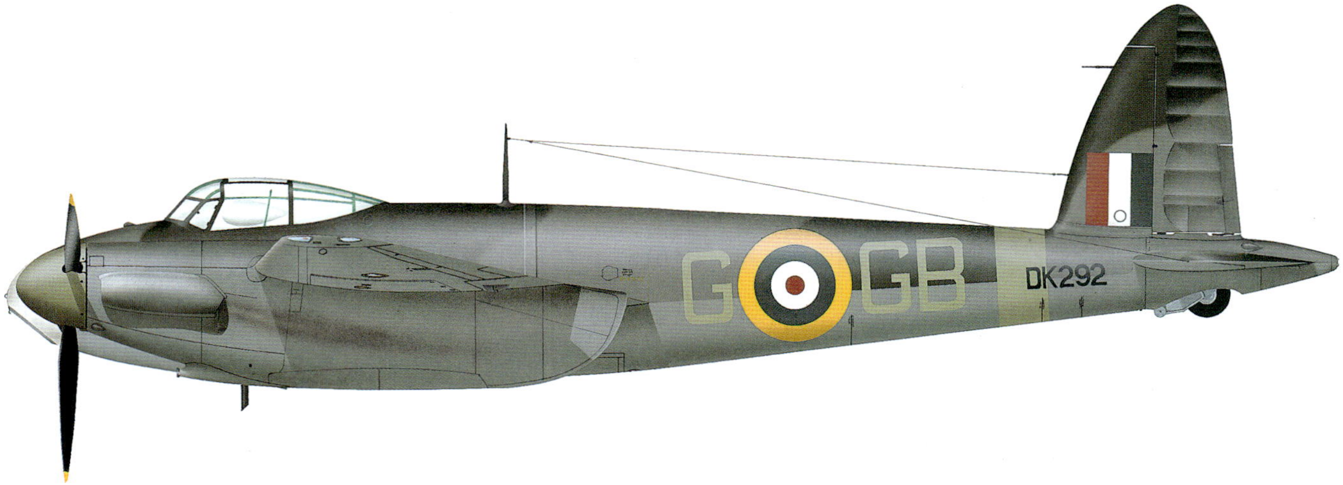

イラスト 2

B.Mk.IVシリーズII　DK292 "GB◎G"
No.2グループ No.105スコードロン 1942年6月 ホーシャム・セントフェイス基地

上面ダークグリーン／オーシャングレイ、下面ミディアムシーグレイ。スピナー、機体コードと胴体後部の帯はスカイ。シリアル・ナンバーは黒。胴体ラウンデルはタイプA1。

Illustration 2

B.Mk.IV Series II, DK292, "GB◎G", No.105 Squadron, No.2 Group, July 1942, Horsham St.Faith.

The upper surfaces are Dark Green and Ocean Grey. The undersurfaces are Medium Sea Grey. The spinners, the codes and the rear fuselage band are Sky. The serials are Black. Note the Type A1 fuselage roundel.

イラスト 3
B.Mk.IVシリーズII　DZ601 "XD◎B" No.2グループ No.139スコードロン 1943年5月　マーラム基地

上面ダークグリーン／オーシャングレイ、下面ミディアムシーグレイ。スピナーはオーシャングレイ。機体コードはスカイ。シリアル・ナンバーは黒。

Illustration 3
B.Mk.IV Series II, DZ601, "XD◎B", No.139 Squadron, No.2 Group, May 1943, Marham.

The upper surfaces are Dark Green and Ocean Grey. The undersurfaces are Medium Sea Grey. The spinners are Ocean Grey. The codes are Sky. The serials are Black.

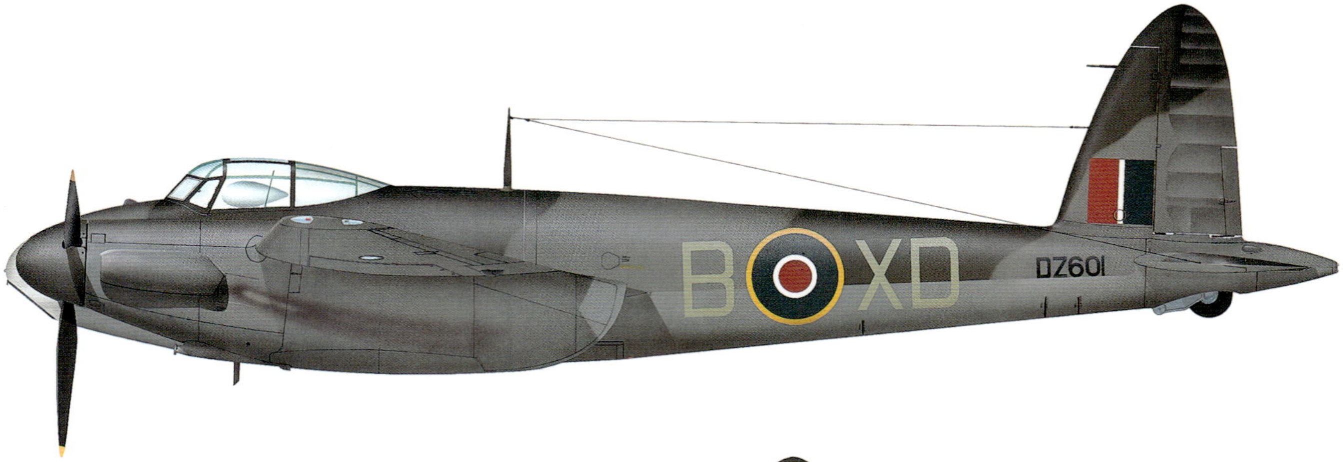

イラスト 4
B.Mk.XVI　ML966 "P3◎P" 第2戦術空軍 No.8（PFF）グループ　No.692スコードロン 1944年3月 グラヴレイ基地

上面ダークグリーン／オーシャングレイ、下面ミディアムシーグレイ。スピナーはオーシャングレイ。機体コードはスムーズナイトDTD308ツヤ消し黒の地に赤（ダルレッド）。シリアル・ナンバーは黒。

Illustration 4
BB.Mk.XVI, ML966, "P3◎P", No.692 Squadron, No.8 (PFF) Group, 2nd TAF, March 1944, Graveley.

The upper surfaces are Dark Green and Ocean Grey. The undersurfaces are Medium Sea Grey. The spinners are Ocean Grey. The codes are Dull Red applied on a Smooth Night DTD308 matt Black background. The serials are Black.

夜間爆撃機の迷彩塗装
Night Bomber Camouflage Scheme

"オーボエ"全天候爆撃／航法システム搭載機のような一部の夜間爆撃機は、上面ダークグリーン／オーシャングレイの基本迷彩に、下面スムーズナイトDTD308を塗装されていた。機体コードは通常赤（ダルレッド）。（イラスト5）

Some night bombers, such as Oboe-equipped aircraft, had a standard finish of Dark Green and Ocean Grey on the upper surfaces, with the undersurfaces painted in Smooth Night DTD308. Codes and serials were usually painted in Dull Red. (**Illustration 5**)

イラスト 5
B.Mk.IVスペシャル DZ650 "P3◎L" 第2戦術空軍　No.8（PFF）グループ　No.692スコードロン　1944年6月 グラヴレイ基地

上面ダークグリーン／オーシャングレイ、胴体側面、テイルコーン、垂直安定板、胴体下面とスピナーはスムーズナイトDTD308ツヤ消し黒。機体コードとシリアル・ナンバーは赤（ダルレッド）。

Illustration 5
B.Mk.IV Special, DZ650, "P3◎L", No.692 Squadron, No.8 (PFF) Group, 2nd TAF, June 1944, Graveley.

The upper surfaces are Dark Green and Ocean Grey. The fuselage sides, the entire tail cone, the vertical stabilizer, the undersurfaces and the spinners are Smooth Night DTD308 matt Black. The codes and the serials are Dull Red.

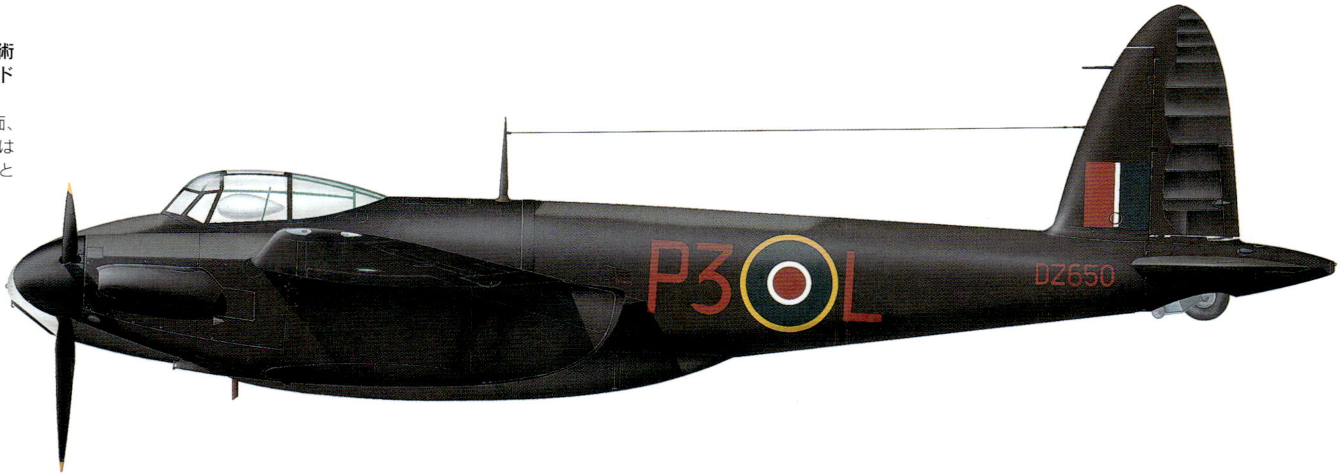

イントルーダーの迷彩塗装　Intruder Camouflage Scheme

敵地へ単機で侵入し、敵飛行場周辺で待ち伏せ攻撃などを行う、いわゆるイントルーダー機の塗装は、1942年10月1日の決定によって上面を昼間戦闘機の基本迷彩であるダークグリーン／オーシャングレイに、下面をスムーズナイトDTD308に塗ることになっていた。シリアル・ナンバーと機体コードは赤（ダルレッド）で記入。しかしこの塗装は迷彩効果の問題から、すぐに改められることになり、同年12月1日に、上面ダークグリーン／ミディアムシーグレイ、下面スムーズナイトDTD308の組みあわせに変更された。垂直安定板、ラダーとスピナーはミディアムシーグレイで塗装され、機体コードは赤（ダルレッド）、シリアル・ナンバーは黒で記載されるようになった。またスピナーをスムーズナイトDTD308に塗装した機体も、しばしばみうけられた。
（イラスト6-9）

The October 1, 1942 directive stated that the finish for intruder aircraft would be the standard day fighter scheme of Dark Green and Ocean Grey upper surfaces but with the undersurfaces painted Smooth Night DTD308. Serials and codes were to be painted in Dull Red. A revision to this scheme soon came into effect from December 1, 1942 when the camouflage was changed to Dark Green and Medium Sea Grey upper surfaces, but retaining the Smooth Night DTD308 undersurfaces. The vertical stabilizer, the rudder and the spinners were to be painted Medium Sea Grey. The codes were Dull Red and the serials were Black. The spinners were often painted Smooth Night DTD308. (Illustrations 6-9)

イラスト 6

F.B.Mk.VIシリーズII HR241 "TH◎M" No.100（爆撃支援）グループ　No.418スコードロン　1944年11月

上面ダークグリーン／ミディアムシーグレイ、胴体下面とスピナーはスムーズナイトDTD308ツヤ消し黒。スピナー先端は白。機体コードは赤（ダルレッド）。シリアル・ナンバーは黒。機首の機関銃は撤去されていることに注目。本機は爆撃機支援用のA.I.レーダー装備機で、機首下面に白で"M"が記入されている。

Illustration 6

F.B.Mk.VI Series II, HR241, "TH◎M", No.418 Squadron, No.100 (Bomber Support) Group, November 1944.

The upper surfaces are Dark Green and Medium Sea Grey. The undersurfaces and the spinners are Smooth Night DTD308 matt Black. Note the tips of the spinners are in White. The codes are Dull Red. The serials are Black. The nose-mounted machine guns are removed. This Mosquito is equipped with A.I. radar for bomber support operations and has an M in White on the underside of the nose.

イラスト 7

F.B.Mk.VIシリーズII　PZ170 "YP◎D" No.100（爆撃支援）グループ　No.23スコードロン　1945年初頭　リトル・スノーリング基地

上面ダークグリーン／ミディアムシーグレイ、胴体下面とスピナーはスムーズナイトDTD308ツヤ消し黒。機体コードは赤（ダルレッド）。シリアル・ナンバーは黒。このイントルーダーは機首にA.I.Mk.XV ASHレーダーを装備。

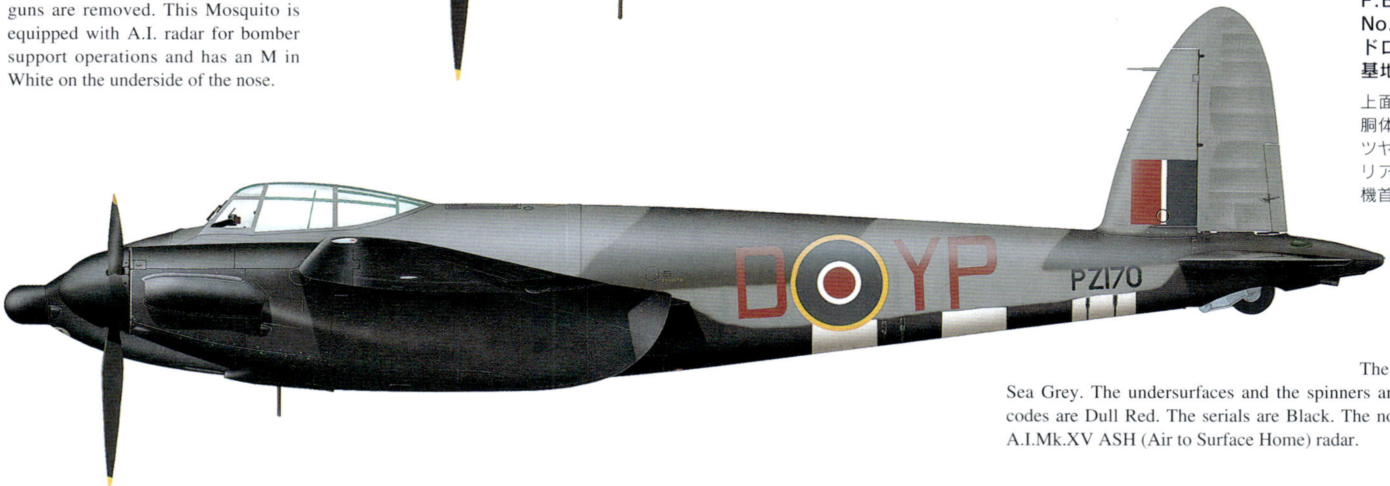

Illustration 7

F.B.Mk.VI Series II, PZ170, "YP◎D", No.23 Squadron, No.100 (Bomber Support) Group, early 1945, Little Snoring.

The upper surfaces are Dark Green and Medium Sea Grey. The undersurfaces and the spinners are Smooth Night DTD308 matt Black. The codes are Dull Red. The serials are Black. The nose of this intruder aircraft is fitted with the A.I.Mk.XV ASH (Air to Surface Home) radar.

イラスト 8

N.F.Mk.II　DZ726 "TW◎Z" No.100（爆撃支援）グループ No.141スコードロン　1944年3月　ウエスト・レイナム基地

上面ダークグリーン／ミディアムシーグレイ。機体下面とスピナーはスムーズナイトDTD308ツヤ消し黒。機体コードは赤（ダルレッド）。シリアル・ナンバーは黒。

Illustration 8

N.F.Mk.II, DZ726, "TW◎Z", No.141 Squadron, No.100 (Bomber Support) Group, March 1944, West Raynham.

The upper surfaces are Dark Green and Medium Sea Grey. The undersurfaces and the spinners are Smooth Night DTD308 matt Black. The codes are Dull Red. The serials are Black.

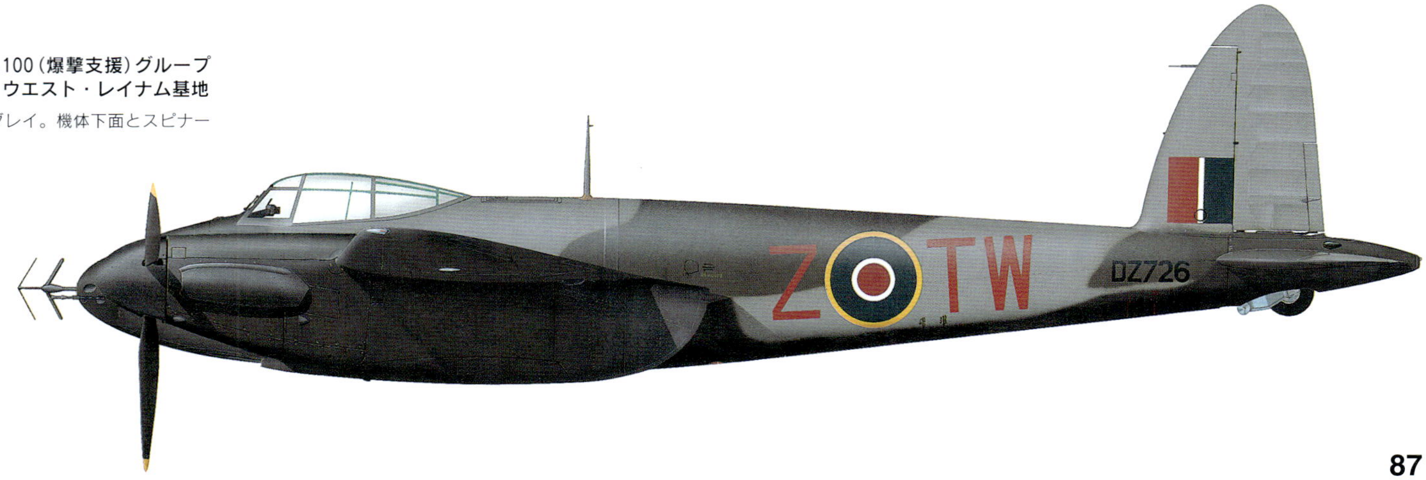

イラスト9

N.F.Mk.30　NT362 "HB◎S" No.100(爆撃支援)グループ　No.239スコードロン　1945年初頭　ウエスト・レイナム基地

上面ダークグリーン／ミディアムシーグレイ。機体下面とスピナーはスムーズナイトDTD308ツヤ消し黒。機体コードは赤（ダルレッド）。シリアル・ナンバーは黒。

Illustration 9

N.F.Mk.30, NT362, "HB◎S", No.239 Squadron, No.100(Bomber Support) Group, early 1945, West Raynham.

The upper surfaces are Dark Green and Medium Sea Grey. The undersurfaces and the spinners are Smooth Night DTD308 matt Black. The codes are Dull Red. The serials are Black.

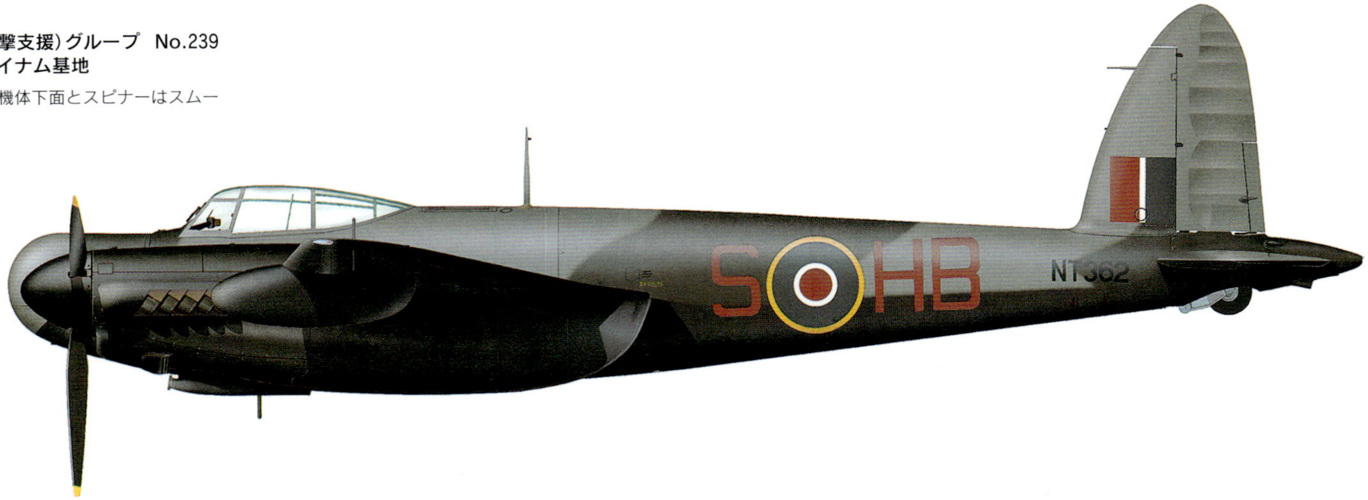

昼間戦闘機の迷彩塗装
Day Fighter Camouflage Scheme

昼間戦闘機の基本迷彩は1942年3月6日の通達により、上面ダークグリーン／オーシャングレイ、下面ミディアムシーグレイ塗装が施されることになった。初期のF.B.Mk.VIは昼間戦闘機の基本迷彩を施され、また戦闘爆撃機型と複式操縦装置付モスキートも胴体後部に18インチ幅のスカイの帯を塗装され、スピナーと機体コードにもスカイが用いられた。しかし、胴体のスカイの帯はF.B.Mk.VIによる昼間低空攻撃には適さないため、戦闘爆撃Sqn.ではただちに廃止され、同じ理由からスピナーのスカイも、ミディアムシーグレイがオーバーペイントされた。また、1942年6月1日の決定により、トレーナーイエローのストライプが敵味方識別色として外翼前縁に塗装されることになり、昼間戦闘機の迷彩のB.Mk.IVと複式操縦装置付モスキートもこの決定に準じた。しかし、同年10月1日付でこのストライプは廃止されている。なお、RAF沿岸航空軍団に配属された戦闘爆撃機型モスキートの塗装は上面ダークシーグレイ、下面スカイであった。**(イラスト10-11)**

The standard day fighter finish of Dark Green and Ocean Grey upper surfaces and Medium Sea Grey undersurfaces was initiated with a directive issued on March 6, 1942. Initially the F.B.Mk.VIs were camouflaged in day fighter scheme, and both fighter-bomber and dual control models were to have the 18in wide Sky band applied to rear fuselage together with Sky-painted spinners and codes. However, the Sky band was quickly removed on the fighter-bomber squadrons as this compromised the camouflage scheme for the low-level daylight raids in which F.B.Mk.VIs were involved. For the same reason the spinners in Sky were overpainted, usually in Medium Sea Grey. A Trainer Yellow identification stripe were added on the leading edge of the outer wings by a directive dated July 1, 1942, encompassing B.Mk.IVs and dual-control Mosquitos in day fighter finish. However, a further order of October 1, 1942, removed this marking. The fighter-bomber Mosquitoes attached to RAF Coastal Command Strike Wing also wore a finish of Dark Sea Grey upper surfaces and Sky undersurfaces. (**Illustrations 10-11**)

イラスト10

F.B.Mk.VIシリーズII　HP927 "SY◎B" No.613スコードロン　1944年3月　レイシャム基地

上面ダークグリーン／オーシャングレイ、下面ミディアムシーグレイ。スピナー、機体コードと胴体後部の帯はスカイ。シリアル・ナンバーは黒。

Illustration 10

F.B.Mk.VI Series II, HP927, "SY◎B", No.613 Squadron, early 1944, Lasham.

The upper surfaces are Dark Green and Ocean Grey. The undersurfaces are Medium Sea Grey. The spinners, the codes and the rear fuselage band are Sky. The serials are Black.

イラスト 11

F.B.Mk.VIシリーズII　HR414 "NE◎L"
No143スコードロン　バンフ・ウイング
1945年初頭　バンフ基地

上面ダークシーグレイ、下面スカイ。スピナーは黄（トレーナーイエロー）。機体コードはスカイ。シリアル・ナンバーは黒。

Illustration 11

F.B.Mk.VI Series II, HR414, "NE◎L", No.143 Squadron, Banff Strike Wing, early 1945, Banff.

The upper surfaces are Dark Sea Grey. The undersurfaces are Sky. The spinners are Trainer Yellow. The codes are Sky. The serials are Black.

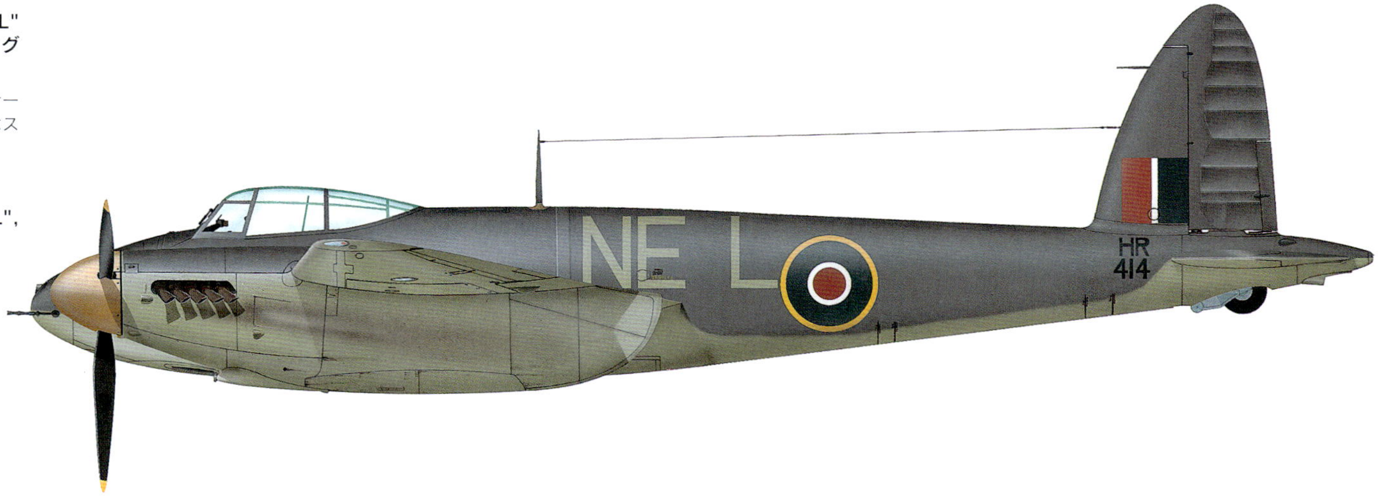

夜間戦闘機の迷彩塗装
Night Fighter Camouflage Scheme

モスキート最初の夜間戦闘機型N.F.Mk.IIの基本塗装は1941年3月24日付の決定に従い、全体をRDM2Aスペシャルナイトと呼ばれる完全ツヤ消しの黒で行われた。ちなみにこのスペシャルナイトRDM2Aは、戦前の夜間迷彩用ツヤ消し黒のナイトDTD308とは異なる塗料で、のちに戦闘機用塗料としてRDM2、スペシャルナイト、もしくはより簡略にナイトと呼称されるようになる。機体コードとシリアル・ナンバーはともに赤（ダルレッド）で記入された。しかしRDM2を塗装した機体と、それまでのDTD308を塗装した機体による比較審査を行ったところ、RDM2Aを塗装した機体には速度の低下がみられた。原因はRDM2Aスペシャルナイトを塗装した機体の表面が荒いツヤ消し仕上げのため、空気抵抗の増大によるものと判明。この試験結果をうけて、1942年4月に夜間戦闘機型N.F.Mk.IIすべての塗装を従来のツヤ消し黒、ナイトDTD308に変更することになった。なお、このとき、色名称による現場での混乱を避けるため、塗り替えに使用される戦前からのナイトDTD308を、スムーズナイトと呼称することにしている。また、この試験の際に、夜間作戦におけるスムーズナイト単色迷彩は、月光に照らされるとかえってシルエットを敵機に発見されやすいことも判明。そこでこの試験結果を受け、1942年10月1日、夜間戦闘機用の新しい迷彩が制定された。それまでの全面スムーズナイト塗装は、全面ミディアムシーグレイに上面のみダークグリーンの雲型迷彩を用いるものに変更、また、スピナー、垂直安定板と方向舵はミディアムシーグレイで塗装された。1943年6月からF.B.MK.VIの生産機のほとんどは配備にあたって夜間戦闘機用迷彩塗装を施されている。機体コードはスカイで記入されたが、スピナーはミディアムシーグレイで塗られた。またN.F.MK.XIIのすべての機体にはこの新しい基本迷彩が採用され、機体コードを（赤ダルレッド）、シリアルナンバーを黒で記入した。機首のレドームははじめツヤ消しで塗装され、のちにはミディアムシーグレイで塗られるようになった。**（イラスト12-16）**

Early night-fighter Mosquito N.F.Mk.IIs were painted with RDM2A Special Night matt Black overall according to directive of March 24, 1941 (DTD Circular No.144), explaining that the night fighter finish was to be RDM2A and not the pre-war Night DTD308. RDM2A Special Night was later abbreviated to "RDM2," "Special Night" or simply "Night Fighter Finish." Both codes and serials were painted in Dull Red. After comparative trials between aircraft painted overall in RDM2A and DTD308, substantial differences in speed performance were noted primarily due to the rough texture of the super matt RDM2A Special Night. Consequently, in April 1942 it was decided to change the scheme of all N.F.Mk.IIs to Night DTD308 matt Black. To avoid possible confusion between RDM2A Special Night and Night DTD308 finishes, the latter was renamed as "Smooth Night" DTD308. During further night tests it was discovered that in certain visibility conditions, especially in strong moonlight, Smooth Night tended to made the aircraft silhouette easily detectable. As a result of these tests a new standard night fighter finish was ordered from October 1, 1942 in which the overall Smooth Night was changed to overall Medium Sea Grey finish with a disruptive pattern of Dark Green on the upper surfaces only. In addition, the spinners, vertical stabilizer and rudder were to be painted Medium Sea Grey. From June 1943 most of the production F.B.Mk.VIs were delivered in the night fighter finish retaining codes in Sky but spinners in Medium Sea Grey, with a few exceptions. All Mosquito N.F.Mk.XIIs were finished in the new two-tone standard night fighter camouflage having Dull Red codes and Black serials. Initially the nose radome was painted matt Black, later restored to Medium Sea Grey. **(Illustrations 12-16)**

イラスト 12

F.B.Mk.VIシリーズII　NS843 "SB◎N"　第2戦術空軍　No.2グループ　No.464スコードロン RAAF　No.140ウイング　1944年9月

全面ミディアムシーグレイに上面のみダークグリーンの雲型迷彩の組みあわせ。スピナーと機体コードはスカイ。シリアル・ナンバーは黒。20インチ（508mm）幅のDデイ・インヴェイジョン・ストライプが機体下面に施され、機体コードとシリアル・ナンバーにオーバーペイントされている。

Illustration 12

F.B.Mk.VI Series II, NS843, "SB◎N", No.464 Squadron, RAAF, No.140 Wing, No.2 Group, 2nd TAF, September 1944.

Medium Sea Grey overall with a disruptive pattern of Dark Green on the upper surfaces. The spinners and the codes are Sky. The serials are Black. Note the 20in (508mm) wide D-Day invasion stripes applied on the underside of the fuselage overlapping the codes and the serials.

イラスト 13
F.B.Mk.XVIII　NT224 "E" No.248スコードロン特別分隊ポートリース・ウイング　1944年11月　ポートリース基地

全面ミディアムシーグレイに上面のみダークグリーンの雲型迷彩の組みあわせ。スピナーはミディアムシーグレイ。機体コードは白フチ付赤（ダルレッド）。シリアル・ナンバーは黒。

Illustration 13
F.B.Mk.XVIII, NT224, "E", No.248 Squadron Special Detachment, Portreath Strike Wing, November 1944, Portreath.

Medium Sea Grey overall with a disruptive pattern of Dark Green on the upper surfaces. The spinners are Medium Sea Grey. The codes are Dull Red with a White border. The serials are Black.

イラスト 14
N.F.Mk.II　W4079 "RS◎F" No.157スコードロン　1942年末　キャッスル・キャンプス基地

全面RDMスペシャルナイト完全ツヤ消し黒。機体コードとシリアル・ナンバーは赤（ダルレッド）。24×27インチのフィン・フラッシュとタイプA1胴体ラウンデル。

Illustration 14
N.F.Mk.II, W4079, "RS◎F", No.157 Squadron, late 1942, Castle Camps.

RDM2A Special Night matt Black overall. The codes and the serials are Dull Red. Aircraft also sports the 24in x 27in fin flash and Type A1 fuselage roundel.

イラスト 15
N.F.Mk.XIII　MM521 "PS◎L" No.264スコードロン　1944年末　フランス、リール／ヴァンデヴィル基地

全面ミディアムシーグレイに上面のみダークグリーンの雲型迷彩の組みあわせ。スピナーと機体コードは赤（ダルレッド）。シリアル・ナンバーは黒。胴体下面のみに施されたDデイ・インヴェイジョン・ストライプはシリアル・ナンバーにオーバーペイント。

Illustration 15
N.F.Mk.XIII, MM521, "PS◎L", No.264 Squadron, late 1944, Lille/Vendeville (France).

Medium Sea Grey overall with a disruptive pattern of Dark Green on the upper surfaces. The spinners and the codes are Dull Red. The serials are Black. Note the D-Day invasion stripes applied only on the underside of the fuselage and overlapping the serials.

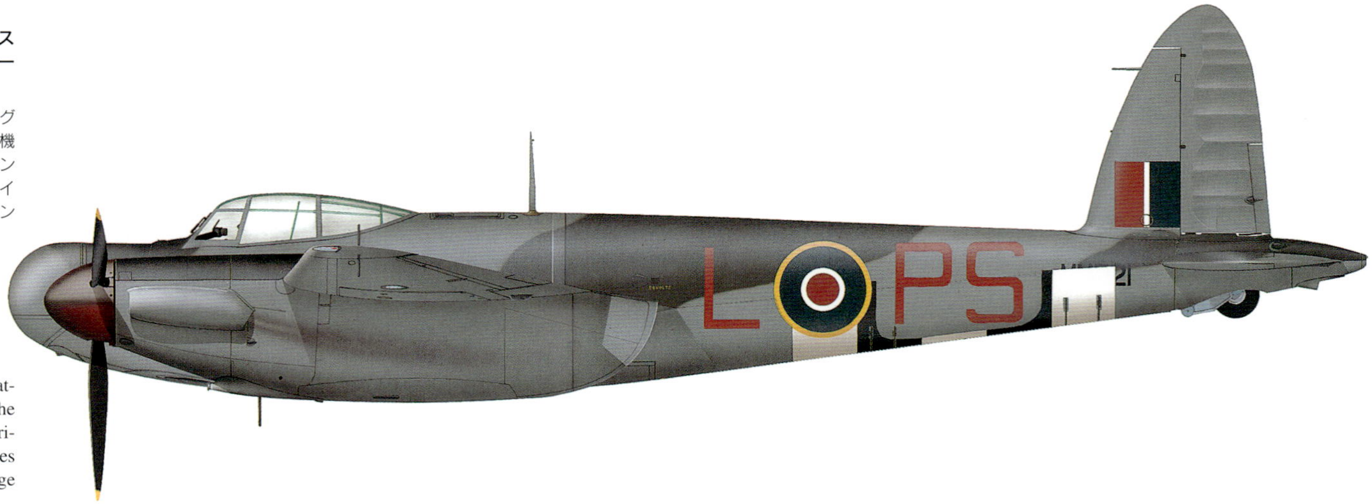

写真偵察ユニット(P.R.U.)所属機の迷彩塗装
P.R.U. Camouflage Scheme

写真偵察機最初の原型機W4051の迷彩試験終了後、最初のモスキート生産型となったP.R.Mk.Iと初期生産分のP.R.Mk.IVには、上面ダークスレートグレイ／スカイグレイ、下面PRUブルーの迷彩が施された。機体上面の迷彩パターンは昼間爆撃機のものに似ており、ダークグリーンにかわってダークスレートグレーが、ダークアース（後にオーシャングレイ）にかわってスカイグレイが置き換えられている。機体コードはスカイ、シリアル・ナンバーは黒で記入された。後期の機体は全面PRUブルー塗装に改められた。機体コードとシリアル・ナンバーは通常のものが用いられ、P.R.Mk.IX、P.R.Mk.XVI、P.R.Mk.34などにシリアルとコードに黒を用いたものがみられる。**(イラスト16-17)**

After camouflage tests on the P.R. prototype W4051, the production P.R. Mk.Is and the early P.R.Mk.IVs had a shaded camouflage of Dark Slate Grey and Sky Grey applied to the upper surfaces with the undersurfaces in PRU Blue. The pattern of the camouflage scheme of the upper surfaces was similar to that of the day bomber scheme, with Dark Green replaced by Dark Slate Grey and Dark Earth (later Ocean Grey) by Sky Grey. The codes were in Sky and the serials in Black. Later the aircraft were painted PRU Blue overall, having usually both serials and codes, when marked, in Black. Examples marked in this way include the P.R.Mk.IX, P.R.Mk.XVI and P.R.Mk.34, with some exceptions. **(Illustrations 16-17)**

イラスト 16

P.R.Mk.I　W4051 "LY◎U" 第１写真偵察大隊　1942年初頭　ベンソン基地

上面ダークスレートグレイとスカイグレー、下面PRUブルー。スピナーとシリアル・ナンバーは黒。機体コードはスカイ。胴体ラウンデルはタイプA1からタイプBに変更されている。

Illustration 16

P.R.Mk.I, W4051,"LY◎U", No.1 P.R.U., early 1942, Benson.

The upper surfaces are Dark Slate Grey and Sky Grey. The undersurfaces are PRU Blue. The spinners and the serials are Black. The codes are Sky. Note the Type A1 fuselage roundel (42in - 1,066.8mm diameter) modified into a Type B roundel.

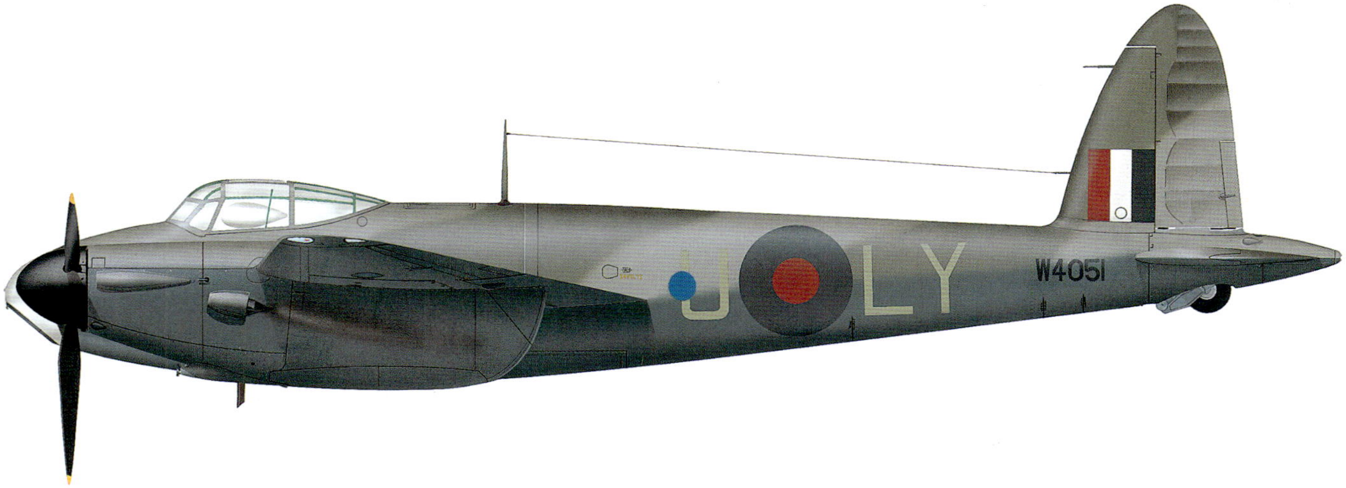

イラスト 17

P.R.Mk.XVI　NS777　No.140スコードロン　1944年12月　ベルギー、メルスブルク基地

全面PRUブルー。スピナーとシリアル・ナンバーは黒。胴体ラウンデルはタイプB、フィン・フラッシュは12 x 12インチタイプ。Dデイ・インヴェイジョン・ストライプは胴体下面のみに塗装。

Illustration 17

P.R.Mk.XVI, NS777, No.140 Squadron, December 1944, Melsbroek (Belgium).

PRU Blue overall. The spinners and the serials are Black. Both the standard Type B fuselage roundel and the 12in x 12in fin flash are applied. The D-Day invasion stripes are painted only on the underside of the fuselage.

その他の迷彩　Other Examples

イラスト 18

F.B.Mk.VIシリーズII　HR559 "UX◎X" No.82スコードロン　1945年初頭　インド、カンビルグラム基地

東南アジア方面に配備され、全面をRAAFアルミニウム・ドープで塗装されていたSEAC（South East Asia Command／東南アジア航空軍団）のモスキート。スピナーは青（ダルブルー）。機体コードとシリアル・ナンバーは黒で、垂直尾翼のフィン・フラッシュ上方にも記入された。ラウンデルの赤（ダルレッド）と白は日本機の日の丸と誤認されやすいため、かわりにアズールブルーが使用されている。主翼ラウンデルは直径30インチ（762mm）、胴体ラウンデルは直径36インチ（914.4mm）。フィン・フラッシュも、ラウンデルと同様日本機に誤認されるのを避けるため、赤（ダルレッド）および白の帯部分をアズールブルーで記入。SEACを示す識別マーキングは主翼、水平安定板および垂直安定板に黒で記入。

Illustration 18

F.B.Mk.VI Series II, HR559, "UX◎X", No.82 Squadron, early 1945, Kumbhirgram (India).

This Mosquito attached to SEAC (South East Asia Command) wears an overall doped RAAF Aluminum finish. The spinners are Dull Blue. The codes and the serials are Black, repeated on the vertical stabilizer just above the fin flash. Dull Red and White are excluded from the roundels and replaced by the Azure Blue to avoid confusion with the bright red "rising sun" *hinomaru* markings applied to Japanese aircraft. The wing roundels are 30in (762mm) diameter, the fuselage roundels are 36in (914.4mm) diameter. In addition, the fin flashes have the Dull Red removed further the White center stripe. The SEAC Black identification markings are applied on the wings, the horizontal and the vertical stabilizers.

イラスト 19

P.R.Mk.XVI "K" アメリカ第8空軍　第492爆撃航空群　1945年　ハリントン基地

胴体前方と主翼上面はPRUブルー。機体下面、胴体後部、尾部およびスピナーは"ジェット"ツヤ有り黒。シリアル・ナンバーは塗りつぶされている。機体識別コードは白。国籍マークはアメリカの標準タイプを記入。

Illustration 19

P.R.Mk.XVI, "K", 8th AF, 492nd Bomb Group, March 1945, Harrington.

The forward fuselage and the wing upper surfaces are PRU Blue. The undersurfaces, the entire rear fuselage, the tail unit and the spinners are "Jet" gloss Black - the serials are painted out. The individual aircraft letter is White. Other standard U.S. national insignia applied.

Illustration 20

F.B.Mk.VI Series II, HR402, "OB◎C", No.45 Squadron, January 1945, Kumbhirgram (India).

The upper surfaces are RAAF Foliage Green and RAAF Earth Brown. The undersurfaces are Medium Sea Grey. The spinners and the serials are Black. The codes are White. Foliage Green applied to this SEAC Mosquito is slightly darker than RAF Dark Green. Dark Green is around FS595A 34096 and Foliage Green FS595A 34092. The latter generally weathered to around FS595A 34097. Earth Brown is around 30099, but some sources quote a color more "yellow," closer to RAF Dark Earth, and others a brown slightly more "red." On January 15, 1945 Japanese Ki-43 Oscars shot down HR402 as it was patrolling over Meiktila airfield.

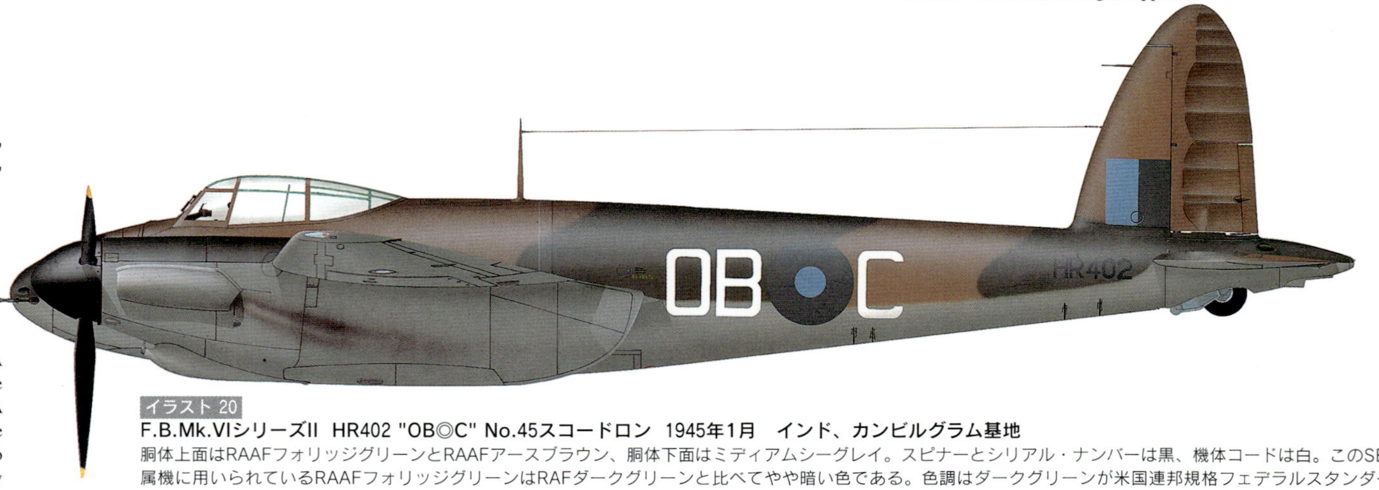

イラスト 20

F.B.Mk.VIシリーズII　HR402 "OB◎C" No.45スコードロン　1945年1月　インド、カンビルグラム基地

胴体上面はRAAFフォリッジグリーンとRAAFアースブラウン、胴体下面はミディアムシーグレイ。スピナーとシリアル・ナンバーは黒、機体コードは白。このSEAC所属機に用いられているRAAFフォリッジグリーンはRAFダークグリーンと比べてやや暗い色である。色調はダークグリーンが米国連邦規格フェデラルスタンダードのFS595A 34096番に近いのに対し、フォリッジグリーンはFS595A 34092に近い。フォリッジグリーンが退色するとFS595A 34097に近い色になる。また、RAAFアースブラウンはFS595A 34099に近い色とされるが、いくつかの資料によると、上記の色調より黄色味がかっていて、RAFダークアースのような色であるとし、また別の資料ではより赤みがかった茶色とする説もある。なお、このHR402は1945年1月15日、メイクティラ飛行場を哨戒中に日本陸軍のキ-43隼4機と遭遇して撃墜された機体。

デ・ハヴィランド・モスキート

写真／航空ファン

Photos : KOKU-FAN

第二次大戦で活躍したイギリス機のなかでも、"木製の驚異"と呼ばれたD.H.98モスキートは、もっとも興味深い軍用機のひとつといえるだろう。

モスキートの起源はデ・ハヴィランド社が独自に開発を開始した高速爆撃機にはじまる。デ・ハヴィランド社は1936年に各航空会社に提示された空軍省仕様P.13／36をうけて、1938年、D.H.91アルバトロス4発旅客機をもとにデザインした双発爆撃機の試案を空軍省へ提出したが、失敗に終った。この失敗から、ハットフィールドの本社で高速爆撃機の開発に着手。開発番号D.H.98はデ・ハヴィランド社のハットフィールド工場内で爆撃機プロジェクトとして知られるようになった。

デ・ハヴィランド社は4発爆撃機から搭乗員や防御武装を減らすなどして機体を軽量化するという発想をつきつめた結果、搭乗員を操縦士と爆撃／航法士の2名に限定し、ロールスロイス・マーリンエンジンを搭載した非武装の小型高速双発爆撃機というアイディアにたどりついた。また、機体の構造材には戦時中に不足する戦略物資と考えられていたアルミニウムなどの金属を節約する目的で、木材を大幅に採用することにした。また、木材の使用には、戦時下で多忙になる金属加工産業のかわりに国内に広く存在する木工産業を利用することで、緊急な増産にも対応できるという利点もあった。

開発の過程で本機の能力から、高速爆撃機に加えて長距離護衛戦闘機と写真偵察機としての可能性も考えられるようになり、まとめられた設計試案は空軍省に提出された。空軍省ではこれをもとにあらたな要求仕様B.1／40をつくり、デ・ハヴィランド社と50機発注の契約を結んだ。この契約書は1940年3月1日にハットフィールドへ届けられた。

1940年10月、ロナルド・E・ビショップを中心とする設計チームはハットフィールドの工場からソールズベリー・ホールへ移動。原型機はここでロールアウトし、デ・ハヴィランド社のクラスB登録番号E0234を与えられた。同年11月25日、原型機はハットフィールドの飛行場で操縦士ジェフリー・デ・ハヴィランド、航法士ジョン・E・ウォーカーにより初飛行に成功、モスキートの歴史がはじまった。

モスキートの生産に関する契約は、当局の無理解により何度も変更されたが、1941年6月に、最初の量産型であるP.R.Mk.I写真偵察機の原型機が初飛行し、同年7月からいよいよ実戦配備が開始された。モスキート最初の実戦参加は同年9月17日、敵地上空に侵入したP.R.Mk.I（W4055）はその高速性能をいかして成功裏のうちに写真を撮影。こうしてヨーロッパはしだいにモスキートの監視下におかれていった。そして、写真偵察機型につづいて爆撃機型、戦闘機／夜間戦闘機型、戦闘爆撃機型が次々と開発され、実戦で戦果をあげていった。モスキートの活躍は多岐におよび、このなかにはV-1飛行爆弾の迎撃、コースタル・コマンド（沿岸航空軍団）による艦船攻撃や、偵察機によるV-2開発基地の写真撮影などもふくまれた。なかでも各種電子機器をそなえた夜間戦闘機型はその重武装と高速性能で、迎撃に上がるドイツ空軍夜間戦闘機隊をなやませている。このモスキートに対してもっとも高い評価を下したのは、ドイツ夜間戦闘機司令部かもしれない。モスキート1機撃墜を、2機分のスコアとして認めたのだ。

モスキートの爆弾搭載量は"空の要塞"と呼ばれたボーイングB-17四発爆撃機と同じで、速度はスピットファイアより30km/hも速く、ベルリンへ向けて1晩に2回のソーティを行うことができた。また、木製のため損傷に対する耐性は高く、修理も金属製の機体に比べて容易だった。

モスキートはヨーロッパにとどまらず、中東、そして極東でも活躍し、その実力を発揮した。特にSEAC（東南アジア航空軍団）に1944年3月から配備されたモスキートF.B.Mk/VIは日本軍を相手に各種の侵攻作戦にあたっている。空戦の機会は少なかったが、1945年1月15日にはモスキートF.B.Mk/VI、3機がメイクティラ上空で隼4機と交戦しそのうちの1機、HR402が撃墜された。なお、この撃墜に際し、日本はただちにモスキート調査団を現地へ派遣している。

第二次大戦中に生産されたモスキートは合計6,710機。そのうち3,054機はデ・ハヴィランド社のハットフィールド工場で作られている。そのほかの内わけはデ・ハヴィランド社のリーヴスデン秘密工場が1,390機、スタンダード・モーターズ社が916機、パーシヴァル・エアクラフト社が198機、エアスピード・エアクラフト社が12機、デ・ハヴィランド・カナダ社が1,032機、デ・ハヴィランド・オーストラリア社が108機であった。デ・ハヴィランド社のハットフィールド工場だけをとってみても、戦時下の約4年間に日産2機のペースで完成させていたことになり、これはモスキートが木製機だからこそ成しえた偉業であろう。

◀タイトな3機編隊を組んで飛行するモスキートF.B.Mk.VI。写真の機体は英空軍指揮下で戦闘任務についたニュージーランド空軍No.487Sqnの所属機。同飛行隊は1943年7月にベンチュラからモスキートに機種改編され、No.21、464SqnとともにNo.140Wing(第140航空団)を編成した。主として昼間精密爆撃任務につき、ナチス占領下のヨーロッパ大陸の重要拠点爆撃に活躍したが、なかでもデンマークのゲシュタポ本部爆撃は今も語り種になっている。とりわけ1944年2月18日のフランス・レジスタンス救出には19機のF.B.Mk.VIが出撃、フランス、アミアン収容所を襲撃、260名のレジスタンスの脱出を成功させた「ジェリコ」作戦は、一連の作戦のハイライトでもある。

Three Mosquito F.B. Mk.VIs seen in a tight formation. These are New Zealand Air Force planes from No. 487 Squadron on fighter duty under RAF command. Making the change from Venturas to Mosquitoes in July of 1943, No. 487 Squadron was part of the No. 140 Wing along with No. 21 and No. 464 Squadron. Primarily involved in precision daylight bombing, they struck crucial targets throughout occupied Europe. Among their noted accomplishments were the bombing of Gestapo headquarters in Denmark, as well as the "Jericho" raid on February 18, 1944 which freed 260 members of the French resistance from prison.

デ・ハヴィランド・モスキート実機写真集 DE HAVILLAND MOSQUITO Photo Album

写真／航空ファン Photos :KOKU-FAN

解説／山田智夫　Commentrary by Tomoo Yamada

◀1940年11月25日午後３時45分、操縦士、ジェフリー・デ・ハヴィランドJr、航法／機関士、ジョン・E・ウォーカー搭乗で30分間の初飛行に成功したモスキート原型第１号機。ソールズベリー・ホールと呼ばれた古館の一郭で組み立てがはじめられた機体は分解されて８キロ離れたハットフィールド飛行場にトレーラーによって運ばれ、ここで再組み立てされて初飛行にのぞんだ。機体はあざやかな黄色に塗られていたが、これは付近の味方対空部隊による誤射を防ぐための措置であった。当初はデ・ハヴィランド社のＢクラス登録番号、E0234を胴体に描いていたが、２回目のテスト飛行終了後、英空軍のシリアル、W4050に描き改められた。スピナー直後にある排気管冷却用のスクープは翌年１月に追加されたもので、同月末には機体上面にグリーンとアースの迷彩が施されている。

The original Mosquito prototype makes its first flight at 3:45pm on November 25, 1940. At the controls for the 30 minute flight were Goeffrey de Havilland Jr. while John E. Walker sat in the right-hand seat. Put together at Salisbury Hall, the aircraft was disassembled, trucked 8 kilometers to Hatfield airfield, reassembled and flown. The bright yellow markings were to prevent the aircraft from being shot at by British AA stations in the vicinity. The aircraft was originally marked with the de Havilland B-class registration number E0234, but this was re-marked to the RAF W4050 following its second test flight. The exhaust pipe cooling intake scoops behind the spinner were added the following January, and the aircraft's upper side was marked in a green and earth scheme at the end of that month.

▶原型機の完成前に英空軍では50機の爆撃機型モスキートを発注していたが、原型機の好性能に気を良くした空軍当局は、この内10機を偵察機型、30機を戦闘機型として生産するよう変更した。偵察機型の原型第１号機、W4051は1941年６月10日に初飛行、量産型、P.R.Mk.Ⅰが翌月にはNo.1PRU(第１写真偵察部隊)に配属され、９月17日、フランスのブレスト軍港を偵察して、モスキート最初の実戦任務を記録した。P.R.Mk.Ⅰは偵察カメラ４基を胴体下に装備、計10機が生産された。写真の機体はその内の１機(W4059)で、ベンソンのNo.1PRUで59回の作戦任務を記録している。機体全面はPRUブルーとよばれる暗いブルーグレイに塗られている。本機は1944年９月20日に事故で損傷、除籍抹消されている。エンジン・ナセルが短いのに注意されたい。

Even before the completion of the Mosquito prototype, the RAF had ordered 50 bomber versions of the plane, but upon seeing the promising performance of the original, they changed the order to include 10 reconnaissance versions and 30 fighters. The first reconnaissance version, W4051, took to the skies for the first time on June 10, 1941. The production version, the P.R. Mk.I was delivered to the No. 1 PRU the following month, and that unit's mission to photograph Brest, France on September 17 marks the first pure military use of the Mosquito in WWII. Carrying four reconnaissance cameras, just ten examples of the Mk.I were completed. One of these, W4059, is seen here, and records show that it completed 59 missions with No. 1 PRU. The entire plane was marked in a dark blue that came to be known as "PRU Blue." W4059 was eventually damaged in an accident on September 20, 1944, and stricken. Note the short engine nacelles.

◀爆撃機型から発注変更されて試作が進められた戦闘機型は、1941年５月15日に原型第１号機(W4052)が初飛行に成功、F.Mk.Ⅱとして量産を開始した。計画当初にはコクピット直後方に7.7mm４連装の回転銃座を装備する予定であったが、機首爆撃窓をソリッドノーズ化して7.7mm機銃４挺、胴体下面に20mm機関砲４門を装備した計画案におちついた。この戦闘機型は合計278機が発注されたが、この内機首の武装を廃し、A.I.Mk.ⅣまたはⅤレーダーを備えた夜間戦闘機型、N.F.Mk.Ⅱが開発され、1941年12月に完成、翌年１月末から実戦配備が開始された。写真のN.F.Mk.Ⅱの機首先端にみえるのがA.I.Mk.Ⅳのレーダーアンテナ。この当時、夜間戦闘機はRDM2Aという完全ツヤ消しブラックで塗られていたが、塗装表面の抵抗が高く、約24km/h近くの速力低下をまねいたとされる。また、月明かりの夜にははっきりとしたシルエットで夜空に浮かび上がってしまうという皮肉な結果のため、ほどなくこの塗装は廃止を余儀なくされた。F./N.F.Mk.Ⅱの合計は467機であった。

The first fighter prototype of the Mosquito (W4052) made its maiden flight on May 15, 1941 and production of the F. Mk.II began almost immediately thereafter. Original plans called for a rotating gunner's position armed with four 7.7mm MGs immediately behind the cockpit, but plans were eventually changed to mounting those four guns in a now-solid nose, and adding the four 20mm Hispano cannon to the underside. 278 examples of the fighter version were ordered, but some of these were eventually produced as N.F. Mk.II night-fighters with A.I. Mk.IV or Mk.V search radars in the nose in place of the guns. The antenna of an A.I. Mk.IV can be seen on the nose of the plane in this photo. Originally, night-fighters were painted entirely in a flat black paint known as RDM2A, but this paint had a rough finish that significantly increased air resistance over the aircraft and slowed it by a remarkable 24km/h. What's more, the pure black aircraft actually stood out against the night sky if there was significant moonlight, and this paint was eventually discarded. At total of 467 F./N.F. Mk.IIs were completed.

▶量産の進むモスキート各型に応じ、搭乗員養成のための訓練用機として登場したのがモスキートT.Mk.Ⅲ。ベースとなったのはF.Mk.Ⅱであるが、武装を全廃し、操縦装置を複操縦装置とした機体で、1942年1月に初飛行、合計343機が量産された。戦後余剰となった機体が実戦型モスキートとあわせて外国空軍に引き渡されたが、写真の機体もそのひとつで、トルコ空軍の国籍標識を機体に記入している。機体全面はアルミニウム・ドープで塗装されている。同空軍では本機の翼下に爆弾架を取付けて爆撃機としても使用したという。このほか、ベルギー、スウェーデン、ノルウェー、イスラエル、フランス、ユーゴスラヴィアなどの各国で使用された。

In order to train pilots for the Mosquitoes rolling off assembly lines, the T. Mk.III trainer version was developed. Based on the F. Mk.II, it eliminated all armament and provided dual flight controls. It first flew in January of 1942, and some 343 examples were produced. Following the war, many of these were sold to other nations as surplus, along with many combat types of the Mosquito as well. The aircraft seen here is one of those, bearing Turkish markings in an all-aluminum dope scheme. The Turkish Air Force is said to have used these trainers as bombers by adding bomb racks under the wings. Other nations receiving surplus Mosquitoes included Belgium, Sweden, Norway, Israel, France and Yugoslavia.

◀写真はB.Mk.Iのシリーズ I の最終号機(W4072)。最初の実戦参加は偵察機型にさらわれたものの、本命の爆撃機型はとりあえず偵察機型P.R.Mk. I のラインから9機が抽出されて生産が進められた。1941年11月に最初の生産機、B.Mk.Ⅳが完成、つづいて残り8機も完成した。これらの機体はP.R.Mk. I と同じくエンジン・ナセルの短い機体であったが、本来の爆撃機の生産ラインから完成した機体はナセルの長い機体で、これを区別するために前者をシリーズ I 、後者をシリーズ II と称した。固定武装は一切なく、爆弾搭載量は合計1,000ポンドであった。原型機当時の面影を残すエンジン・ナセル周辺や、ダークアース、ダークグリーンの迷彩、黄色フチの太いタイプA1ラウンデルにも注意されたい。下面はスカイ。本機はのちにNo.105、627Sqnなどに配属されたが、1944年1月9日、フランクフルト爆撃の帰途、墜落して失われている。

The first combat sorties by Mosquitoes were by P.R. versions, but production of the "main" bomber version moved forward, too, with the first examples coming off the production lines for recon. versions. There were only nine such B. Mk.IV aircraft, the first being produced in November of 1941. These, like the P.R. aircraft from the same line, had the shorter engine nacelles. To distinguish them from the standard-production bombers, the short-nacelle type was designated Series I, while the longer version was Series II. The photo depicts the final production example of the Series I planes, W4072. It carried no fixed armament and could haul 1,000 lbs. of bombs. Note the prototype-style engine nacelles, the dark earth/dark green marking scheme, and the A1 roundels with the wide yellow borders. The underside was marked in sky. This plane was delivered to No. 627 Squadron, but lost while returning from a raid on Frankfurt on January 9, 1944.

◀雲海をバックに飛行するモスキートB.Mk.Ⅳシリーズ II (DZ353:GB-E)。爆撃機型モスキートを最初に配備したNo.105Sqn所属機で、マーラムを基地としていた。機体に搭乗しているのはロイ・ラルストン中佐／シド・クレイトン大尉で、機首爆撃窓横には爆弾のシルエットで表わした出撃マークが４個描かれている。当初爆弾搭載量は1,000ポンドとされていたが、爆弾のフィンを縮めると合計2,000ポンド搭載可能と判断され、以後の爆撃機型ではこれが標準となった。なお、爆撃機型モスキートの初出撃は1942年５月31日のケルン市の爆撃戦果の偵察および爆撃であった。塗装はオーシャングレイ／ダークグリーンの昼間爆撃機迷彩が採用されている。長く後ろに伸びたエンジン・ナセルにも注意。B.Mk.Ⅳはシリーズ I 、II あわせて300機が生産されている。

A B. Mk.IV Series II (DZ353:GB-E) makes way against a sea of clouds. This is a plane from the first unit to take deliveries of bomber models, No. 105 Squadron based at Marham. Flying the plane in the photo were Wing Commander (Wg Cdr) Roy Ralston and Flight Lieutenant (Flt Lt) Sid Clayton. Note the four bombs painted on the nose, indicating four successful sorties. Originally, the Mosquito's bomb load was thought to be just 1,000 lbs., but when it was discovered that up to 2,000 lbs. could be mounted by simply creating a bomb with smaller fins this became the standard payload. The first action by bomber versions of the plane was the reconnaissance and bombing of Köln on May 31, 1942. Markings were the standard daytime bombing ocean gray/dark green scheme. Note the long engine nacelles. 300 B. Mk.IVs were produced including both Series I and Series II types.

▶スマートな機影をみせるモスキートF.B.Mk.Ⅵ。F.B.Mk.Ⅵは戦闘爆撃機型としてF.Mk.IIから発達した機体で、固定武装は変わらず、20mm機関砲点検扉後部を開閉式として爆弾倉とし、500ポンド爆弾を２発搭載、さらに主翼を補強してラックを増設、同じく500ポンド爆弾計２発を搭載できるようにした。生産機数は数あるモスキート・シリーズ最多の2,584機にのぼった。初飛行は1943年２月で、実戦配備は翌月からはじまった。写真はNo.487Sqn所属機(MM417:EG-T)で、主翼下面のラックには250ポンド爆弾が吊下されている。機体塗装はオーシャングレイ／ダークグリーンの機体上面、シーグレイミディアムの下面という昼間迷彩が施されているが、上面迷彩が胴体下方にまで延びている。またパターンも夜間戦闘機のものが使われているという不思議な塗装。

An F.B. Mk.VI shows off its extremely clean lines. The Mk.VI fighter-bomber was derived from the Mk.II fighter and mounted the same fixed armament, but was modified to convert the area behind the cannon to a bomb bay capable of carrying two 500 lb. bombs. In addition, the wings were strengthened and racks were added to allow the aircraft to carry two 500-lb. bombs under the wings. This was the most-produced version of all Mosquito variants, with no less than 2,584 examples being produced. First flying in February of 1943, it began combat operations the following month. The photo shows MM417:EG-T of No. 487 Squadron carrying 250 lb. bombs under its wings. This particular aircraft has a unique marking scheme. It's marked in the daylight bombing scheme of an ocean gray/dark green upper side and a sea gray medium underside, but the top side markings extend down lower than normal on this example. What's more, the camouflage pattern is that normally seen on night fighters, not fighter-bombers.

▶P.R.Mk.Ⅰに続いて登場した写真偵察機型がこのP.R.Mk.Ⅳ。爆撃機型B.Mk.Ⅳをベースに爆弾倉に大型の燃料タンクを装備し、航続距離を増加させた機体。1942年４月に初飛行した。最初はベンソンのNo.1PRUで２機のB.Mk.Ⅳが写真偵察機型に改造されたのを皮切りに、生産ラインから抽出されたB.Mk.Ⅳが27機、P.R.Mk.Ⅳ仕様に改造された。写真の機体も生産ラインで改造されたうちの１機で、全面PRUブルー、国籍標識は白い部分を省略したタイプBが用いられている。国籍標識前方に偵察カメラ撮影用の窓がみえる。

The P.R. Mk.IV was the successor to the P.R. Mk.I. Flying for the first time in April of 1942, it was based on the B. Mk.IV, but mounted a large additional fuel tank in the bomb bay. The first two examples were B. Mk.IVs of No. 1 PRU that were converted to the recon. configuration. Ultimately, 27 B. Mk.IV aircraft were converted on the assembly line to P.R. Mk. IVs. The aircraft in the photo is one of these, painted entirely in PRU blue and with the "Type B" national insignia that dropped the white part of the marking. The camera window is visible forward of the national insignia.

◀A.I.Mk.Ⅳレーダー装備のN.F.Mk.Ⅱにかわって A.I.Mk.Ⅷセンチメートル波レーダーを搭載した夜間戦闘機として1942年８月に初飛行したのが、このモスキートN.F.Mk.ⅩⅡ。機首レーダーアンテナはシンブル(指抜き)と呼ばれるファイバー製のレドームに収納されている。機体はすべてN.F.Mk.Ⅱからの改造で、新造機は一切ない。合計97機が完成し、No.85Sqnに1943年２月末に配属されている。写真の機体は全面シーグレイミディアムに上面のみダークグリーンの迷彩が施されているが、この当時の夜間戦闘機の標準迷彩で、パターンも独自のものが使われている。

The Mosquito N.F. Mk. XII, which succeeded the N.F. Mk.II by replacing it's A.I. Mk.IV radar with an A.I. Mk.VIII centimeter-band air search radar, first flew in August of 1942. The radar antenna is concealed under the "thimble" fiberglass radome on the nose. All examples of this mark were converted from N.F. Mk.II aircraft; there was no new production. A total of 97 such conversions were made with deliveries beginning to No. 85 Squadron in February of 1943. This plane has an all-sea gray medium with dark green on the upper side scheme, the standard night fighter markings of the time, but the pattern appears to be unique.

▶F.B.Mk.Ⅵの機体にA.I.Mk.Ⅷセンチメートル波レーダーを搭載したのがN.F.Mk.ⅩⅢ。1943年３月に初飛行し、こちらはすべて新造機としてラインオフしている。レーダーは写真のようにシンブル・ノーズのものと、ブルノーズ(牛の鼻)とも呼ばれたユニバーサル・ノーズの２種類のレドームに収容されていた。武装は胴体下の20mm機関砲４門のみ。写真の機体塗装は夜間戦闘機迷彩塗装に、下面をブラックで塗った、いわゆるイントルーダー迷彩で、1942年12月１日より適用された。ブラックも以前のRDM2Aではなく塗装表面が平滑になったスムーズナイトと呼ばれる塗料が使われている。写真の機体(HK476?)は判然としないが、シリアルが正しければ、No.410Sqn(RA-Q)所属で、1944年６月14～15日にかけて海峡上空でミステル親子飛行爆弾を撃墜している。なお、N.F.Mk.ⅩⅢは翼下に増加タンクが吊下でき、航続距離も向上している。総生産機数は270機。

The N.F. Mk. XIII was derived by mounting the A.I. Mk.VIII centimeter-band air search radar in the F.B. Mk.VI. This type first flew in March 1943, and these were all examples were created on the assembly line as new builds. Thimble-type radomes and bull-nose (universal) radomes were both seen in this variant. The four 20mm Hispano cannon were the only armament. This plane is marked in a night-fighter scheme with black on the underside in the so-called "intruder" scheme. The black paint is not the troublesome RDM2A described above, but a glossier finish known as "smooth night." Though difficult to confirm in this photo, if the s/n "HK476" is correct, than this plane belongs to No. 410 Squadron (RA-Q) and shot down a German Mistel on the night of June 14-15, 1944. The N.F. Mk.XIII could mount drop tanks under the wings for added range. 270 were produced.

◀1942年中頃から高高度で侵入、写真偵察を行なうドイツ空軍のJu86Pを要撃するために単発、双発を問わず高高度要撃戦闘機の開発が進められた。N.F.Mk.XVはモスキートB.Mk.Ⅳをベースに、エンジンを二段二速過給器付マーリン77に換装、コクピットを与圧式とし、防弾鋼板を外すなど機体重量の軽減化が図られ、さらに主翼が伸長され、全幅は16.45メートルから18メートルとなった。機首のシンブル・ノーズにはA.I.Mk.Ⅷを搭載、武装は胴体下面に新たに設けたブリスターに7.7mm機銃４挺のみであった。写真はその試作第１号機(MP469)で、４枚に増加したプロペラも特長のひとつ。この他４機が改造され、1943年３月よりNo.85Sqnに全機配備されたが、肝心の高高度偵察機の侵入もなく、せっかくの機体も空振りに終わってしまった。なお、夜戦エースとして有名なJ.カニンガム中佐(20機撃墜)が本機を駆って高度43,500フィート(13,268メートル)に到達したといわれている。

Development of a plane which could intercept and destroy the ultra-high altitude German Ju86P recon. plane began in mid-1942. To this end, a B. Mk.IV was used as the base for the aircraft, which was powered by two-stage, two-speed supercharger-equipped Merlin 77 engines, had a pressurized cockpit, removed all armor to lighten the plane, and had its wings extended from a span of 16.45m to 18m. A thimble nose housing the A.I. Mk. VIII was added, and a new gun blister housing four 7.7MGs was added to the belly. The photo is the prototype of the N.F. Mk.XV, MP469, and four-bladed propellers are another of its unique traits. Four other examples were built as conversions of existing planes. All of these planes were delivered to No. 85 Squadron, but the feared ultra-high altitude recon. intrusions never occurred and this Mosquito never had a chance to prove itself. The famous night-fighter ace Wg Cdr J. Cunningham (20 kills) is said to have taken one of these planes up to 43,500 feet (13,268m).

▶側面よりみたN.F.Mk.ＸⅤ。爆撃機型から改造された機体のため、戦闘機ながら爆撃機用のキャノピーを装着している。塗装は高高度作戦機に採用されたディープスカイという色が使われているそうだが、カラーチャートをみるとかなり暗いブルー(ごく少量グレイがかっている)で、FS15102に気持ち黒を加えたような色なのだが、この写真をみる限りそんなダークな印象が感じられない。使用されたフィルムの結果そのようになったのかもしれない。とにかくディープスカイという色はPRUブルーと明度的には近似した難解な色調ではある。

A side view of an N.F. Mk.XV. Though a fighter, it was a modification of a bomber version of the plane, hence the bomber-style canopy. According to available resources, it was marked in “deep sky” which, on color charts, if a very dark blue that appears to have a bit of gray in it (something like FS15102 with a touch of black added), but the color suggested by this photo is not that dark. Perhaps it has something to do with film aging or other factors. In any event, PRU blue and deep sky are difficult to distinguish in photos.

◀爆撃機型B.Mk.Ⅳの性能、とりわけ高空性能向上型として二段二速過給器付マーリン72/73(1,680hp)装備のB.Mk.Ⅸが1943年４月に就役、計54機が生産されたが、これに前後して同じくマーリン72/73装備の写真偵察機型、P.R.Mk.Ⅸが開発され、合計90機が生産された。つづいて、より高高度性能を高めた写真偵察機型、P.R.Mk.ＸⅥが1943年７月に初飛行に成功、合計432機もの多数が量産され、同年12月には部隊配備がはじめられた。本機は与圧キャビンを装備し、爆弾倉に相当する機内に大型の燃料タンクを装備して、ヨーロッパ全域での偵察任務に大活躍した。なお、後期の機体ではエンジンをマーリン76/77に換装している。写真のP.R.Mk.XVI(MM341)はマーリン72/73装備機である。胴体下面の写真偵察窓に注意。

The B. Mk.IX was a derivative of the B. Mk.IV, created primarily to improve its high-altitude performance. To that end, the plane’s engines were replaced with two-stage, two-speed turbocharger-equipped Merlin 72/73s (1,680hp) beginning in April of 1943, and 54 were produced. At the same time, 90 P.R. Mk.IX variants that also mounted the Merlin 72/73 were produced. To boost high-altitude performance even further, the P.R. Mk. XVI was produced beginning in July 1943, and unit placements begin in December of the same year. Production of this successful variant was much higher, with 432 examples being completed. The Mk.XVI boasted a pressurized cockpit and a bomb bay-sized fuselage fuel tank that gave it the ability to spy on virtually any location in Europe. Later models mounted the Merlin 76/77 engine. The plane in the photo, MM341, is powered by the Merlin 72/73. Note the camera windows on the underside of the fuselage.

▶機体全面PRUブルーに、白／黒のインヴェイジョン・ストライプ(敵味方識別帯)を描いた、ベンソン基地のNo.544Sqn所属のモスキートP.R.Mk.XVI(NS502/M)。果てしない雲海をバックに、美しくかつ力強いショットだが、時あたかも連合軍によるノルマンディ上陸作戦たけなわのころで、この雲海の下では凄惨な戦いが絶え間なく繰り広げられているはずだ。コクピットのキャノピー頂部には天測ドームが取り付けられている。本機は翌年2月21日に不時着、損傷し、修理後の1947年11月7日、こんどは英海軍に引き渡されている。

Here's NS502:M from No. 544 Squadron, a P.R. Mk.XVI based at Benson, sporting an all-PRU blue scheme with invasion stripes. Note the astrodome on the top of the canopy. This aircraft was damaged in a landing accident on February 21, 1945. Following repairs, it was transferred to the Royal Navy on November 7, 1947.

◀P.R.Mk.XVIの開発と相前後して、爆撃機型でもB.Mk.IXの後継として、より高高度性能の向上を企図したモスキートB.Mk.XVIが開発された。これは与圧キャビンをもち、実用上昇限度を1万メートルとした機体で、最初の12機をのぞいてすべて大型爆弾倉を標準装備しており、4,000ポンド爆弾の運用が可能となった。また、B.Mk.IXと同じく主翼はユニバーサルウィングを採用、爆弾または増加タンクの懸吊することができた。後期の機体では出力向上したマーリン76/77(1,710hp)が搭載されている。402機が生産された。写真は逆光で詳細の把握にはふさわしくはないが、有名な写真家、チャールズ・E・ブラウンの作品だけあって味わいある一葉である。本機(ML963/8K-K)はNo.571Sqn所属機で、1945年4月10、11日にかけてのベルリン爆撃の際に失われてしまった。なお、本文中のB.Mk.35(TA634)のマーキングは本機のそれを模したものである。

About the same time as development of the P.R. Mk. XVI took place, the B. Mk.XVI was created to follow-up the B. Mk.IX with even better performance at altitude. With a pressurized cabin, this variant could reach 10,000m. With the exception of the first 12 aircraft, all had the enlarged bomb bay that could handle 4,000-lb. ordnance. Like the B. Mk.IX, it mounted the universal wing that could handle drop tanks or additional ordnance. Later models were powered by the 1,710hp Merlin 76/77. 402 examples of the type were produced. Though the back lighting makes details difficult to discern, this photo is nevertheless the work of the renowned Charles E. Brown. The plane seen here, ML963:8K-K of No. 571 Squadron was lost while bombing Berlin on April 10-11, 1945. The B. Mk.35 in this book (TA634) is marked as this aircraft.

◀わずか27機が生産されただけの対艦船攻撃機型モスキートがこのF.B.Mk.XⅧ。F.B.Mk.Ⅵをもとに改造された機体で、機首武装の7.7mm機銃を２梃に減じ、胴体下面の20mm機関砲４門の代わりに６ポンド対戦車砲(口径57mm)１門を搭載、コクピット、エンジン周囲には装甲が施されていた。６ポンド砲の装弾数は25発で、これを20秒間で発射した。なお、機首の7.7mm機銃は前方掃射用というよりも照準用に使われたのかもしれない。写真の機体(NT255/O)はコースタル・コマンド(沿岸航空機軍団)隷下No.248Sqn(本来のコードDM-)所属機で、主翼、胴体にはインヴェイジョン・ストライプが巻かれている。同飛行隊はF.B.Mk.Ⅵとの混成で、本機が艦船攻撃、F.B.Mk.Ⅵがエアカバーを担当した。Ｄデイ以後６月中に対艦、対潜、そして対空砲火制圧にと活躍、Ｕボート１隻撃沈、１隻撃破、そして撃墜数機を記録している。胴体下面の６ポンド砲にあやかってか、本機はモスキート(蚊)ならぬ「ツェツェ蝿」の別称を賜った。

The anti-shipping F.B. Mk.XVIII seen here was a modification of the F.B. Mk.VI. Only 27 examples were completed. Nose armament was reduced to just two 7.7mm guns, and the belly cannon were replaced with a single six-pounder (57mm) cannon. Armor plate was also added around the cockpit and engines. The six-pounder held 25 rounds and could fire them all in a span of 20 seconds via an auto-loader mechanism. The nose guns were primarily used to aim the main cannon. The plane in this photo (NT255:0) is of No. 248 Squadron, assigned to Coastal Command. Note the invasion stripes. No. 248 was a mixture of Mk.XVIII and Mk.VI aircraft, with the former attacking shipping while the latter flew cover. During the month of June, 1944 alone, the squadron flew scores of sorties against shipping, submarines and anti-aircraft emplacements, and was credited with one U-boat sunk, one damaged, and several aircraft downed. In some circles, Mosquitoes armed with the six-pounder were referred to as "tsetse flies."

▶1944年５月に初飛行した夜間戦闘機であるN.F.Mk.XⅨは最初からブルノーズとよばれたユニバーサル・ノーズを装備した機体で、装備レーダーは英国製のA.I.Mk.Ⅷ、または米国製SCR720ー英国名A.I.Mk.Ｘのどちらかが搭載できた。本機は主として中高度以下での戦闘を主眼としており、エンジンもマーリン25が搭載されていた。おそらく、同年始めから開始されたドイツ空軍による英本土爆撃作戦「シュタインボック」に対応するために急遽夜間戦闘機の整備、充実が求められたためであろう。もっとも本機の初飛行に前後してドイツ空軍の作戦も先細りになってしまったのは皮肉ではあったが。とはいえ、合計220機が生産されている。

First flying in May of 1944, the N.F. Mk.XIX was the first Mosquito to sport the "bull nose" universal nose. Underneath that cover was either a British A.I. Mk.VIII radar unit, or an American SCR720 (known as the A.I. Mk.X in British use). This aircraft was primarily a fighter in mid to high altitudes and was powered by the Merlin 25. Certainly one of the reasons for its rapid development was the "Steinbock" bombing campaign launched by Germany against the British Isles early in the same year. 220 examples were produced.

▶ブルノーズの愛称がぴったりのレドームを突き出して基地に憩うモスキートN.F.Mk.30(MM687)。二段二速マーリン装備が標準化しつつあったモスキート各型にあって、夜間戦闘機にも遅ればせながらその波が寄せてきた。機首レドームには最新鋭のA.I.Mk.Ⅹレーダーを装備、強力な電子機器への更新にくわえ、エンジンはマーリン72/73、76/77が搭載され、文字通り最強の夜間戦闘機としてデビューした。初期の機体では排気管とそのシュラウドは写真のような旧来のタイプが継承されたが、エンジン出力向上に適応しなかったのか、トラブルに悩まされた。結局、新たに開発したルーバー型の新排気管シュラウドの採用によってこのトラブルは克服された。実戦配備は1944年６月ころからはじまり、英本土の防空のほか、夜間爆撃機部隊のエスコートとしてドイツ本土にも侵攻し、迎え撃つドイツ夜戦と終戦まで死闘を演じた。本機は計530機もの多数が生産されたが、結局、本型が大戦に参加した最後のモスキートになった。

An N.F. Mk.30 (MM687) showing off its "bull nose" radome. Within that resides a state-of-the-art A.I. Mk.X radar system. The wave of conversions to the two-stage, two-speed supercharger-equipped engine reached the night-fighting Mosquitoes with this model. Powered by the Merlin 72/73 or 76/77, it was perhaps the best night-fighter of the war. Early models of the plane had exhaust pipes and shrouds like that of the plane seen here, but this configuration caused trouble and was eventually converted to the louver-type pipes and shrouds. The N.F. Mk.30 first saw combat in June of 1944 in missions to protect the night skies of Britain as well as escorting night bombing missions to Germany. 530 were completed. This was the final version of the Mosquito introduced in WWII.

◣B.Mk.ⅩⅥによって採用された大型爆弾倉に燃料タンクを増設して航続距離を大幅に向上させた写真偵察機がP.R.Mk.34。機内燃料だけでも3,600リットルにも達し、さらに主翼下面のラックも最大900リットル増加タンク２個が懸吊でき、これにより最大航続距離は実に5,600kmを記録した。エンジンはマーリン76/77または113/114(1,690hp)を搭載、最大速度684km/h、上昇限度12,000mという高性能を発揮した。また、与圧キャビンを備え、偵察カメラは計５基を装備するなど、モスキート写真偵察機型の掉尾を飾るにふさわしい機体として大きく期待された。しかし、初飛行は1944年12月４日と大戦も終局段階にあり、実戦配備も翌年４月で、ヨーロッパでのデビューには時期を逸し、さらに極東に派遣されたのも日本が降伏する直前で、結局、さしもの高性能を発揮できないまま終戦を迎えた。本機は計50機が生産された。英空軍におけるモスキートの最終任務は1955年12月15日、RAFセレターに駐留するNo.81Sqnのモスキート P.R.Mk.34(RG314)による偵察任務だとされている

Developed from the B. Mk.XVI, the P.R. Mk.34 was an ultra long-range version of the Mosquito. With a huge fuel tank mounted in the bomb bay the plane could carry a remarkable 3,400 liters of fuel, and could also add a total of 900 liters in tanks mounted under the wings. With this much fuel, the plane had a range of 5,600km. However, range was not the only remarkable point in its performance. Powered by the Merlin 76/77 or 113/114 engine (1,690hp), it could reach 684km/h and climb to 12,000m. Boasting a pressurized cockpit and five reconnaissance cameras as well, it was truly the ultimate reconnaissance Mosquito. However, it first flew in December of 1944, and did not reach units until April of the following year when the war in Europe was in its final stages. Likewise, by the time deliveries were switched to the Pacific, the conflict there was winding down as well, and the type never saw combat. 50 examples were built. The final mission flown by any RAF Mosquito is said to have been carried out by a P.R. Mk.34 of No. 81 Squadron at Seletar on December 15, 1955.

◀爆撃機型モスキートのフィナーレを飾ったB.Mk.35。エンジンはマーリン113/114で、大型爆弾倉を標準装備、与圧キャビンを備えていた。しかしながら本機も初飛行が1945年３月で、結局、P.R.Mk.34と同じく実戦デビューを果たすことなく終わった。生産機数は276機。戦後の英空軍部隊に配属されたほか、民間にも多くの機体が払い下げられ気象観測や連絡などにも使われた。その他、偵察カメラを搭載、P.R.Mk.35として使われたり、本誌で紹介されている機体のように標的曳航機に改造された例も多い。

The B. Mk.35 was the final bomber version of the Mosquito. Powered by the Merlin 113/114, it carried the enlarged bomb bay and a pressurized canopy as standard equipment. However, its first flight did not come until March of 1945, and like the P.R. Mk.34 it was too late to see combat. 276 examples were produced. Many continued in RAF service after the war, and many others were sold into the private sector, being used meteorological observations and other roles. Some were also converted into recon. planes with the addition of cameras, or like the example covered in this book, used as target tugs.

▶モスキートの高性能とそのめざましい活躍に注目していた英海軍では本機の艦上機化を計画していたが、それがこのシーモスキートT.R.33(Torpedo Reconnaissance-fighter:雷撃／偵察・戦闘)である。本機は既存のF.B.MkⅥの機体を利用して、機首にASH捜索レーダー、降着装置も通常の油圧式に改め、艦上機として主翼折畳み、着艦フック追加などが施され、離艦距離短縮のため、RATO取り付けも可能であった。20mm機関砲４門の固定武装のほか、胴体下面に900kg魚雷または爆弾を吊下できた。本機は当初極東での作戦を想定して開発されたが、実機が配備されたのは戦後の1946年８月、翌年７月には唯一の運用飛行隊No.811Sqnも解散されてしまった。写真は50機改造生産されたシーモスキートT.R.33の１機(LR387)で４枚プロペラ、胴体下の着艦フック、機首レドームなどの特長がよくわかる。

Impressed by the Mosquito's performance, the Royal Navy sought to deploy the plane aboard aircraft carriers. The Mosquito T.R. 33 (Torpedo Reconnaissance-Fighter) was the result of these plans. Built from modified F.B. Mk.VIs, the modifications included adding a nose-mounted ASH radar, converting the landing gear to standard oleos, adding folding wings and a landing hook, and mounting RATO gear to shorten the takeoff run. In addition to four 20mm cannons, the plane could carry a 900kg torpedo or bombs under the fuselage. Conceived for operations in the Far East, the plane did not arrive on scene until August of 1946, and the only squadron which operated it (No. 811) was disbanded the following July. This photo, of Sea Mosquito T.R. 33 (LR387) shows off its four-blade propeller, landing hook, nose radome and other unique characteristics well.

◀同じく大戦終結によって余剰となった機体はさまざまな後方任務用に供せられたが、本機もそのひとつ。B.Mk.ⅩⅥの機体をベースに、標的曳航機として改造された機体で、英海軍の発注によって26機が生産され、シーモスキートT.T.(Target Tug:標的曳航)Mk.39と呼称された。写真の機体(PF576)からもおわかりのように機首に大振りなグラスノーズを備え、胴体下面には標的曳航装置を装備していた。写真では機体に隠れて見えないが、胴体後上方には標的観測員用のキューポラが設置されている。流麗なシルエットが魅力のモスキートではあったが、いくら任務からの必然的帰結とはいえ、本機のスタイルにモスキートの老醜をみてしまったような気がしてならない。

Another Mosquito that was relegated to support duties due to the end of the war was the Sea Mosquito T.T. Mk.39 (target tug). Developed from the B. Mk.XVI, the Royal Navy ordered 26 of these aircraft. As can be seen from this photo of PF576, the T.T. Mk.39 had a large glass nose and of course mounted towing gear under the fuselage. Though it is not visible in the photo, there was also a cupola at the rear of the fuselage for a crewmember to observe the towed target.

▶デ・ハヴィビランド社でのモスキート生産開始とほぼ時期を同じくして、カナダの子会社、デ・ハヴィランド・カナダでも本機の委託生産が1941年9月より開始された。モスキートB.Mk.Ⅳに相当するモスキートB.Mk.Ⅶ(パッカード・マーリン31、1,300hp)が25機、B.Mk.ⅩⅩ(パッカード・マーリン33、1,460hp)が245機、さらにB.Mk.25(パッカード・マーリン225、1,635hp)が400機生産された他、F.B.Mk.Ⅵに相当するモスキートF.B.Mk.26(パッカード・マーリン225)が337機生産された。写真の機体(KB162)はモスキートB.Mk.ⅩⅩで、1943年8月にはじめて大西洋を渡って英本土に到着した最初のカナダ製モスキートである。本機は同年末にNo.139 Sqn(XD-J)に配属されたが、1944年10月14日、ウォーボーイズを離陸直後、エンジン故障で墜落し、失われた。本機にとって80回目の出撃であった。胴体後部には本機を戦闘機にみせかけるためのスカイ色の帯による欺瞞塗装が施されているのに注意。

At about the same time that de Havilland began Mosquito production in Britain, their subsidiary in Canada (de Havilland Canada) began license production in September 1941. The company produced 25 examples of the B. Mk.VII (equivalent to the B. Mk.IV, powered by the 1,300hp Packard Merlin 31), 245 examples of the B. Mk.XX (Packard Merlin 33, 1,460hp), 400 examples of the B. Mk.25 (Packard Merlin 225, 1,635hp) and 337 examples of the F.B. Mk. 26 (equivalent to the F.B. Mk.VI, powered by the Packard Merlin 225). The Mosquito in this photo, KB162, is a B. Mk.XX and in August of 1943 became the first Canadian-built Mosquito to cross the Atlantic and reach England. Late that year, the plane was placed with No. 139 Squadron (XD-J) and stationed at Warboys but on October 14, 1944, it was destroyed when it crashed due to engine trouble shortly after take-off. The plane had marked 80 sorties before that. Note the sky color marking on the rear of the fuselage as part of a camouflage scheme designed to make it look like a fighter.

◀カナダと同じく、当時英連邦の一員であったオーストラリアの、デ・ハヴィランド・オーストラリア社でもモスキートの生産が準備された。生産された機体はF.B.MkⅥに相当するF.B.Mk.40で、エンジンはパッカード・マーリン31ないし33で1943年7月より178機、写真偵察機P.R.Mk.40が6機、練習機型T.Mk.43が11機(その他、11機がF.B.Mk.40からのちに改造)であった。これに加えてエンジンを二段二速のパッカード・マーリン69(1,720hp)搭載のモスキートMk.41が17機が生産された。本機はP.R.Mk.41と表記され、あたかも軍用写真偵察機と同一視されている向きもあるが、実際は地形測量や調査目的の機体である。この他オーストラリア空軍では英国より76機(F.B.Mk.Ⅵ/T.Mk.Ⅲ/P.R.Mk.ⅩⅥ)が購入された。写真はF.B.Mk.40のラインナップ。機体全面フォリッジグリーン(FS34092が近似色)塗装。

Like Canada, de Havilland's subsidiary in Australia also manufactured the aircraft. Planes built were 178 examples of the F.B. Mk.40 (equivalent to the F.B. Mk.VI, powered by Packard Merlin 31 or 33) by July of 1943, six examples of the P.R. Mk.40, eleven examples of the T. Mk.43 trainer (later, eleven more were produced by converting F.B. Mk.40s). In addition, they produced 17 examples of the Mosquito Mk.41, which was powered by the two-stage, two-speed supercharger-equipped Packard Merlin 69 (1,720hp). Though this sounds like the name of a military-use aircraft, it was actually developed for geographic survey work. In addition, Australia imported 76 Mosquitoes from Britain, including F.B. Mk.VI, T. Mk.III and P.R. Mk.XVI aircraft. The photo shows a line-up of F.B. Mk.40s. The are marked in all-foliage green (very close to FS34092) schemes.

作図／カルミネ・ディ・ナポリ　解説／ラファエレ・マンチーニ、ステファニア・グイドリッツィ
Illustrations by Carmine Di Napoli, Commentary by Raffaele Mancini & Stefania Guidorizzi

左エンジン・ナセル外側の変遷
Left Engine Nacelle Variations (outboard)

Short, one-stage engine nacelle fitted to P.R.Mk.I and B.Mk.IV Series I (P.R./Bomber Conversion Type)

P.R.Mk.IとB.Mk.IVシリーズ I（写真偵察機型から爆撃機型への改修タイプ）の1速過給器付エンジン・ナセル（ナセル延長前のタイプ）

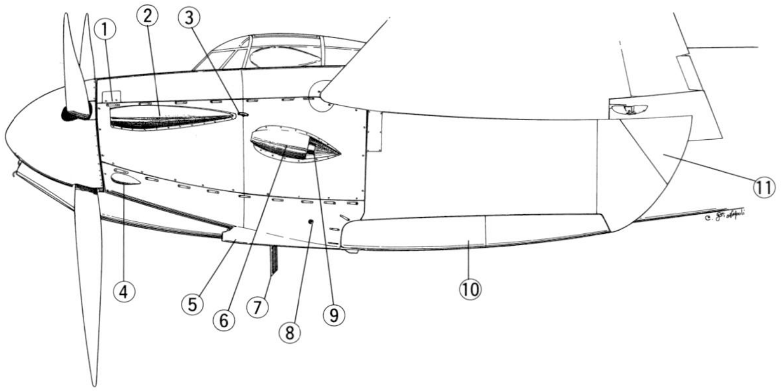

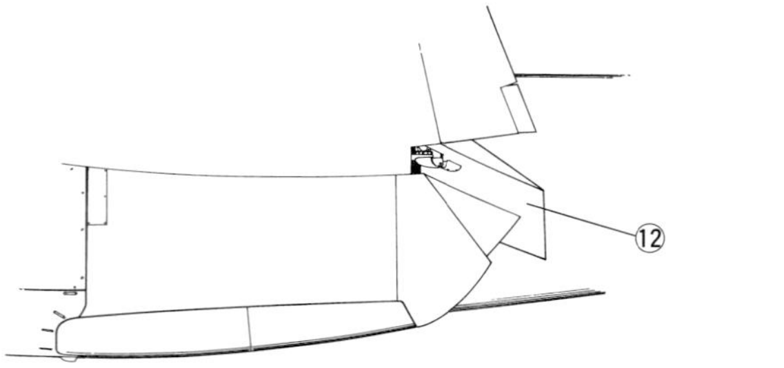

① 冷却液注入ハッチ
② 排気管フェアリングの空気取入口
③ 排気ベンチュリ
④ 冷却液パイプのバルジ（ナセル左側のみ）
⑤ 極初期型の気化器空気取入れ口フェアリング
⑥ 極初期型の排気口フェアリング
⑦ 曳航式アンテナ用フェアリング
⑧ 燃料ポンプ・ドレイン（ナセル左側のみ）
⑨ 排気管
⑩ 極初期型の主脚収容ドア
⑪ 極初期型のナセル後部フェアリング（取り外し式）
⑫ 1枚式フラップ下げ位置

① Coolant fill hatch
② Exhaust air intake fairing
③ Exhaust venturi
④ Coolant pipe bulge (left side only)
⑤ Carburetor air intake fairing (very early type)
⑥ Exhaust port fairing (very early type)
⑦ Trailing antenna fairlead fairing
⑧ Fuel pump drain (left side only)
⑨ Exhaust pipe
⑩ Main wheel door (very early type)
⑪ Movable nacelle tail fairing (very early type)
⑫ One-piece flap in down position

Lengthened one-stage engine nacelle early-model

ナセル延長後の1速過給器付エンジン・ナセル（初期タイプ）

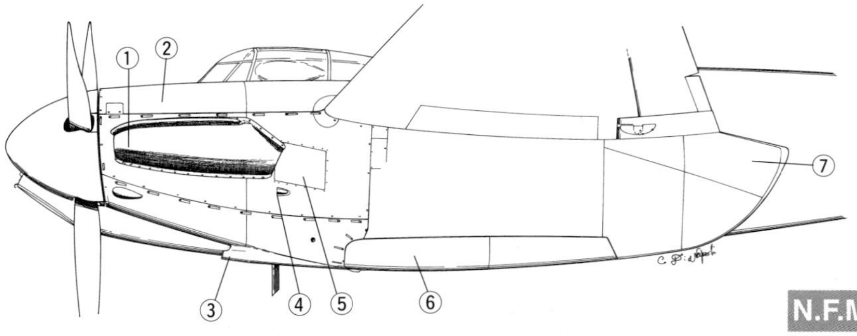

① 排気管消焔シュラウド
② 排気マニフォールド・フランジ冷却空気ダクト
③ 大型化された初期型の気化器空気取入口フェアリング
④ 燃料ポンプ冷却空気取入口
⑤ 排気用耐熱パネル
⑥ 初期型の主脚収納ドア
⑦ 延長されたナセル後部フェアリング

① Exhaust flame damper shroud
② Cooling spouts air duct fairing for exhaust manifold flanges
③ Carburetor air intake fairing (enlarged, early type)
④ Fuel pump cooling air intake
⑤ Heat resistant panel
⑥ Main wheel door (early type)
⑦ Fixed fairing to trailing edge of wing

Lengthened one-stage engine nacelle late model

ナセル延長後の1速過給器付エンジン・ナセル（後期タイプ）

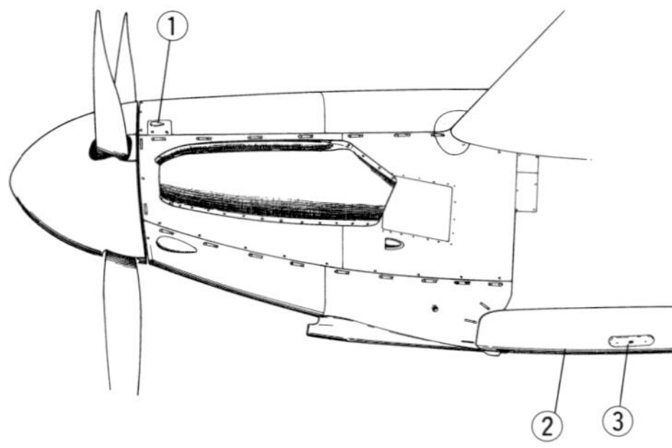

① 冷却液注入ハッチのバルジ（マーリン25装備機のみ）
② 後期型の主脚収納ドア
③ 主脚収納ドアロック用ラッチ点検ハッチ

① Coolant fill hatch bulge (Merlin 25-equipped only)
② Main wheel door (late type)
③ Wheel door locking latch inspection hatch

N.F.Mk.XIX and F.B.Mk.VI Series II engine nacelle late-model

N.F.Mk.XIXとF.B. Mk.VIシリーズIIのエンジン・ナセル（後期タイプ）

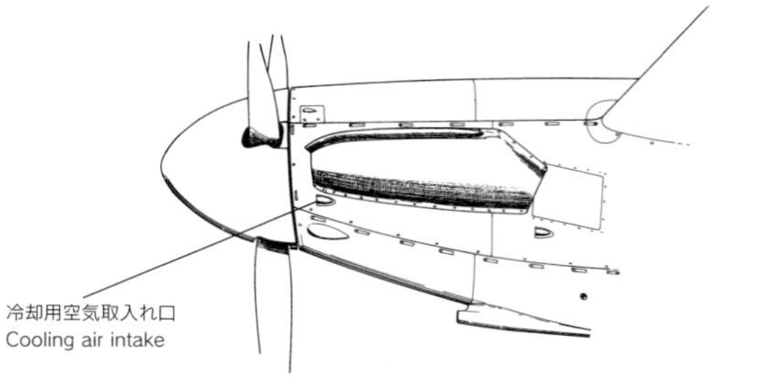

Two-stage engine nacelle fitted to B.Mk.IX, P.R.Mk.IX, N.F.Mk.XV, B.Mk.XVI, P.R.Mk.XVI, P.R.Mk.34, B.Mk.35

B.Mk.IX、P.R.Mk.IX、N.F.Mk.XV、B.Mk.XVI、P.R.Mk.XVI、P.R.Mk.34、B.Mk.35の2段過給器付エンジン・ナセル

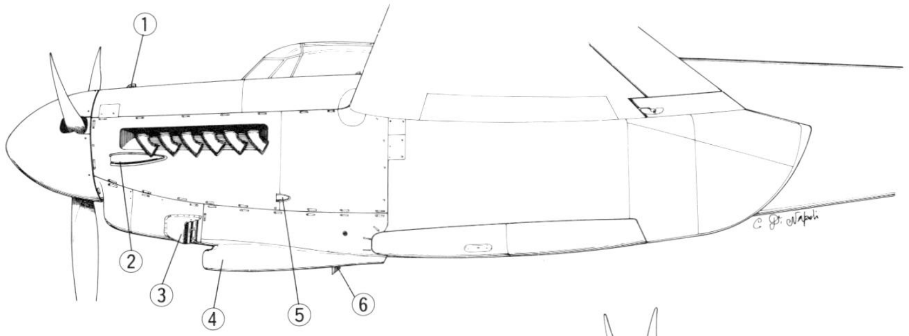

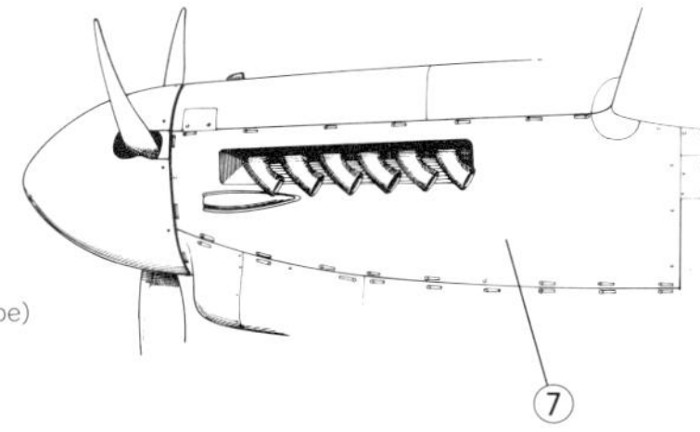

① 冷却液タンク・ベント
② 磁石発電機冷却用空気取入口
③ 中間冷却器用空気排出口
④ 後期型の気化器空気取入口フェアリング
⑤ 燃料ポンプ冷却空気取入口
⑥ 曳航式アンテナ用フェアリング
⑦ 後期タイプのカウリング側面パネル

① Coolant header tank vent
② Magneto cooling air intake
③ Intercooler duct outlet
④ Carburetor air intake fairing (late type)
⑤ Fuel pump cooling air intake
⑥ Trailing antenna fairlead fairing
⑦ Cowling side panel (late type)

P.R.Mk.34 and B.Mk.35 engine nacelle (late model) fitted with longer spinner (overall length 41ft-6in)

延長型スピナーを装備したP.R.Mk.34、B.Mk.35後期タイプのエンジン・ナセル（全長41ft-6in）

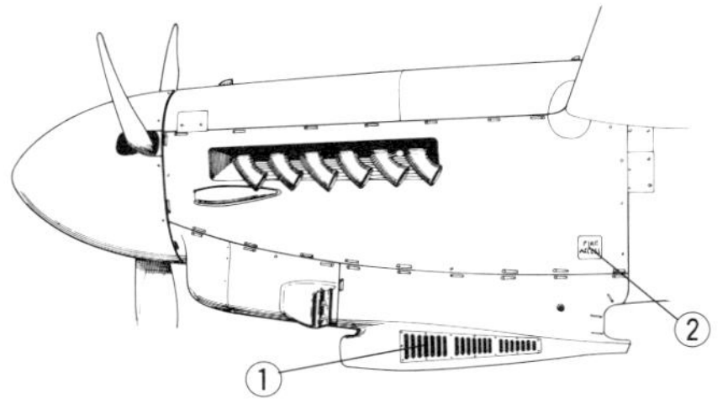

① 冷却用ルーバー付きの気化器空気取入口フェアリング（B.Mk.XVIの一部にもこのタイプを装備した機体あり）
② エンジン始動用カートリッジスターター用点検ハッチ

① Carburetor air intake fitted with cooling louver plate (some B.Mk.XVIs were also equipped with this fitting)
② Fire access hatch

Engine nacelle of B.Mk.35 converted into T.T.Mk.35 and fitted with longer spinner (overall length 41ft-6in)

B.Mk.35を改修したT.T.Mk.35の延長型スピナーとエンジン・ナセル（全長41ft-6in）

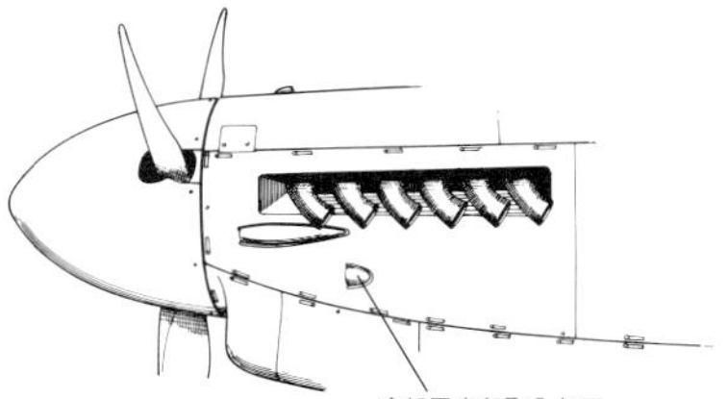

冷却用空気取入れ口
Cooling air intake

N.F.Mk.30 two-stage engine nacelles

N.F.Mk.30の2段過給器付エンジン・ナセル

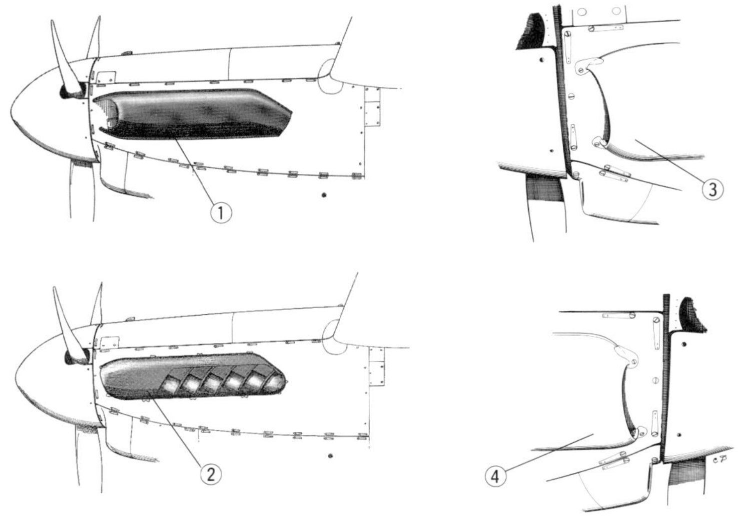

① 初期型の排気管消焔シュラウド
② 後期型の排気管消焔シュラウド
③④ 後期型の排気管消焔シュラウド前方空気取入口は左右で形状が異なる

① Early-type exhaust flame damper shroud
② Late-type exhaust flame damper shroud
③ Late-type front exhaust shroud intake fairing fitted on the left side of both nacelles
④ Differing shape of the front of the late-type exhaust shroud intake fairing fitted on the right side of both nacelles

エンジンマウント
ENGINE MOUNT

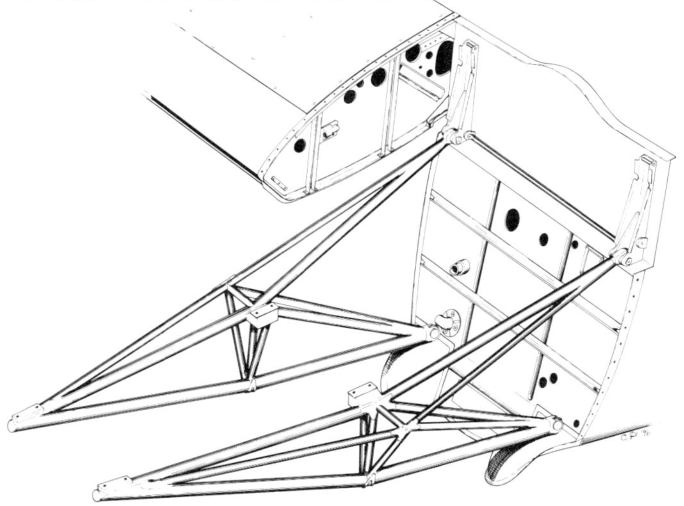

Merlin engine mount

マーリンエンジンのマウント部

翼端のバリエーション Wingtip Variations

Very early-type
極初期型

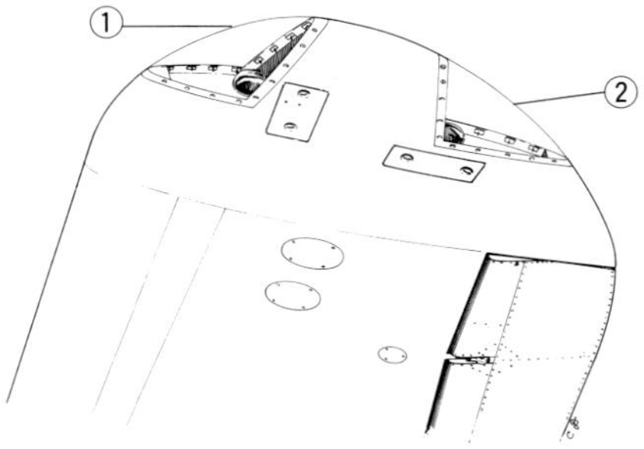

▲P.R.Mk.I、B.Mk.IVシリーズI（写真偵察機型から爆撃機型への改修タイプ）、B.Mk.IVシリーズII初期生産分とF./N.F.Mk.II初期生産分の翼端。
Very early-type fitted to P.R.Mk.I, B.Mk.IV Series I (P.R.U./Bomber Conversion Type), B.Mk.IV Series II early production aircraft, F./N.F.Mk.II early production aircraft.

Early-type
初期型

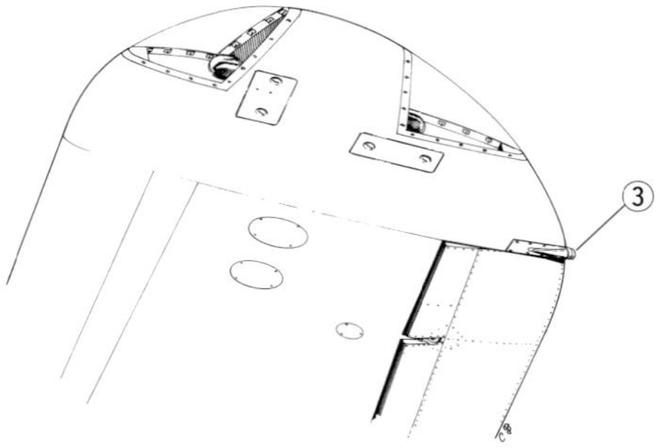

▲レジンランプ付きの初期型翼端。B.Mk.IVシリーズII中期生産分にみられる。
Early-type equipped also with a resin lamp, fitted to B.Mk.IV Series II mid-production aircraft.

① 航法灯
② 編隊灯（白）
③ レジンランプ

① Navigation light
② Formation light (white)
③ Resin lamp

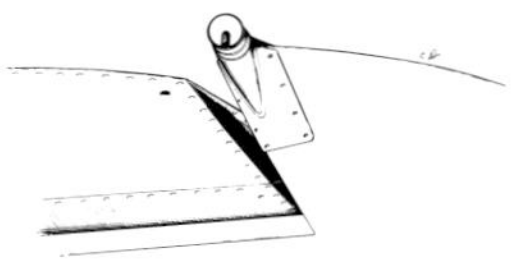

翼端後部のレジンランプ詳細
The rearward resin lamp

Standard late-type
後期型の標準的翼端

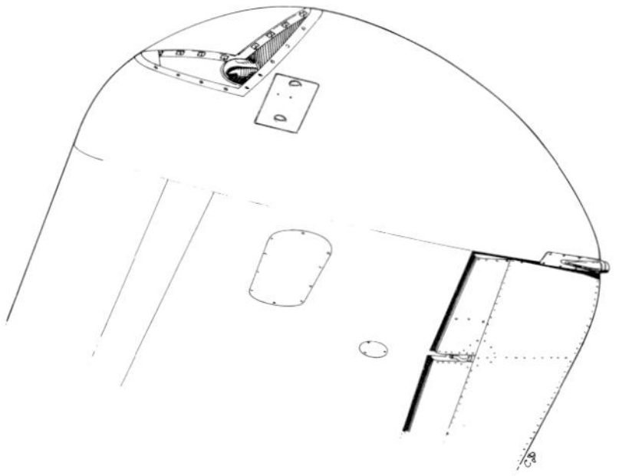

▲航法灯１灯とレジンランプ１灯を取り付けた後期型の標準的翼端。F./N.F.Mk.II後期生産分と後期の戦闘機型バリエーション、B.Mk.IVシリーズIIの後期、B.Mk.IXと後期の爆撃機型バリエーション、P.R.Mk.Iと後期の写真偵察機型バリエーション、F.B.Mk.VI、F.B.Mk.XVIIIにみられる。

Standard late-type equipped with one navigation light and one resin lamp, fitted to F./N.F.Mk.II late production aircraft and later fighter variants, B.Mk.IV Series II late production aircraft, B.Mk.IX and later bomber variants, P.R.Mk.IX and later photo-reconnaissance variants, F.B.Mk.VI, F.B.Mk.XVIII.

60ポンドロケット弾発射レール
RP SINGLE RAILS

※図は左主翼下面に60ポンドロケット弾4発を搭載した状態と、その発射レールを示す
※Arrangement of four underwing single rails equipped with 60-lb RPs.

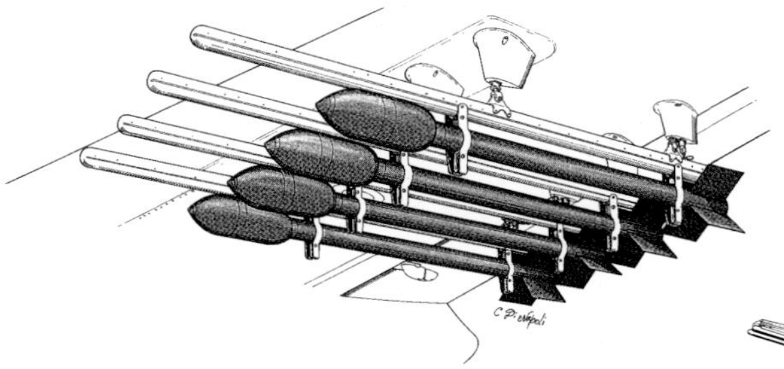

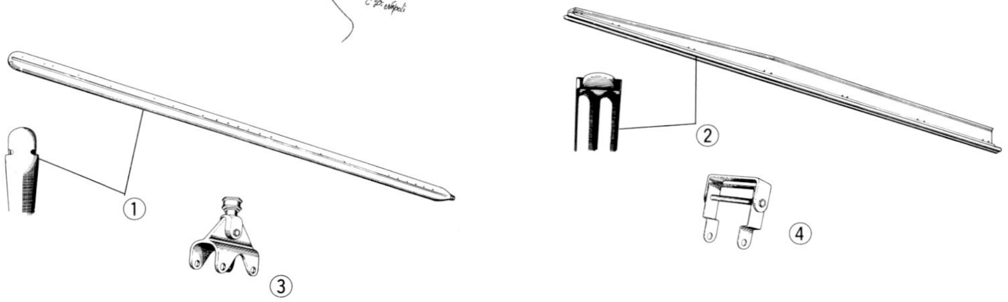

① Mk.IBロケット弾発射レール
② Mk.IIIAロケット弾発射レール
③ Mk.IBの発射レイル固定アタッチメント
④ Mk.IIIAの発射レイル固定アタッチメント

① Mk.IB model rail
② Mk.IIIA model rail
③ Mk.IB rail fork attachment
④ Mk.IIIA rail attachment

N.F.Mk.XVの翼端
N.F.Mk.XV wingtip

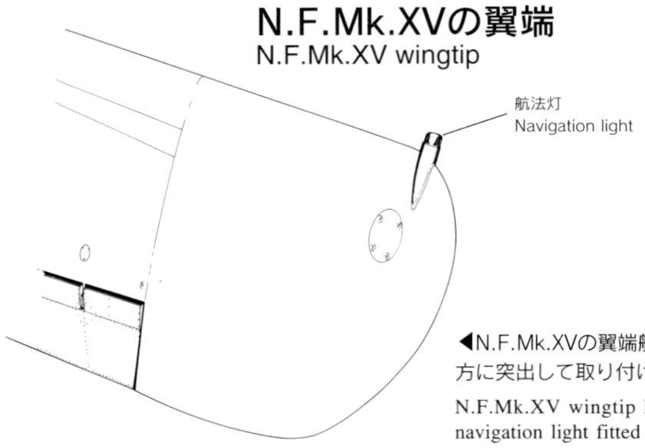

◀N.F.Mk.XVの翼端航法灯は前方に突出して取り付けられる。
N.F.Mk.XV wingtip having the navigation light fitted on a small protuberance.

爆弾架
Underwing Bomb Carrier

※図は外翼下面のラックに落下式増槽タンクか爆弾架のどちらかを懸吊可能な"ユニバーサル・ウイング"で使用する爆弾架を示す。
※Bomb carrier and fairing fitted to "universal wing"

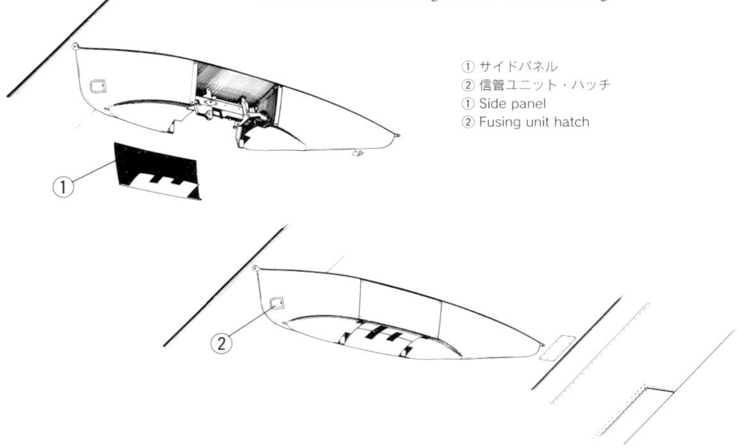

昇降舵ホーンバランス
Elevator Horn Balances

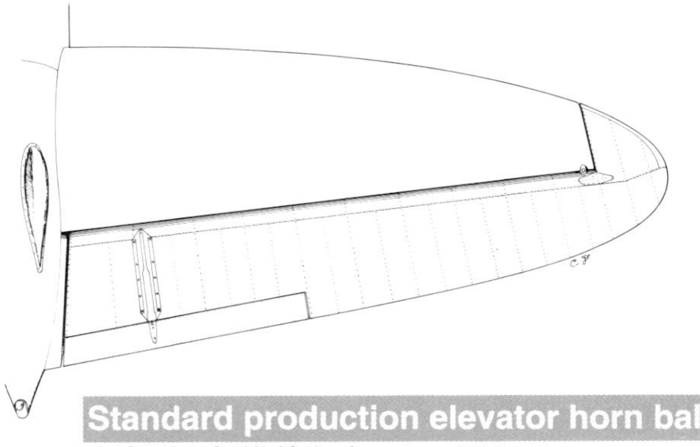

Standard production elevator horn balance
生産型の標準的なホーンバランス

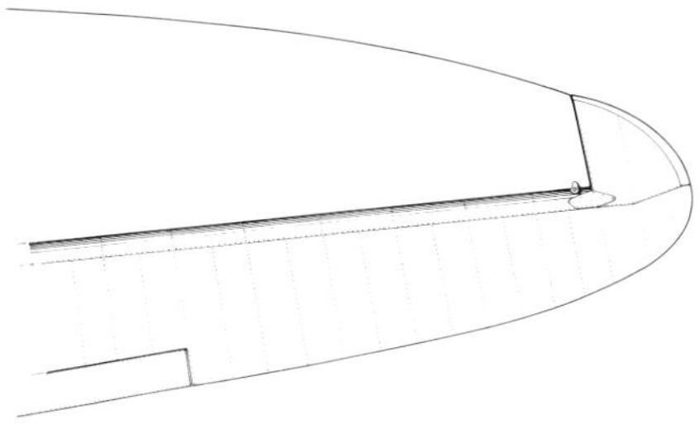

Enlarged horn balance fitted to B.Mk.IV Special and P.R.Mk.34
B.Mk.IVスペシャルとP.R.Mk.34の増積されたホーンバランス

落下増槽
Underwing Drop Tanks

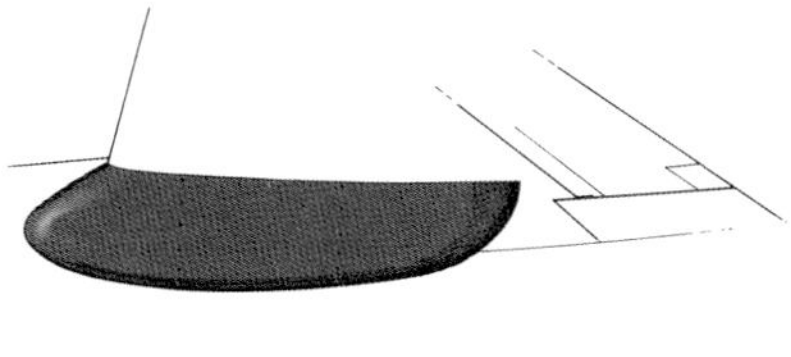

50gal underwing drop tank
50ガロンタイプの落下式増槽タンク

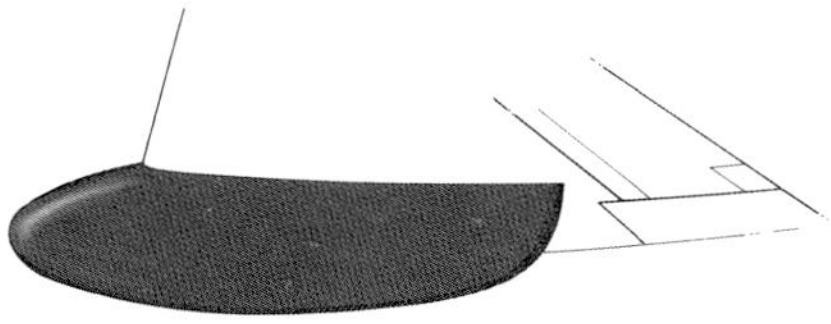

100gal underwing drop tank
100ガロンタイプの落下式増槽タンク

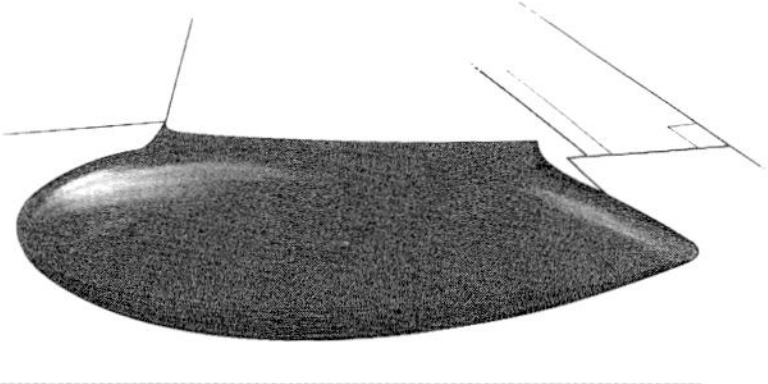

200gal underwing drop tank
200ガロンタイプの落下式増槽タンク

増槽ラック
Slip release gear

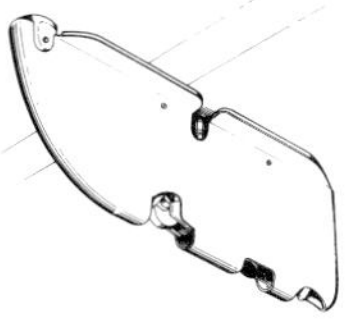

プロペラのバリエーション
Propeller Variations

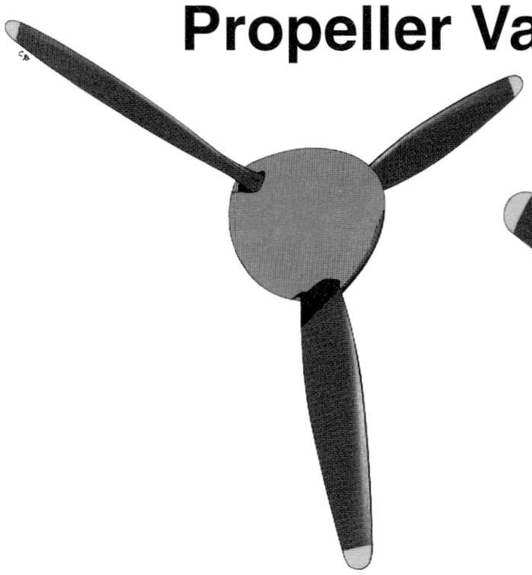

De Havilland Hydromatic needle blade propeller (three-blade, early-type, 12ft - 3.658m diameter)
デ・ハヴィランド・油圧式3翅プロペラ（ニードルブレードタイプの初期型、直径12ft／3.658m）

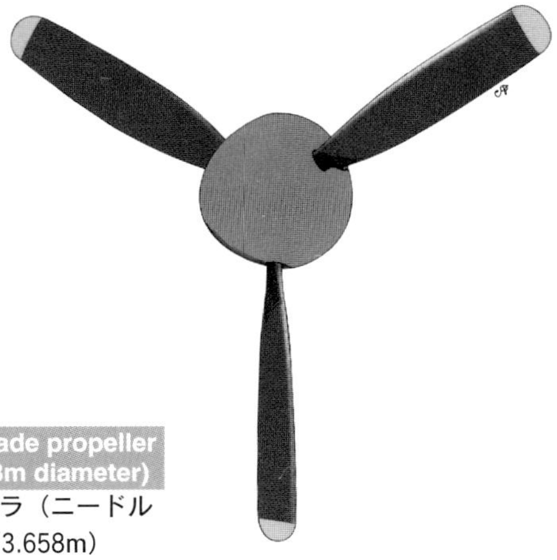

De Havilland Hydromatic paddle blade propeller (three-blade, late-type, 12ft-6in - 3.810m diameter)
デ・ハヴィランド・油圧式3翅プロペラ（パドルブレードタイプの後期型、直径12ft-6in／3.810m）

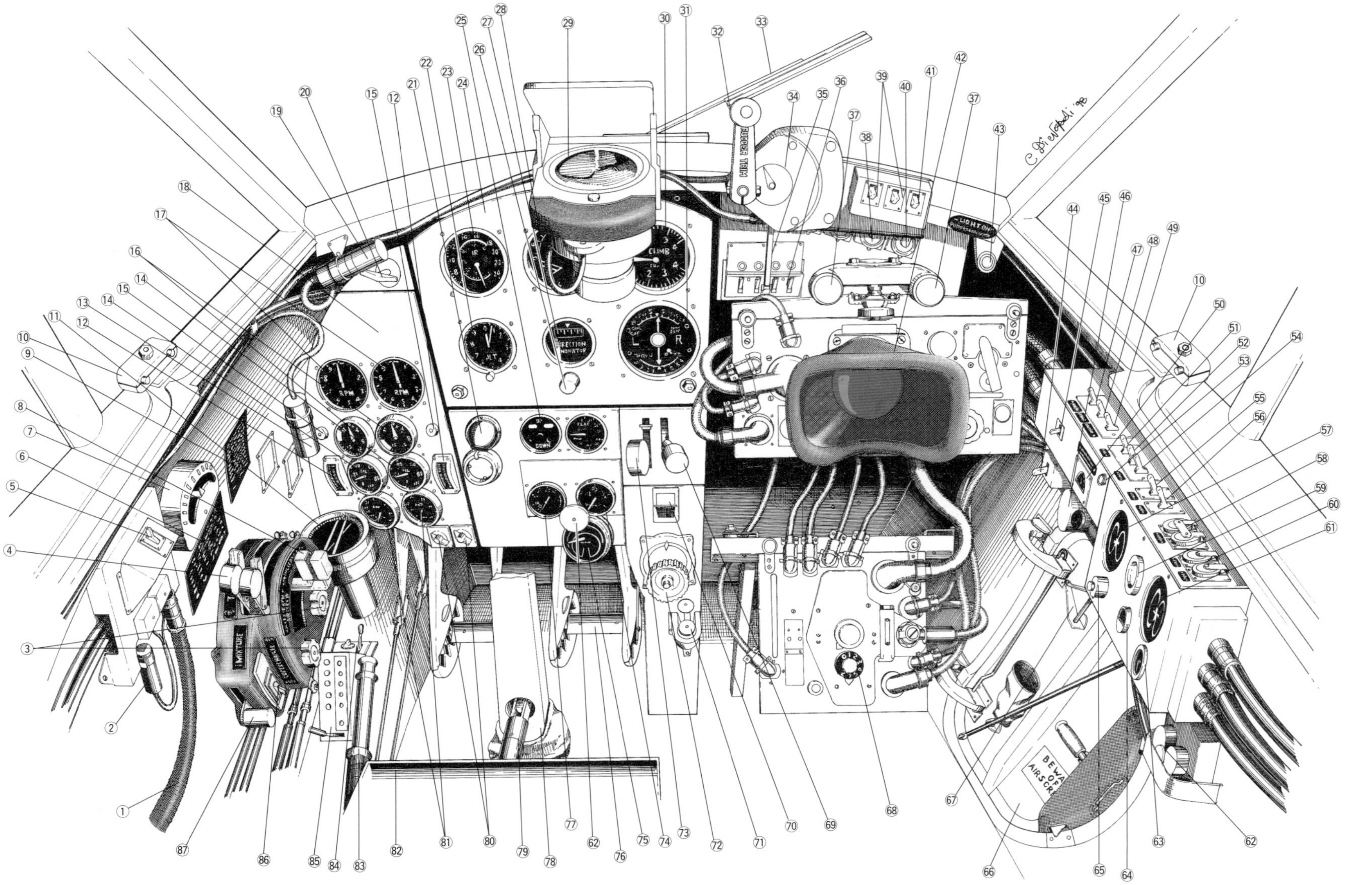

コクピット　N.F.Mk.IIのコクピット内部／N.F.Mk.II COCKPIT LAYOUT

1. パイロット用酸素パイプ
2. インターカム・ジャック
3. 摩擦調整器
4. スロットル・レバー
5. ビーム・アプローチ・スイッチ
6. 緊急時主脚操作手順指示プレート
7. 昇降舵トリムタブ指示計
8. プロペラ調速レバー
9. エンジン制限データ・プレート
10. 側面窓ボール・スライド
11. コンパス
12. 油圧計
13. コンパス灯
14. 油温計
15. 燃圧警告灯
16. ブースト圧計
17. 回転指示計
18. R.I.リピーターコンパス格納部
19. 照明灯
20. ブースト・コントロール・スイッチ
21. パイロット用ベンチレーター
22. 高度計
23. 速度計
24. 飛行計器盤
25. 降着装置位置指示計
26. フラップ位置指示計
27. 方向指示計
28. 人工水平儀
29. 光像式射撃照準器
30. 昇降計
31. 旋回計
32. 方向舵トリムタブ・コンロトール・ハンドル
33. 前部風防ワイパー
34. 方向舵トリムタブ指示計
35. 電装系スイッチ
36. 磁石式発電機スイッチ
37. プロペラ・フェザリング・ボタン
38. エンジン始動スイッチ
39. ラジエター・フラップ用スイッチ
40. ブースター・コイル・スイッチ
41. スペア・スイッチ
42. A.I.（空中要撃）レーダー用ディスプレイ
43. 増槽燃料ポンプ警告灯
44. 編隊灯点灯スイッチ
45. 敵味方識別灯切換えスイッチ
46. 敵味方識別灯コントロール
47. 射撃カメラ・スイッチ
48. 航法灯点灯スイッチ
49. 紫外線ランプ・メインスイッチ
50. 電圧計
51. ピトー管暖房スイッチ
52. 増槽タンク用燃料ポンプ・スイッチ（増槽懸吊時に使用）
53. 発電機警告灯
54. 発電機スイッチ（未使用時）
55. 前方航法灯点灯スイッチ
56. I.F.F.（敵味方識別装置）マスタースイッチ
57. 内翼タンク燃料計
58. I.F.F.破壊ボタン
59. 増槽タンク燃料計（増槽タンク使用機）
60. 消火器スイッチ
61. 外翼タンク燃料計
62. Mk.VIIIc酸素調節器
63. 航空時計
64. 外気温度計
65. 前部風防ワイパー・コントロール
66. 搭乗員乗降ドア
67. 非常時ドア投棄ハンドル
68. A.R.I. 5093受信器
69. フラップ切換レバー
70. 機銃マスタースイッチ
71. 風防防氷ポンプ
72. 補助翼トリムタブ・コントロールと補助翼トリムタブ指示器
73. 降着装置切換レバー
74. 酸素調節器ゲージ
75. 3重圧力計
76. フットペダル
77. 酸素量指示計
78. 操縦桿（図では基部以外を省略）
79. 昇降舵操作ロッド
80. 着陸灯スイッチ
81. 冷却温度計
82. 操縦士席
83. 座席調節レバー
84. 無線選択スイッチ
85. 無線器コントロール・ユニット
86. 過給器ギア変換スイッチ
87. 気化器混合機調整レバー

1. Pilot's oxygen line
2. Intercom jack
3. Friction adjusting knobs
4. Throttle levers
5. Beam approach switch
6. Landing gear emergency plate
7. Elevator trim tab indicator
8. Propeller speed control levers
9. Engine limitations data plate
10. Bottom ball slide
11. Compass
12. Oil pressure gauges
13. Compass floodlight
14. Oil temperature gauge
15. Fuel pressure warning light
16. Boost pressure gauges
17. R.P.M. indicators
18. Stowage for R.I. compass repeater
19. Instrument light
20. Boost control cut-out
21. Pilot's ventilator
22. Altimeter
23. Airspeed indicator
24. Instrument flying panel
25. Landing gear position indicator
26. Flap position indicator
27. Direction indicator
28. Artificial horizon
29. Reflecting gunsight
30. Climb indicator
31. Turn indicator
32. Rudder trim tab control handle
33. Windscreen wiper
34. Rudder trim tab indicator
35. Electrical services switches
36. Magneto switches
37. Propeller feathering buttons
38. Engine electric starter switches
39. Radiator flap switches
40. Booster coil switches
41. Spare switch
42. A.I. radar CRT display
43. Immersed fuel pump warning light
44. Formation lights switch
45. Identification lights selector switch
46. Identification lights control
47. Camera gun switch
48. Navigation lights switch
49. U.V. lighting switch
50. Voltmeter
51. Pitot tube heating switch
52. Immersed fuel pump switch (for long-range tanks when mounted)
53. Generator warning light
54. Generator switch (not in use)
55. Forward navigation light switch
56. I.F.F. master switch
57. Inner wing tanks fuel gauge
58. I.F.F. self-destruct controls
59. Fuel contents gauge (for long-range tanks when mounted)
60. Fire extinguisher switches
61. Outer wing tanks fuel gauge
62. Mk.VIIIc oxygen regulator
63. Clock
64. Air temperature gauge
65. Windscreen wiper control
66. Cockpit entry door
67. Emergency door-jettison handle
68. A.R.I. 5093 receiver
69. Flaps selector lever
70. Gun master switch
71. Windscreen deicing pump
72. Aileron trim tab control and indicator
73. Landing gear control
74. Oxygen contents gauge
75. Triple pressure gauge
76. Rudder pedals
77. Oxygen flow indicator
78. Control column (deleted for clarity)
79. Elevator connecting rod
80. Landing lights switches
81. Coolant temperature gauges
82. Pilot's seat
83. Seat control lever
84. Radio set-selector switch
85. Radio control unit
86. Supercharger gear change switch
87. Engine mixture control lever

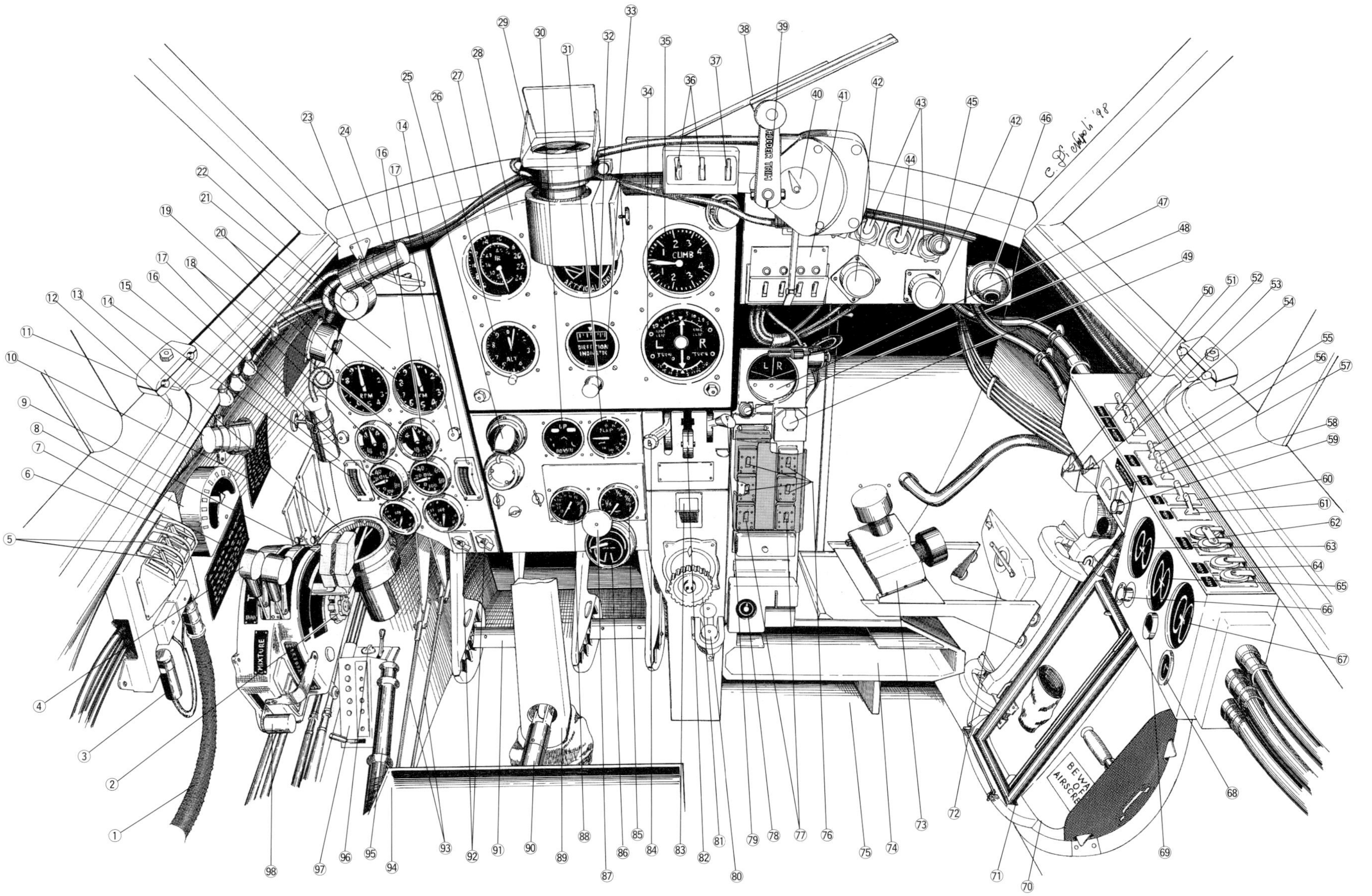

F.B.Mk.VIのコクピット内部／F.B.Mk.VI COCKPIT LAYOUT

1 パイロット用酸素パイプ
2 摩擦調整器
3 インターカム・ジャック
4 スロットル・レバー
5 R.I.コンパス・スイッチ
6 ビーム・アプローチ・スイッチ
7 緊急時主脚操作手順指示プレート
8 昇降舵トリムタブ指示計
9 プロペラ調速レバー
10 エンジン制限データ・プレート
11 側面窓ボール・スライド
12 紫外線ランプ
13 コンパス
14 油圧計
15 コンパス灯
16 油温計
17 燃圧警告灯
18 ブースト圧計
19 照明灯照度調節用加減抵抗器
20 回転指示計
21 R.I.リピーターコンパス格納部
22 紫外線ランプ発光ボタン
23 照明灯
24 ブースト・コントロール・スイッチ
25 パイロット用ベンチレーター
26 高度計
27 速度計
28 飛行計器盤
29 光像式射撃照準器
30 降着装置位置指示計
31 フラップ位置指示計
32 人工水平儀
33 方向指示計
34 旋回計
35 昇降計
36 ラジエター・フラップ用スイッチ
37 空気取入口フィルター・スイッチ
38 前部風防ワイパー
39 方向舵トリムタブ・コンロトール・ハンドル
40 方向舵トリムタブ指示計
41 磁石式発電機スイッチ
42 プロペラ・フェザリング・ボタン
43 ブースター・コイル・スイッチ
44 エンジン始動スイッチ
45 増槽燃料ポンプ警告灯
46 航法士用ベンチレーター
47 ビームアプローチ指示計
48 爆弾倉ドア警告灯
49 爆弾および主翼増槽タンク投下スイッチ
50 カメラ用マスタースイッチ
51 レジンランプ点灯スイッチ
52 航法灯点灯スイッチ
53 紫外線ランプ・メインスイッチ
54 機体下面敵味方識別灯点灯／点灯色切換えスイッチ
55 ピトー管暖房スイッチ
56 増槽タンク用燃料ポンプ・スイッチ（増槽懸吊時に使用）
57 光像式反射照準器スイッチ
58 電圧計
59 発電機警告灯
60 前方航法灯点灯スイッチ
61 I.F.F.（敵味方識別装置）マスタースイッチ
62 I.F.F.破壊ボタン
63 内翼タンク燃料計
64 消火器スイッチ
65 中央タンクおよび増槽タンク燃料計
66 前部風防ワイパー・コントロール
67 外翼タンク燃料計
68 航空時計
69 外気温度計
70 搭乗員乗降ドア
71 乗員昇降用ハシゴ
72 ドリフト・レコーダー・ドア
73 ドリフト・レコーダー
74 チャートボードと地図入れ
75 航法士用パラシュート収納部
76 爆弾切換スイッチ
77 爆弾信管セット・スイッチ
78 距離計
79 フラップ切換レバー
80 風防防氷ポンプ
81 補助翼トリムタブ・コントロールと補助翼トリムタブ指示器
82 降着装置切換レバー
83 機銃マスタースイッチ
84 爆弾倉ドア切換スイッチ
85 酸素調節器ゲージ
86 3重圧力計
87 Mk.VIIIc酸素調節器
88 酸素量指示計
89 操縦桿（図では基部以外を省略）
90 昇降舵操作ロッド
91 フットペダル
92 着陸灯スイッチ
93 冷却温度計
94 操縦士席
95 座席調節レバー
96 無線選択スイッチ
97 無線器コントロール・ユニット
98 気化器混合機調整レバー

1 Pilot's oxygen line
2 Friction controls
3 Intercom jack
4 Throttle levers
5 R.I. compass switches
6 Beam approach switch
7 Landing gear emergency plate
8 Elevator trim tab indicator
9 Propeller speed control levers
10 Engine limitations data plate
11 Bottom ball slide
12 U.V. light
13 Compass
14 Oil pressure gauges
15 Compass floodlight
16 Oil temperature gauge
17 Fuel pressure warning light
18 Boost pressure gauges
19 Floodlight rheostat
20 R.P.M. indicators
21 Stowage for R.I. compass repeater
22 Exciter button for U.V. lighting
23 Instrument light
24 Boost control cut-out
25 Pilot's ventilator
26 Altimeter
27 Airspeed indicator
28 Instrument flying panel
29 Reflecting gunsight
30 Landing gear position indicator
31 Flap position indicator
32 Artificial horizon
33 Direction indicator
34 Turn indicator
35 Climb indicator
36 Radiator flap switches
37 Air intake filter switch
38 Windscreen wiper
39 Rudder trim tab control handle
40 Rudder trim tab indicator
41 Magneto switches
42 Propeller feathering buttons
43 Booster coil switches
44 Engine electric starter switches
45 Immersed fuel pump warning light
46 Observer's ventilator
47 Beam approach visual indicator
48 Bomb doors warning light
49 Bomb containers and wing drop tanks jettison switch
50 Cine-camera master switch
51 Resin lamp switch
52 Navigation lights switch
53 U.V. lighting switch
54 Downward identification lights color control
55 Pitot tube heating switch
56 Immersed fuel pump switch (for long-range tanks when mounted)
57 Reflecting gunsight switch
58 Voltmeter
59 Generator warning light
60 Forward navigation light switch
61 I.F.F. master switch
62 I.F.F. self-destruct controls
63 Inner wing tanks fuel gauge
64 Fire extinguisher switches
65 Fuel gauge - center tank and long-range tank (when mounted)
66 Windscreen wiper control
67 Outer wing tanks fuel gauge
68 Clock
69 Air temperature gauge
70 Cockpit entry door
71 Crew access ladder
72 Drift recorder door
73 Drift recorder
74 Chart board and map stowage
75 Observer's parachute stowage
76 Bomb selector switches
77 Bomb fusing switches
78 Footage indicator
79 Flaps selector lever
80 Windscreen deicing pump
81 Aileron trim tab control and indicator
82 Landing gear selector level
83 Gun master switch
84 Bomb doors selector
85 Oxygen contents gauge
86 Triple pressure gauge
87 Mk.VIIIc oxygen regulator
88 Oxygen flow indicator
89 Control column (deleted for clarity)
90 Elevator connecting rod
91 Rudder pedals
92 Landing lights switches
93 Coolant temperature gauges
94 Pilot's seat
95 Seat control lever
96 Radio set-selector switch
97 Radio control unit
98 Engine mixture control lever

モスキート各型変遷 Variations of the Mosquito

作図／カルミネ・ディ・ナポリ　解説／ラファエレ・マンチーニ
Illustrations by Carmine Di Napoli, Commentary by Raffaele Mancini

（1／72スケール）
(All diagrams are 1/72 scale)

P.R.Mk.I Left view
左側面図

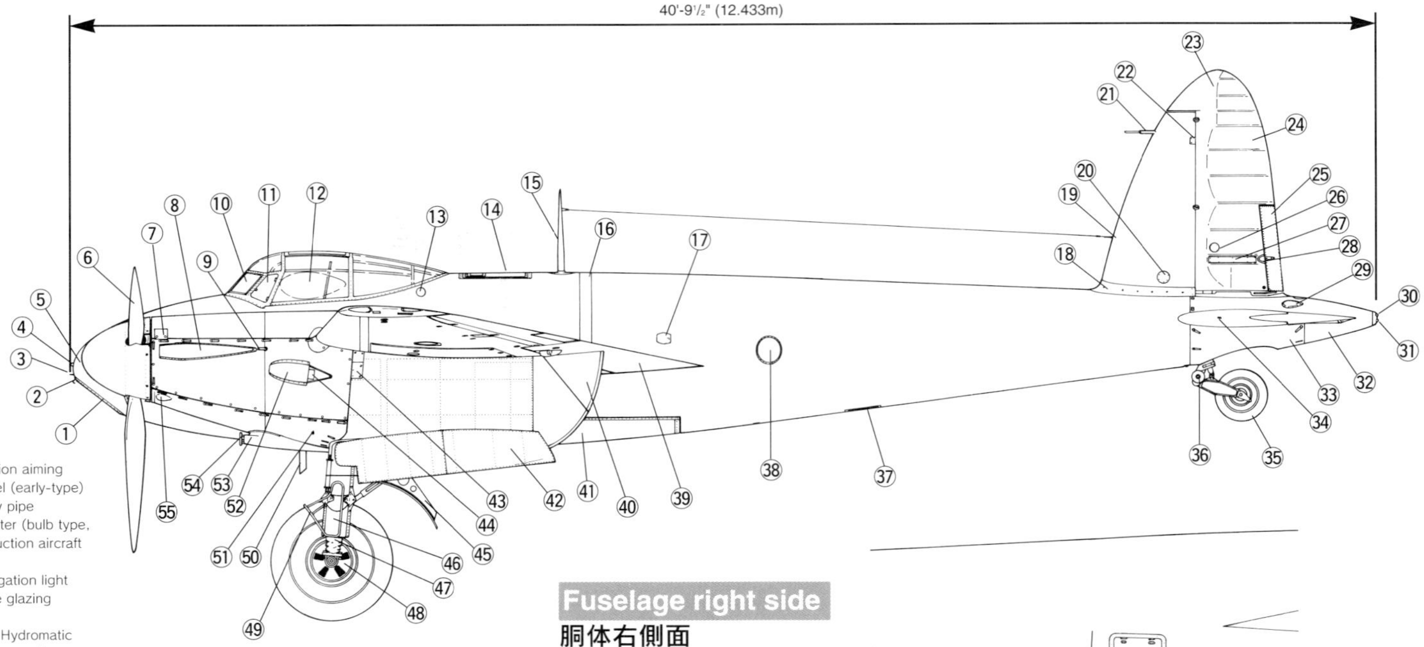

① 初期型の偵察／観測用平面ガラス
② 防氷スプレーパイプ
③ 極初期型のみにみられるバルブタイプの外気温度計
④ 航法灯
⑤ 初期型の機首透明ガラス
⑥ デ・ハヴィランド・油圧式3翅プロペラ（ニードルブレードタイプの初期型、直径12ft／3.658m）
⑦ 冷却液注入ハッチ（左右ナセル左側のみ）
⑧ 排気管フェアリングの空気取入口
⑨ 排気ベンチュリ

① Flat observation aiming Perspex panel (early-type)
② Deicing spray pipe
③ Air thermometer (bulb type, earliest production aircraft only)
④ Forward navigation light
⑤ Perspex nose glazing (early-type)
⑥ De Havilland Hydromatic needle blade propeller (three-blade, early-type, 12ft - 3.658m diameter)

Fuselage right side
胴体右側面

P.R.Mk.Iの乗降ハッチまわり
P.R.Mk.I Entry/access door

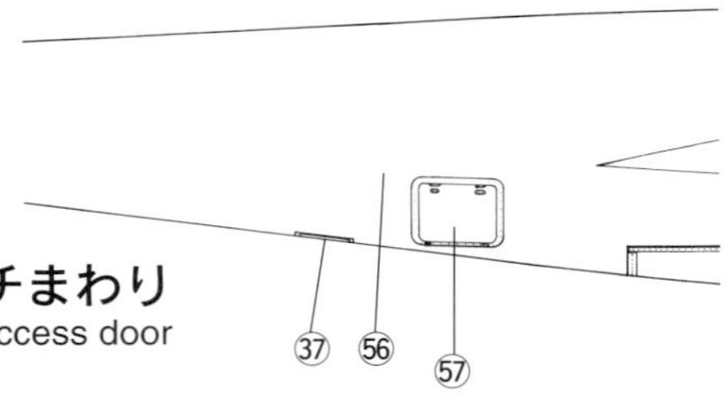

モスキートP.R.Mk.I

写真偵察型P.R.Mk.Iの原型機は、モスキートのプロトタイプ1号機であるW4050号機が初期試験を終了したのち、1941年6月10日に初飛行した。同年7月13日にベンソンの第1写真偵察大隊に配備を開始。その後9機が生産され、9月から10月にかけて部隊配備された。生産型9機のシリアルナンバーはW4054からW4056とW4058からW4063。1941年9月17日、ドイツ占領下のフランスへP.R.Mk.Iによる偵察飛行が行われた。これが第二次大戦におけるモスキート最初の作戦飛行である。P.R.Mk.Iはロールスロイス・マーリン21エンジンを搭載し、プロペラはデ・ハヴィランド初期型ニードルブレード定速プロペラを装備。W4050号機を踏襲したエンジン・ナセルと水平尾翼はのちのタイプにくらべて短く、水平尾翼の翼長は19フィート$5\frac{1}{2}$インチ(5.931m)であった。W4050号機ともっとも異なる点は20インチ（0.508m）延長された主翼と偵察用カメラである。W4050号機の主翼スパン52フィート6インチ(16.002m)に対し、P.R.Mk.Iは翼端の形状を変更したスパン54フィート2インチ(16.510m)の主翼を採用。偵察用カメラは前方2基＋後方1基のセットによる胴体下面の垂直撮影用カメラ計3基と、胴体左側面の斜め下方撮影用カメラ1基の計4基を装備していた。また、シングル・ドラムブレーキが右側のみでドラム左側にスロットのある初期型の主車輪、初期型の主脚収容ドアガイド、初期型尾脚ストラットも、P.R.Mk.I生産型の特徴である。なお胴体下面の敵味方識別灯と主翼下面の着陸灯は、まだ装備されていない。燃料搭載量は通常690ガロン(3,137リッター)であった。

Mosquito P.R.Mk.I

After the early tests of the first Mosquito prototype W4050, the P.R.Mk.I photo-reconnaissance prototype, serialled W4051, made its maiden flight on June 10, 1941, and it was later delivered to the No.1 Photographic Reconnaissance Unit at Benson on July 13, 1941. In all nine production aircraft were manufactured (W4054 to W4056, W4058 to W4063), and they were delivered between September and October 1941. The first operational mission carried out by a Mosquito during WWII was mounted by a P.R.Mk.I on September 17, 1941, flying over French and German territory. This model, equipped with 1,480hp Rolls-Royce Merlin 21 engines driving the early-type de Havilland Hydromatic needle blade constant speed propellers, retained the short engine nacelles and the short horizontal tail planes (19ft-$5\frac{1}{2}$in, 5.931m) of the original prototype W4050. The major differences from W4050 were the longer wings 54ft-2in (16.510m) instead of 52ft-6in (16.002m), with new wingtips, and the camera installation which consisted of three vertical cameras (two forward and one rear) and one oblique camera fitted on the left fuselage underside. The production aircraft were equipped with the very early-type exhaust ports, the early-type main wheels having single brake drums on the right side and slots on the left side, the early-type main wheel door guides and the early-type tail wheel strut construction equipped with the one-track tire. The downward identification code lights on the fuselage underside and the underwing landing lights were not fitted. The fuel capacity was normally 690gal (3,137 liters).

⑩ V字タイプの初期型前部風防
⑪ 内側開閉式の前部風防側面窓
⑫ 側面窓のブリスター
⑬ 燃料注入ハッチ
⑭ 救命ボート収納部パネル（投棄式）
⑮ アンテナ支柱
⑯ 胴体補強用帯金
⑰ 地上スターターソケットハッチ
⑱ 垂直尾翼基部フェアリング
⑲ 初期型のアンテナ空中戦接続部
⑳ 垂直尾翼基部点検ハッチ
㉑ ピトー管
㉒ 外部電源接続部パネル
㉓ 方向舵ホーンバランス
㉔ 方向舵（羽布張り、作動角26°-1°+2°）
㉕ 方向舵トリムタブ（作動角16°±2°30"）
㉖ 方向舵操舵装置点検ハッチ
㉗ 方向舵トリムタブ操作ロッド・アクセスパネル
㉘ 方向舵トリムタブ操作ロッド・フェアリング
㉙ 昇降舵操舵リンケージ・フェアリング
㉚ 航法灯
㉛ 編隊灯
㉜ 初期型のテイルコーン・フェアリング（取り外し式）
㉝ 引込式尾脚アクセスパネル
㉞ アンテナ空中線接続部（アンテナ空中線はない）
㉟ 初期型タイヤ付き引込式尾輪
㊱ 初期型の尾輪ストラット
㊲ 胴体下面カメラ窓
㊳ 斜め下方撮影用カメラ窓
㊴ 極初期型の1枚式フラップ（合板仕上げ、作動角45°±2°）
㊵ 極初期型のナセル後部フェアリング（取外し式）
㊶ 胴体下面コンパートメント・ドア
㊷ 極初期型の主脚収容ドア
㊸ ジャッキ用アクセスハッチ（両ナセル外側のみ）
㊹ 排気管
㊺ マッドガード

⑦ Coolant fill hatch (left side only of both nacelles)
⑧ Exhaust air intake fairing
⑨ Exhaust Venturi
⑩ Perspex V-windscreen (early-type)
⑪ Direct vision panel (opens inwards)
⑫ Side panel blister
⑬ Fuel fill hatch
⑭ Dinghy access panel (jettisonable)
⑮ Antenna mast
⑯ Reinforcing strip
⑰ Ground starter socket hatch
⑱ Vertical stabilizer root fairing
⑲ Antenna wire mount (early position)
⑳ Root fitting inspection hatch
㉑ Pitot tube
㉒ Electrical connection access hatch
㉓ Rudder horn balance
㉔ Rudder (fabric-covered, 26° -1° +2° arc of movement)
㉕ Rudder trim tab (16° ±2°30" arc of movement)
㉖ Rudder controls inspection hatch
㉗ Rudder trim tab control rod access panel
㉘ Rudder trim tab control rod fairing
㉙ Elevator operating linkage fairing
㉚ Tail navigation light
㉛ Tail formation light
㉜ Detachable tail cone fairing (early-type)
㉝ Retracting gear access panel
㉞ Antenna wire mount (antenna wire not fitted)
㉟ Rearward-retracting tail wheel with one-track tire (early-type)
㊱ Tail wheel strut construction (early-type)
㊲ Rear vertical camera window
㊳ Oblique camera window
㊴ One-piece flap (ply-covered, 45°±2° arc of movement, very early-type)
㊵ Movable nacelle tail fairing

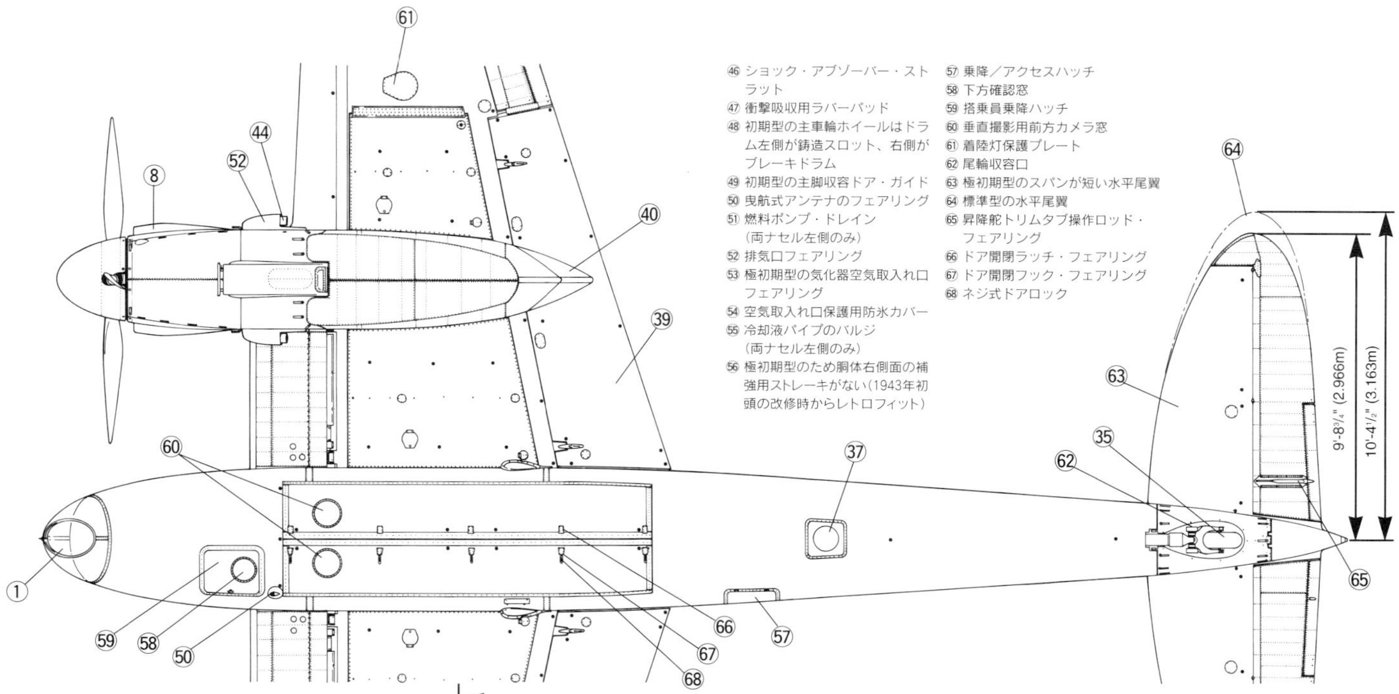

46 ショック・アブゾーバー・ストラット
47 衝撃吸収用ラバーパッド
48 初期型の主車輪ホイールはドラム左側が鋳造スロット、右側がブレーキドラム
49 初期型の主脚収容ドア・ガイド
50 曳航式アンテナのフェアリング
51 燃料ポンプ・ドレイン（両ナセル左側のみ）
52 排気口フェアリング
53 極初期型の気化器空気取入れ口フェアリング
54 空気取入れ口保護用防氷カバー
55 冷却液パイプのバルジ（両ナセル左側のみ）
56 極初期型のため胴体右側面の補強用ストレーキがない（1943年初頭の改修時からレトロフィット）
57 乗降／アクセスハッチ
58 下方確認窓
59 搭乗員乗降ハッチ
60 垂直撮影用前方カメラ窓
61 着陸灯保護プレート
62 尾輪収容口
63 極初期型のスパンが短い水平尾翼
64 標準型の水平尾翼
65 昇降舵トリムタブ操作ロッド・フェアリング
66 ドア開閉ラッチ・フェアリング
67 ドア開閉フック・フェアリング
68 ネジ式ドアロック

41 Ventral compartment door (fitted with door locks)
42 Main wheel door (very early-type)
43 Access hatch for jacking (outboard side only of both nacelles)
44 Exhaust pipe
45 Mudguard
46 Shock-absorber strut
47 Rubber shock pads
48 Main wheel with slots cast on the left side and single brake drums on the right side (early-type)
49 Main wheel door guide (early-type)
50 Trailing antenna fairlead fairing
51 Fuel pump drain (left side only of both nacelles)
52 Exhaust port fairing
53 Carburetor air intake fairing (very early-type)
54 Air intake guard
55 Coolant pipe bulge (left side only of both nacelles)
56 No reinforcing strake (very early-type, retro-fitted as modification from early 1943)
57 Entry/access door
58 Drift sight window
59 Crew entry door
60 Front vertical camera window
61 Landing light covering plate
62 Tail wheel bay
63 Short horizontal tail plane (very early-type)
64 Standard horizontal tail plane
65 Elevator trim tab control rod fairing
66 Lock catch mount fairing
67 Lock hook screw operated mount fairing
68 Screwdriver opening for door locking

P.R.Mk.I Bottom view

P.R.MK. I 下面図

モスキートB.Mk.IVシリーズ I（写真偵察機から爆撃型への改修タイプ）

モスキート爆撃機型初期の生産タイプ。P.R.Mk.IのシリアルナンバーW4064から4072を改修して9機製作された。爆撃機型の部隊配備は最初に1941年11月15日、スワントン・モーレーに基地をおくNo.105Sqnに行われ、配備は1942年2月に完了。翌月31日、B.Mk.IVによる初空襲がドイツ本土のケルンに対して実行された。B.Mk.IVの観測／爆撃照準用機首透明ガラスはP.R.Mk.Iと同じ初期型のものが取り付けられ、エンジンナセルも長さが後のタイプに比べて短く、排気管まわりも異なる極初期型のタイプである。また、水平尾翼も同様にP.R.Mk.Iの短いタイプである。エンジンはロールスロイス・マーリン21、1段過給器エンジンを搭載し、ニードルブレード・プロペラを装備。主脚と尾脚も初期型のままで、尾輪はワン・トラック・タイヤである。P.R.Mk.Iとの主な相違点は胴体下面の敵味方識別灯と主翼下面の着陸灯の装備。機首平面観測窓左内側に装備されたMk.XIV爆撃照準器、また250ポンド（113.4kg）爆弾4発を搭載可能にした点である。のちには安定翼（フィン）を短くして全長を切りつめた500ポンド（226.8kg）爆弾4発も搭載可能になった。爆弾装扉の前後には、あらたに4個の油圧式ジャッキが取り付けられた。燃料搭載量はすべての内部タンクを使用した場合で690ガロン（3,137リッター）、通常は539ガロン（2,450リッター）であった。

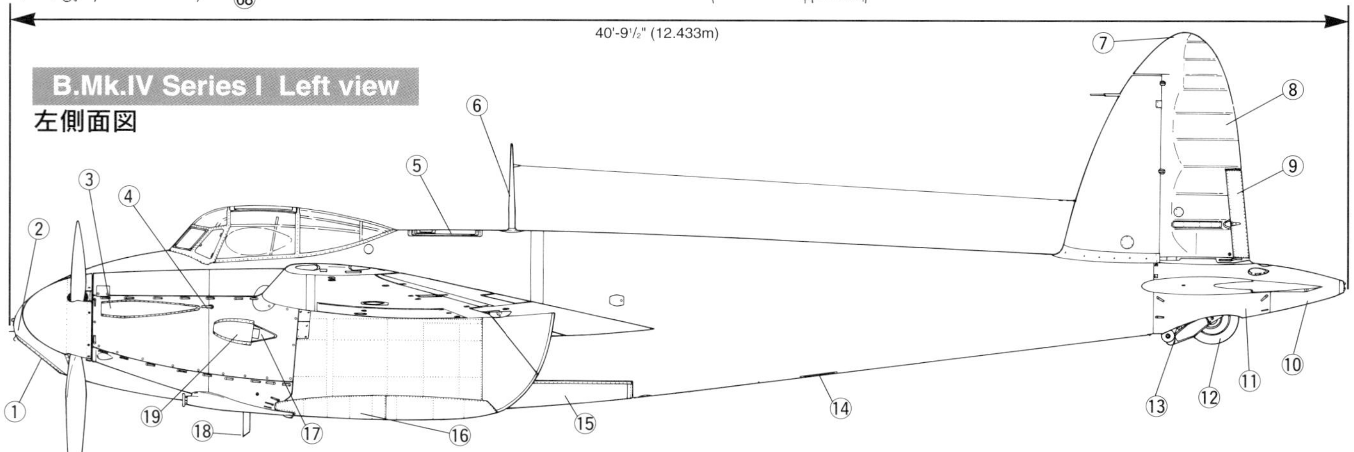

B.Mk.IV Series I Left view

左側面図

① 初期型の偵察／爆撃照準用平面ガラス
② 初期型の機首透明ガラス
③ 排気管フェアリングの空気取入れ口
④ 排気ベンチュリ
⑤ 救命ボート収納部パネル（投棄式）
⑥ アンテナ支柱
⑦ 方向舵ホーンバランス
⑧ 方向舵（羽布張り、作動角26°-1°+2°）
⑨ 方向舵トリムタブ（作動角16°±2°30"）
⑩ 初期型のテイルコーン・フェアリング（取り外し式）
⑪ 引込式尾脚アクセスパネル
⑫ 初期型タイヤ付き引込式尾輪
⑬ 初期型の尾輪ストラット
⑭ 胴体下面カメラ窓
⑮ 油圧式ジャッキ付爆弾装扉
⑯ 初期型の主脚収容ドア
⑰ 排気管
⑱ 曳航式アンテナ・フェアリング
⑲ 排気口フェアリング

① Flat observation/bomb aiming Perspex panel (early-type)
② Perspex nose glazing (early-type)
③ Exhaust air intake fairing
④ Exhaust Venturi
⑤ Dinghy access panel (jettisonable)
⑥ Antenna mast
⑦ Rudder horn balance
⑧ Rudder (fabric-covered, 26° -1° +2° arc of movement)
⑨ Rudder trim tab (16° ±2°30" arc of movement)
⑩ Detachable tail cone fairing (early-type)
⑪ Retracting gear access panel
⑫ Rearward-retracting tail wheel with one-track tire (early-type)
⑬ Tail wheel strut construction (early-type)
⑭ Vertical camera window
⑮ Bomb bay door (fitted with hydraulic jacks)
⑯ Main wheel door (very early-type)
⑰ Exhaust pipe
⑱ Trailing antenna fairlead fairing
⑲ Exhaust port fairing

Mosquito B.Mk.IV Series I (P.R./Bomber Conversion Type)

This was the early production Mosquito bomber of which nine aircraft were built, serialled W4064 to W4072, converted from P.R.Mk.I model. The first aircraft for the Bomber Command was delivered on November 15, 1941, to No.105 Squadron based at Swanton Morley. The deliveries were completed in February 1942. The first WWII air raid carried out by the B.Mk.IV Series I was mounted on May 31, 1942, and Cologne (Germany) was the bombing target. The model retained the early-type glazed observer/bomb aimer's nose and both the very early-type short engine nacelles with the exhaust ports and the short horizontal tail planes fitted to the P.R.Mk.I variant. Powered by the Rolls-Royce Merlin 21 one-stage supercharger engine driving needle blade propeller, the B.Mk.IV Series I was fitted with the early-type main wheels and the early-type tail wheel strut construction equipped with one-track tire. Major changes from P.R.Mk.I included the fitting of the downward identification code lights on the fuselage underside and the underwing landing lights, the Mk.XIV bombsight unit mounted on a bracket on the left side of the nose behind the flat observation panel, and the bomb bay that was initially capable of carrying four 250-lb (113.4kg) bombs, and later four short-finned 500-lb (226.8kg) bombs, doubling the original bomb load. The bomb bay doors were modified with the fitting of front and rear hydraulic jacks in order to be operated in flight. The model could carry a maximum fuel load of 690gal (3,137 liters) with all internal long-range tanks installed, but normally it carried 539gal (2,450 liters) of fuel with the maximum useful load.

B.Mk.IV Series II Left view
左側面図

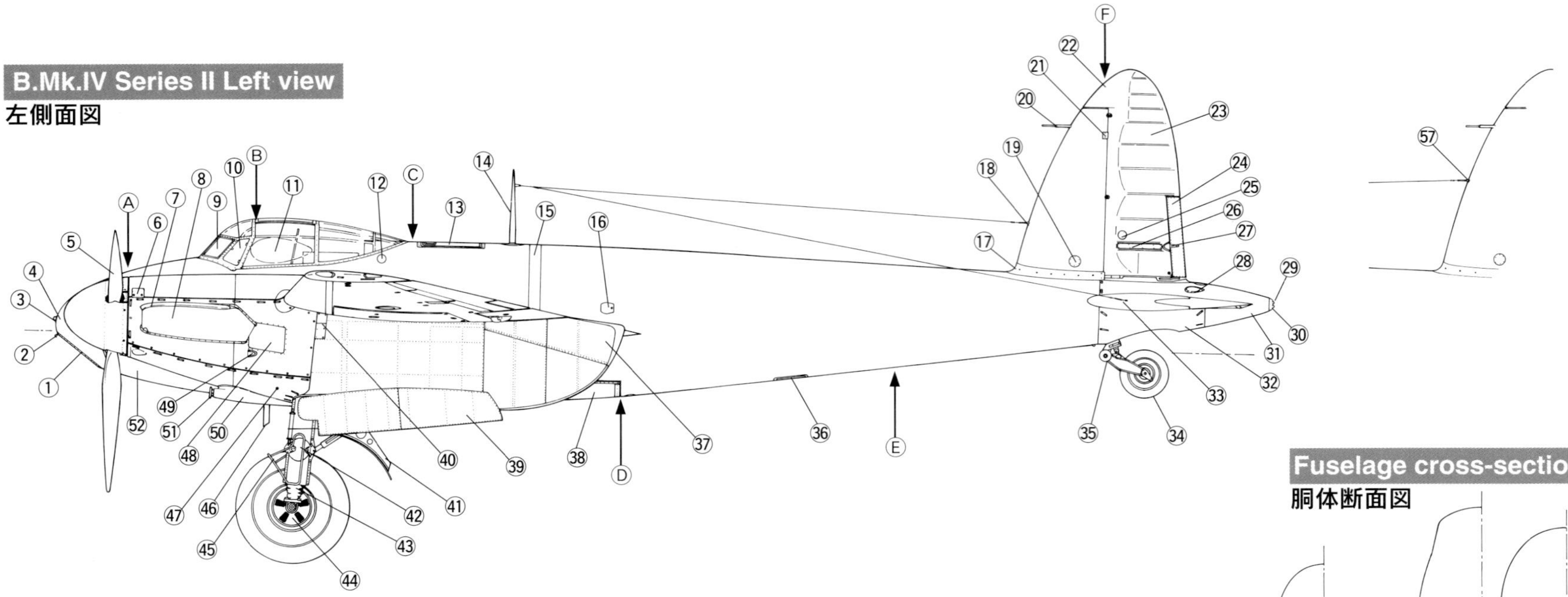

Fuselage cross-section
胴体断面図

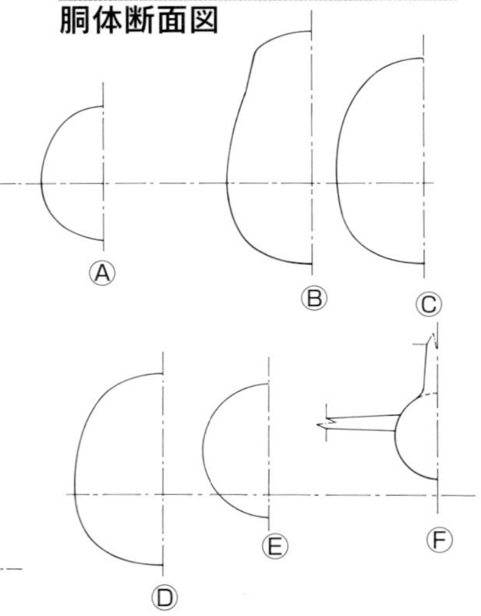

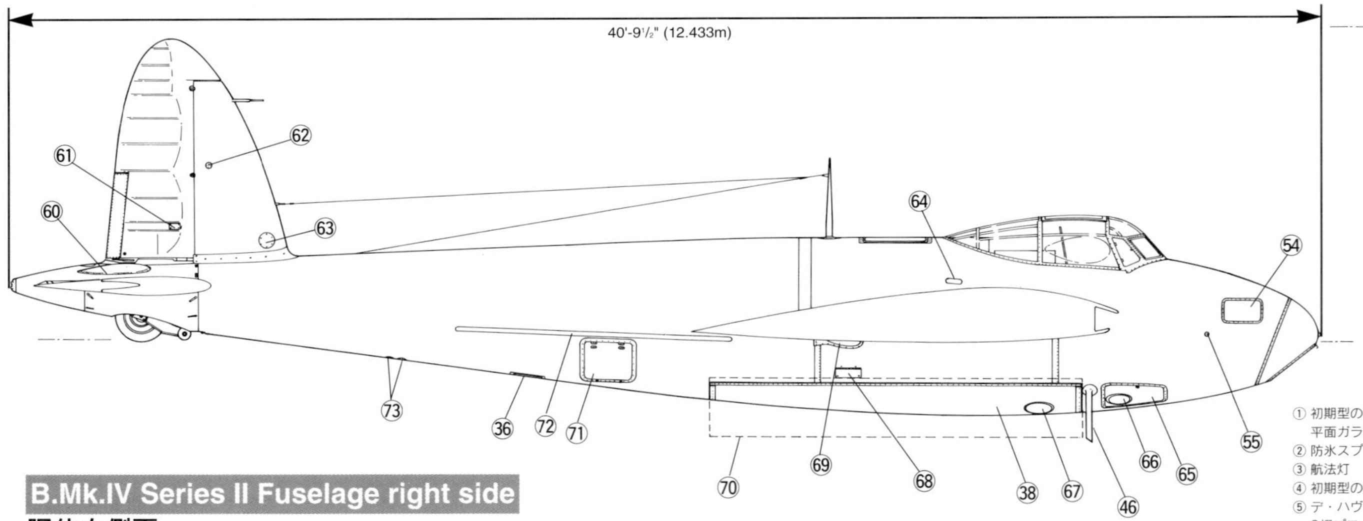

B.Mk.IV Series II Fuselage right side
胴体右側面

① 初期型の偵察／爆撃照準用平面ガラス
② 防氷スプレーパイプ
③ 航法灯
④ 初期型の機首透明ガラス
⑤ デ・ハヴィランド・油圧式3翅プロペラ（ニードルブレードタイプの初期型）
⑥ 冷却液注入ハッチ（両ナセル左側のみ）

① Flat observation/bomb aiming Perspex panel (early-type)
② Deicing spray pipe
③ Forward navigation light
④ Perspex nose glazing (early-type)
⑤ De Havilland Hydromatic needle blade propeller (three-blade, early-type)
⑥ Coolant fill hatch (left side only of both nacelles)

モスキートB.Mk.IVシリーズII

モスキートB.Mk.IVシリーズIIは1段過給器エンジンを搭載した爆撃型の標準タイプである。生産型の初飛行は1942年3月、偵察型P.R.シリーズに改修された機体を含む291機を生産、配備は1942年4月から翌年8月にかけて行われた。ほとんどの機体はロールスロイス・マーリン21エンジンを搭載したが、一部の機体はマーリン23エンジンを搭載、プロペラは両タイプとも、ニードルブレード・タイプを装備していた。シリーズIからの大きな改修点はエンジン・ナセル後部と、昇向舵の延長である。後方に延長されたエンジン・ナセルは主翼後縁より突き出るかたちになり、以前の型では1枚だったフラップが、ナセル後端をはさんだ2枚式に変更された。この改良にともない、主翼面積が改修前の41.806m²から40.469m²になっている。延長された水平尾翼のスパンは20フィート-9インチ（6.325m）になった。ほかにも主脚収納ドアの大型化、排気管まわりの形状変更、カウリング・パネルの一部変更、消焔ダンパー・シュラウドの追加、やや大型化された空気取入口のフェアリングなど、各部に改修が加えられている。なお後期生産型の排気系には、初期型の消焔ダンパー・シュラウド付サキソフォン形排気管にかわって速度向上のために5本の推力式単排気管が採用されている。昼間行動作戦に従事した爆撃型のうち、一部の機体では消焔ダンパー・シュラウドを取り外して、この推力式単排気管をむき出しにしたものもあった。また、爆撃行動範囲を拡大するため主翼構造に補強が加えられ、スリッパ型の50ガロン（227.3リッター）入落下増槽タンクを装備可能にした。爆弾は全長を切りつめた500ポンド（226.8kg）爆弾４発か、1,000ポンド（453.6kg）爆弾2発を搭載可能であった。また、主翼両端は前方に編隊灯。後方に航法灯のついたものに変更され、主脚も後期型タイプになった。尾脚には中央が滑車のようにくぼんだシミーダンパー付"マーストランド"タイヤが採用された。また、低空飛行中に鳥が衝突することへの対策として、風防に耐弾ガラスが採用された。そして胴体右側面の乗降／アクセスハッチのすぐ上に胴体補強用のストレーキが追加された。この補強ストレーキはシリーズII以前に製造されたすべての機体にレトロフィットされている。垂直尾翼のアンテナ空中線取付位置も上方に変更され、以後、これが標準となる。また、後期生産分ではアンテナ支柱が撤去され、ウィップアンテナにおきかえられた。全天候爆撃航法システム"オーボエ"を装備したモスキート B.Mk.IVシリーズIIの初出撃は1942年12月20日、No.109Sqnによる作戦であった。燃料の標準搭載量は539ガロン（2,450リッターであった）。

B.Mk.IV Series II Front view
正面図

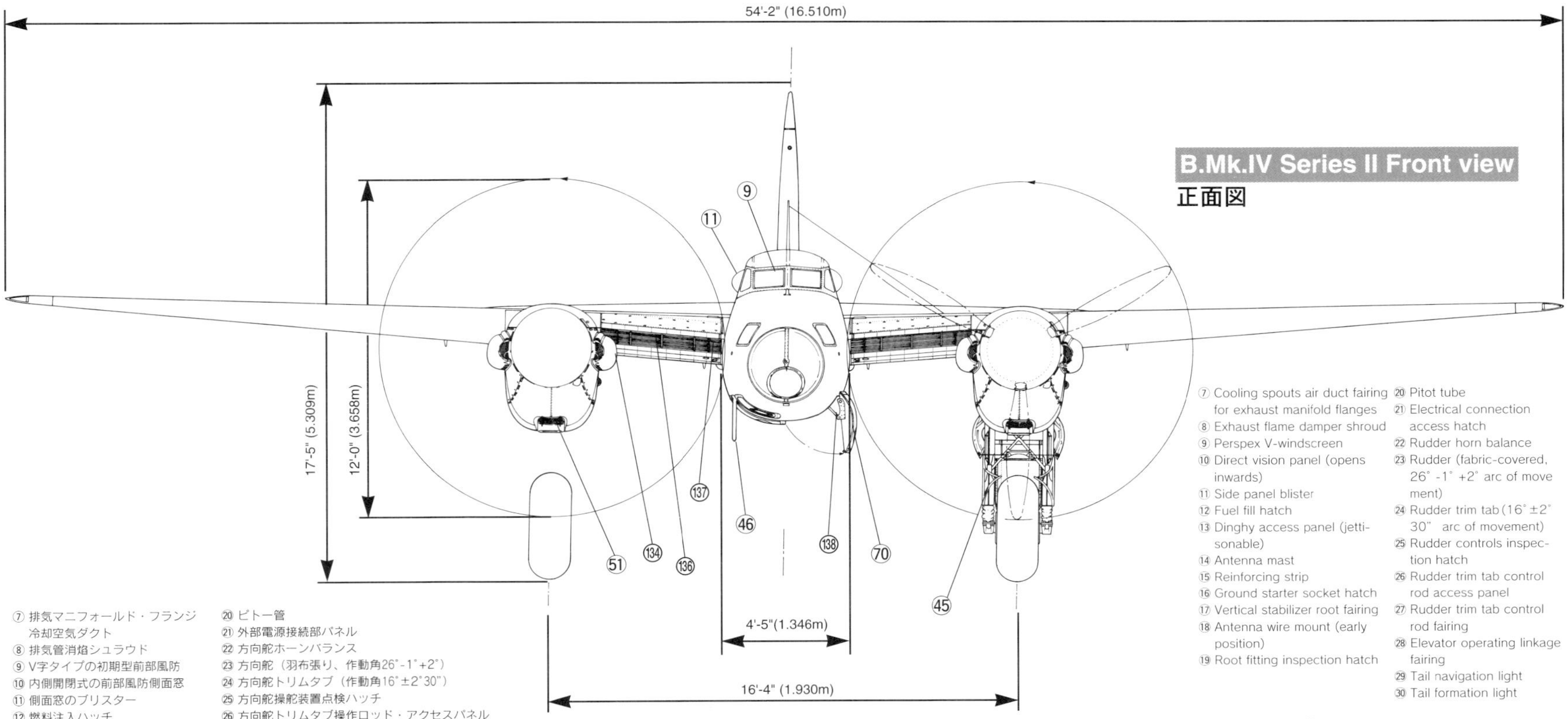

⑦ Cooling spouts air duct fairing for exhaust manifold flanges
⑧ Exhaust flame damper shroud
⑨ Perspex V-windscreen
⑩ Direct vision panel (opens inwards)
⑪ Side panel blister
⑫ Fuel fill hatch
⑬ Dinghy access panel (jettisonable)
⑭ Antenna mast
⑮ Reinforcing strip
⑯ Ground starter socket hatch
⑰ Vertical stabilizer root fairing
⑱ Antenna wire mount (early position)
⑲ Root fitting inspection hatch
⑳ Pitot tube
㉑ Electrical connection access hatch
㉒ Rudder horn balance
㉓ Rudder (fabric-covered, 26° -1° +2° arc of movement)
㉔ Rudder trim tab (16° ±2° 30" arc of movement)
㉕ Rudder controls inspection hatch
㉖ Rudder trim tab control rod access panel
㉗ Rudder trim tab control rod fairing
㉘ Elevator operating linkage fairing
㉙ Tail navigation light
㉚ Tail formation light

⑦ 排気マニフォールド・フランジ冷却空気ダクト
⑧ 排気管消焔シュラウド
⑨ V字タイプの初期型前部風防
⑩ 内側開閉式の前部風防側面窓
⑪ 側面窓のブリスター
⑫ 燃料注入ハッチ
⑬ 救命ボート収納部パネル(投棄式)
⑭ アンテナ支柱
⑮ 胴体補強用帯金
⑯ 地上スターターソケットハッチ
⑰ 垂直尾翼基部フェアリング
⑱ 初期型のアンテナ空中線接続部
⑲ 垂直尾翼基部点検ハッチ
⑳ ピトー管
㉑ 外部電源接続部パネル
㉒ 方向舵ホーンバランス
㉓ 方向舵（羽布張り、作動角26°-1°+2°）
㉔ 方向舵トリムタブ（作動角16°±2°30"）
㉕ 方向舵操舵装置点検ハッチ
㉖ 方向舵トリムタブ操作ロッド・アクセスパネル
㉗ 方向舵トリムタブ操作ロッド・フェアリング
㉘ 昇降舵操舵リンケージ・フェアリング
㉙ 航法灯
㉚ 編隊灯
㉛ 初期型のテイルコーン・フェアリング（取り外し式）
㉜ 引込式尾脚アクセスパネル

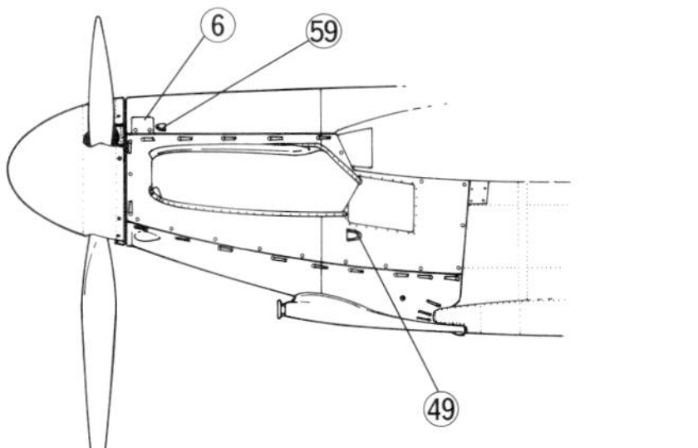

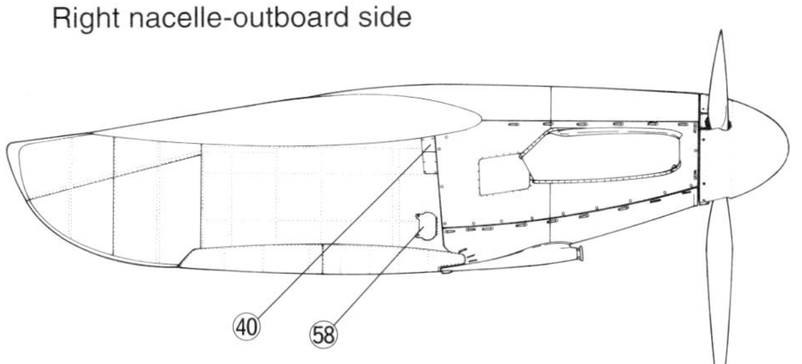

右ナセル外側
Right nacelle-outboard side

右ナセル内側
Right nacelle-inboard side

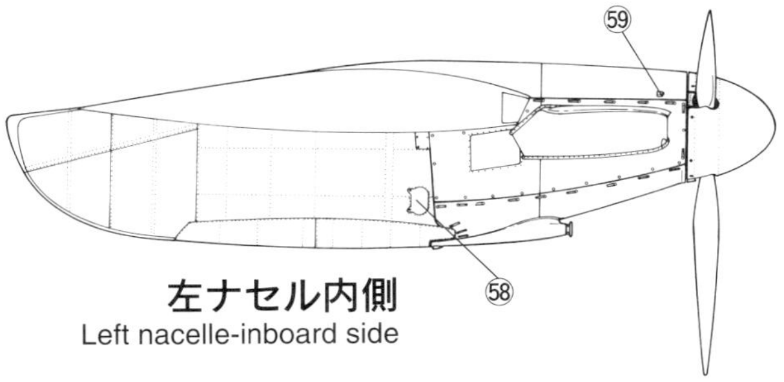

左ナセル内側
Left nacelle-inboard side

Mosquito B.Mk.IV Series II

The Mosquito B.Mk.IV Series II was the standard production bomber equipped with one-stage supercharger engines. The first production aircraft made its maiden flight in March 1942, and about 291 B.Mk.IV Series IIs, including a batch converted to the P.R. variant, were delivered from April 1942 to August 1943. Most had Rolls-Royce Merlin 21 engines, though a few had Merlin 23s, with both driving ordinary needle blade propellers. Major changes from Series I aircraft were the modified engine nacelles and the standard enlarged tail plane elevators. This model was fitted with lengthened engine nacelles extending beyond the wing trailing edge, dividing the original one-piece wing flap in two and reducing the wing area to 40.469m^2 from the original 41.806m^2. New enlarged standard horizontal tail planes having a span of 20ft-9in (6.325m) were fitted. The main wheel doors were also revised, the exhaust ports and some cowling panels were modified, exhaust flame damper shrouds and slightly enlarged carburetor air intake fairings were fitted. On late production aircraft the exhaust system shifted to five stub exhaust pipes which gave an increase in speed compared with the standard early-type "saxophone" exhausts fitted with flame damper shrouds. However, for daylight operations, some bombers had the shroud removed from the five stub ejector exhausts. To further improve the bomber's range, the wings were internally reinforced for the fitting of two underwing 50gal (227.3-liter) slipper-type drop tanks. The bomb bay could carry four short-finned 500-lb (226.8kg) or two 1,000-lb (453.6kg) bombs. In the course of production, the Series IIs were fitted with late-type main wheels, being plan-type on both sides, a new wingtip with a rearward resin lamp for air-to-air recognition added initially to a pair of navigation and formation lights and later to only one navigation light, the late-type tail wheel strut construction with Marstrand two-track anti-shimmy tire having a groove in the center, and a bulletproof windscreen to protect against bird strikes during low-level operations. A reinforcing strake on the right side of the rear fuselage, just above the entry/access door, was fitted as a modification to all aircraft and the antenna wire mount on the leading edge of the vertical stabilizer was moved to a new position higher up. Late production aircraft had the antenna mast removed and replaced by a whip antenna. No.109 Squadron operated the Oboe-equipped Mosquito B.Mk.IV Series II for the first time on December 20, 1942. This bomber normally carried 539gal (2,450 liters) of fuel.

B.Mk.IV Series II Top view

上面図

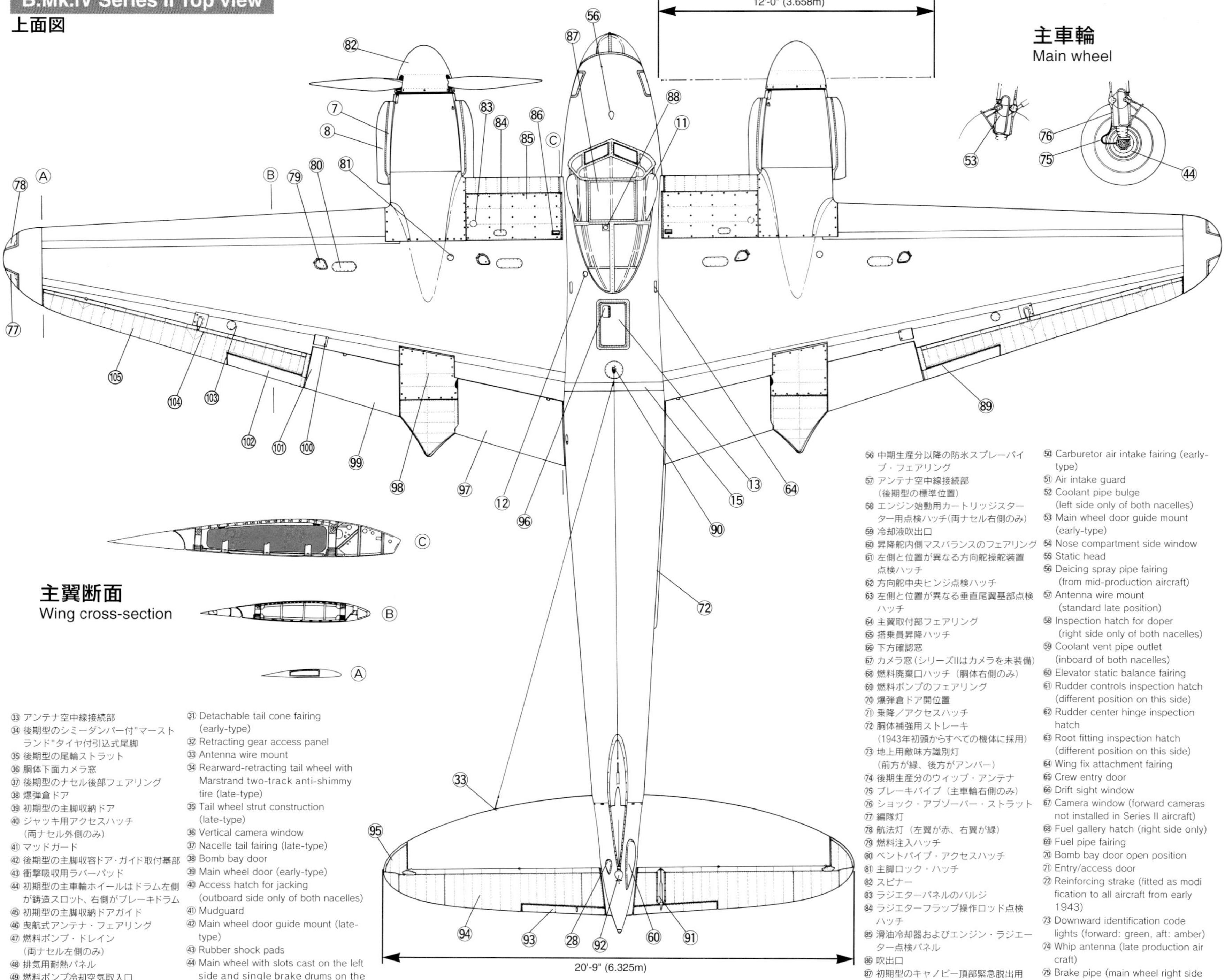

主車輪
Main wheel

主翼断面
Wing cross-section

33 アンテナ空中線接続部
34 後期型のシミーダンパー付"マーストランド"タイヤ付引込式尾脚
35 後期型の尾輪ストラット
36 胴体下面カメラ窓
37 後期型のナセル後部フェアリング
38 爆弾倉ドア
39 初期型の主脚収納ドア
40 ジャッキ用アクセスハッチ（両ナセル外側のみ）
41 マッドガード
42 後期型の主脚収容ドア・ガイド取付基部
43 衝撃吸収用ラバーパッド
44 初期型の主車輪ホイールはドラム左側が鋳造スロット、右側がブレーキドラム
45 初期型の主脚収納ドアガイド
46 曳航式アンテナ・フェアリング
47 燃料ポンプ・ドレイン（両ナセル左側のみ）
48 排気用耐熱パネル
49 燃料ポンプ冷却空気取入口
50 初期型の気化器空気取入口フェアリング
51 空気取入れ口保護用防氷カバー
52 冷却液パイプのバルジ（両ナセル左側のみ）
53 初期型の主脚収容ドア・ガイド取付基部
54 機首側面窓
55 ピトー管静圧ヘッド

31 Detachable tail cone fairing (early-type)
32 Retracting gear access panel
33 Antenna wire mount
34 Rearward-retracting tail wheel with Marstrand two-track anti-shimmy tire (late-type)
35 Tail wheel strut construction (late-type)
36 Vertical camera window
37 Nacelle tail fairing (late-type)
38 Bomb bay door
39 Main wheel door (early-type)
40 Access hatch for jacking (outboard side only of both nacelles)
41 Mudguard
42 Main wheel door guide mount (late-type)
43 Rubber shock pads
44 Main wheel with slots cast on the left side and single brake drums on the right side (early-type)
45 Main wheel door guide (early-type)
46 Trailing antenna fairlead fairing
47 Fuel pump drain (left side only of both nacelles)
48 Heat resistant panel
49 Fuel pump cooling air intake

56 中期生産分以降の防氷スプレーパイプ・フェアリング
57 アンテナ空中線接続部（後期型の標準位置）
58 エンジン始動用カートリッジスターター用点検ハッチ（両ナセル右側のみ）
59 冷却液吹出口
60 昇降舵内側マスバランスのフェアリング
61 左側と位置が異なる方向舵操舵装置点検ハッチ
62 方向舵中央ヒンジ点検ハッチ
63 左側と位置が異なる垂直尾翼基部点検ハッチ
64 主翼取付部フェアリング
65 搭乗員昇降ハッチ
66 下方確認窓
67 カメラ窓（シリーズIIはカメラを未装備）
68 燃料廃棄口ハッチ（胴体右側のみ）
69 燃料ポンプのフェアリング
70 爆弾倉ドア開位置
71 乗降／アクセスハッチ
72 胴体補強用ストレーキ（1943年初頭からすべての機体に採用）
73 地上用敵味方識別灯（前方が緑、後方がアンバー）
74 後期生産分のウィップ・アンテナ
75 ブレーキパイプ（主車輪右側のみ）
76 ショック・アブゾーバー・ストラット
77 編隊灯
78 航法灯（左翼が赤、右翼が緑）
79 燃料注入ハッチ
80 ベントパイプ・アクセスハッチ
81 主脚ロック・ハッチ
82 スピナー
83 ラジエターパネルのバルジ
84 ラジエーターフラップ操作ロッド点検ハッチ
85 滑油冷却器およびエンジン・ラジエーター点検パネル
86 吹出口
87 初期型のキャノピー頂部緊急脱出用ハッチ
88 信号弾発射口
89 補助翼バランスタブ（右翼のみ）
90 ベークライト製アンテナ保護板
91 昇降舵トリムタブ操作ロッド・フェアリング（右昇降舵は上面のみ）
92 昇降舵左右中央ヒンジ点検ハッチ

50 Carburetor air intake fairing (early-type)
51 Air intake guard
52 Coolant pipe bulge (left side only of both nacelles)
53 Main wheel door guide mount (early-type)
54 Nose compartment side window
55 Static head
56 Deicing spray pipe fairing (from mid-production aircraft)
57 Antenna wire mount (standard late position)
58 Inspection hatch for doper (right side only of both nacelles)
59 Coolant vent pipe outlet (inboard of both nacelles)
60 Elevator static balance fairing
61 Rudder controls inspection hatch (different position on this side)
62 Rudder center hinge inspection hatch
63 Root fitting inspection hatch (different position on this side)
64 Wing fix attachment fairing
65 Crew entry door
66 Drift sight window
67 Camera window (forward cameras not installed in Series II aircraft)
68 Fuel gallery hatch (right side only)
69 Fuel pipe fairing
70 Bomb bay door open position
71 Entry/access door
72 Reinforcing strake (fitted as modification to all aircraft from early 1943)
73 Downward identification code lights (forward: green, aft: amber)
74 Whip antenna (late production aircraft)
75 Brake pipe (main wheel right side only)
76 Shock-absorber strut
77 Formation light
78 Navigation light (left: red, right: green)
79 Fuel fill hatch
80 Vent pipe access hatch
81 Landing gear uplock hatch

B.Mk.IV Series II Bottom view
下面図

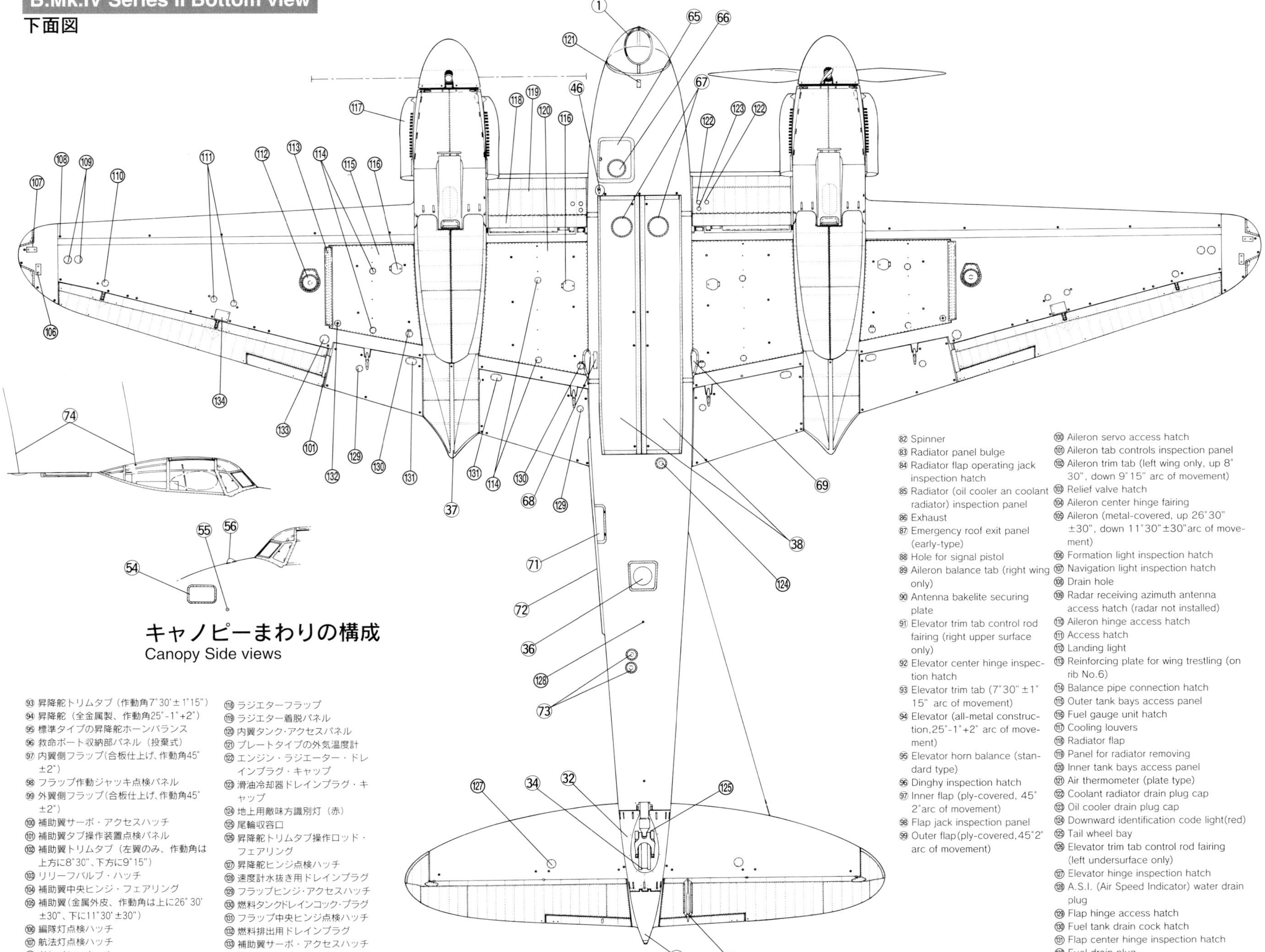

キャノピーまわりの構成
Canopy Side views

93 昇降舵トリムタブ（作動角7°30'±1°15"）
94 昇降舵（全金属製、作動角25°-1°+2°）
95 標準タイプの昇降舵ホーンバランス
96 救命ボート収納部パネル（投棄式）
97 内翼側フラップ(合板仕上げ、作動角45°±2°)
98 フラップ作動ジャッキ点検パネル
99 外翼側フラップ(合板仕上げ、作動角45°±2°)
100 補助翼サーボ・アクセスハッチ
101 補助翼タブ操作装置点検パネル
102 補助翼トリムタブ（左翼のみ、作動角は上方に8°30"、下方に9°15"）
103 リリーフバルブ・ハッチ
104 補助翼中央ヒンジ・フェアリング
105 補助翼(金属外皮、作動角は上に26°30'±30"、下に11°30"±30")
106 編隊灯点検ハッチ
107 航法灯点検ハッチ
108 ドレイン・ホール
109 方位受信レーダー・アクセスパネル（レーダーは未搭載）
110 補助翼ヒンジ・アクセスハッチ
111 アクセスハッチ
112 着陸灯
113 主翼トレッスル用補強板（第6リブ）
114 バランスパイプ連結ハッチ
115 外翼タンク・アクセスパネル
116 燃料計ユニット・アクセスハッチ
117 冷却用ルーバー
118 ラジエターフラップ
119 ラジエター着脱パネル
120 内翼タンク・アクセスパネル
121 プレートタイプの外気温度計
122 エンジン・ラジエーター・ドレインプラグ・キャップ
123 滑油冷却器ドレインプラグ・キャップ
124 地上用敵味方識別灯（赤）
125 尾輪収容口
126 昇降舵トリムタブ操作ロッド・フェアリング
127 昇降舵ヒンジ点検ハッチ
128 速度計水抜き用ドレインプラグ
129 フラップヒンジ・アクセスハッチ
130 燃料タンクドレインコック・プラグ
131 フラップ中央ヒンジ点検ハッチ
132 燃料排出用ドレインプラグ
133 補助翼サーボ・アクセスハッチ
134 補助翼ヒンジおよび左右連動調節部点検ハッチ
135 初期型の滑油冷却器
136 エンジン・ラジエター
137 キャビン暖気ユニット
138 爆弾倉ドア前方ジャッキ

82 Spinner
83 Radiator panel bulge
84 Radiator flap operating jack inspection hatch
85 Radiator (oil cooler an coolant radiator) inspection panel
86 Exhaust
87 Emergency roof exit panel (early-type)
88 Hole for signal pistol
89 Aileron balance tab (right wing only)
90 Antenna bakelite securing plate
91 Elevator trim tab control rod fairing (right upper surface only)
92 Elevator center hinge inspection hatch
93 Elevator trim tab (7°30" ±1° 15" arc of movement)
94 Elevator (all-metal construction,25°-1°+2° arc of movement)
95 Elevator horn balance (standard type)
96 Dinghy inspection hatch
97 Inner flap (ply-covered, 45° 2°arc of movement)
98 Flap jack inspection panel
99 Outer flap(ply-covered,45°2° arc of movement)
100 Aileron servo access hatch
101 Aileron tab controls inspection panel
102 Aileron trim tab (left wing only, up 8° 30", down 9°15" arc of movement)
103 Relief valve hatch
104 Aileron center hinge fairing
105 Aileron (metal-covered, up 26°30" ±30", down 11°30"±30"arc of movement)
106 Formation light inspection hatch
107 Navigation light inspection hatch
108 Drain hole
109 Radar receiving azimuth antenna access hatch (radar not installed)
110 Aileron hinge access hatch
111 Access hatch
112 Landing light
113 Reinforcing plate for wing trestling (on rib No.6)
114 Balance pipe connection hatch
115 Outer tank bays access panel
116 Fuel gauge unit hatch
117 Cooling louvers
118 Radiator flap
119 Panel for radiator removing
120 Inner tank bays access panel
121 Air thermometer (plate type)
122 Coolant radiator drain plug cap
123 Oil cooler drain plug cap
124 Downward identification code light(red)
125 Tail wheel bay
126 Elevator trim tab control rod fairing (left undersurface only)
127 Elevator hinge inspection hatch
128 A.S.I. (Air Speed Indicator) water drain plug
129 Flap hinge access hatch
130 Fuel tank drain cock hatch
131 Flap center hinge inspection hatch
132 Fuel drain plug
133 Aileron servo access hatch
134 Aileron center hinge and differential control inspection hatch
135 Oil cooler (early-type)
136 Coolant radiator (early-type)
137 Cabin heater unit
138 Front bomb bay door jack

B.Mk.IV Special Left view
左側面図

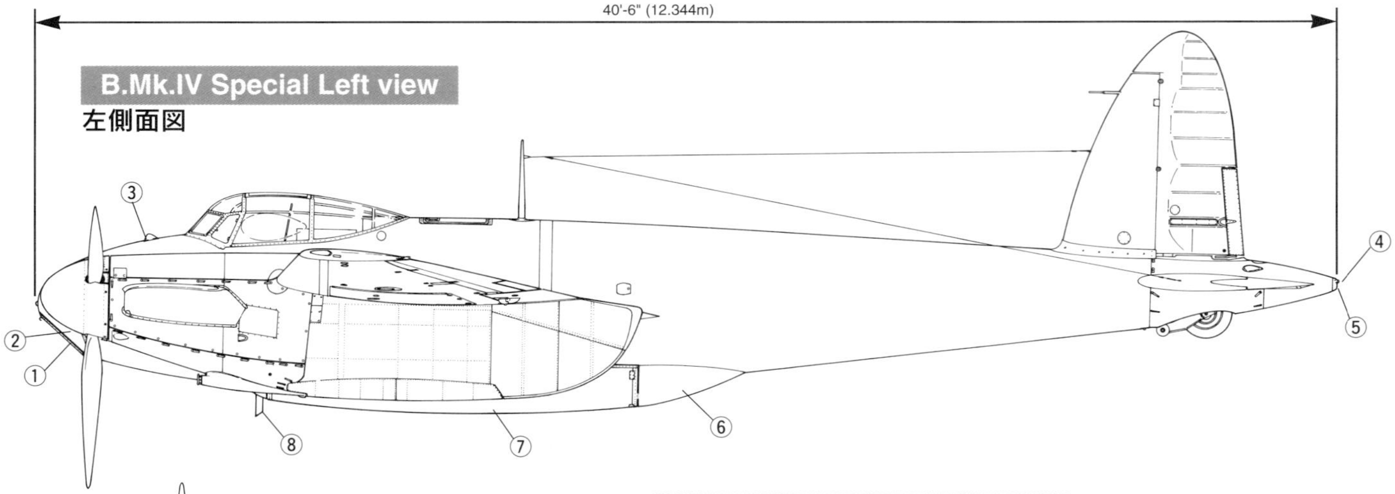

B.Mk.IV Special Bottom view
下面図

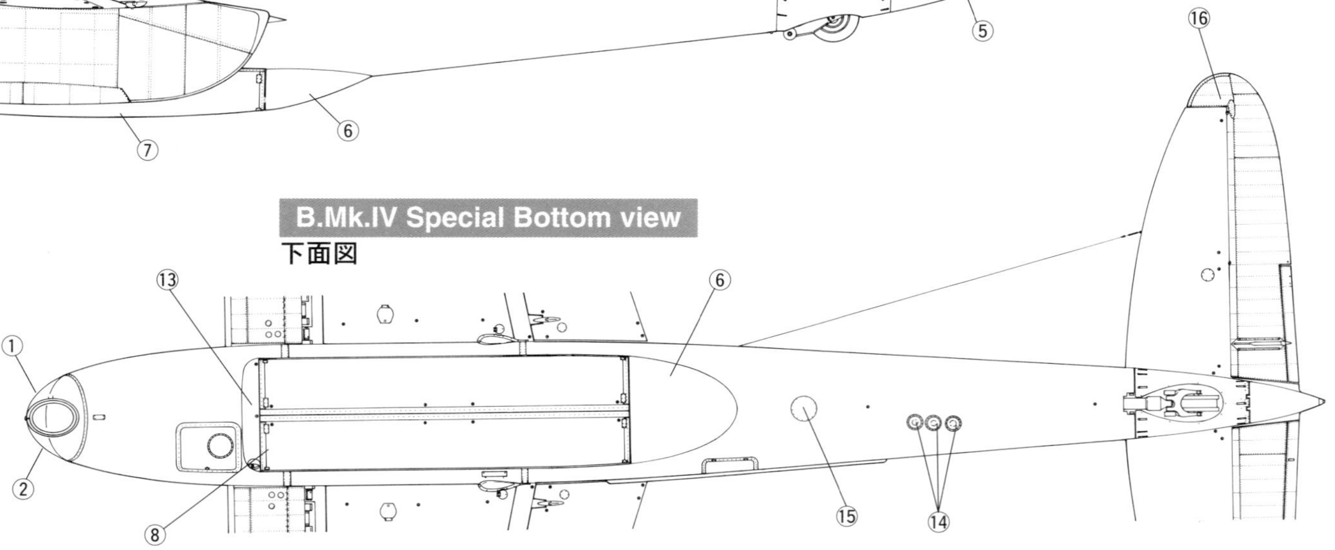

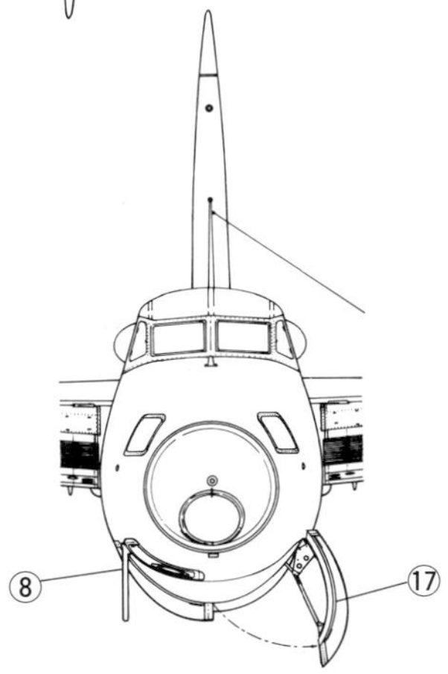

B.Mk.IV Special Front view
正面図

B.Mk.IVスペシャルの爆弾倉左側面

B.Mk.IV Special
Bomb bay Left side

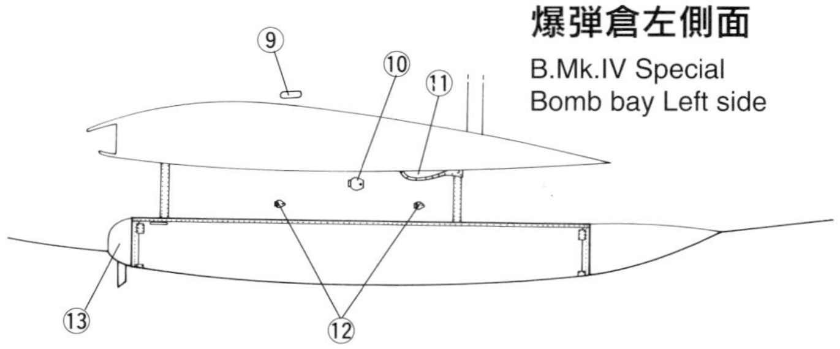

モスキート B.Mk.IVスペシャル

B.Mk.IVスペシャルは、B.Mk.IVに4,000ポンド（1,814.4kg）爆弾を搭載できるようにしたタイプで、1943-44年の間にB.Mk.IVの後期生産分から23機が改造された。最初のモデルはNo.8グループに引き渡され、同年2月23日から24日にかけての夜間出撃で、No.692SqnのシリアルDZ535、DZ637、DZ647号機がドイツのデュッセルドルフに"ブロックバスター"爆弾を投下したのが実戦デビューとなった。通常型のB.Mk.IVシリーズIIとの相違点は、バルジ型の爆弾倉ドア、増積された昇降舵のホーンバランス、そして中央フレームのない中空合わせガラス式偵察／爆撃照準用平面ガラスを採用した、後期型の機首透明ガラスである。機首透明ガラスが更新されたのにともない、全長は40フィート6インチ（12.344m）になっている。胴体下面のカメラは撤去されて撮影窓が塞がれ、機首には機体重心位置調整のために重さ60ポンド（27.2kg）のバラストが積まれた。一部には後期型パドルブレードタイプのプロペラを装備した機体もあった。通常、本機の作戦時における燃料搭載量は50ガロン（227.3リッター）入落下タンク２個を含む497ガロン(2,259リッター)であった。

Mosquito B.Mk.IV Special

During 1943-44 twenty-three late production B.Mk.IV Series IIs were transformed into B.Mk.IV special aircraft capable of carrying a 4,000-lb (1,814.4kg) bomb. The first model was delivered to No.8 Group in early 1944 and the operational debut was carried out by No.692 Squadron on the night of 23-24 February when three aircraft, serialled DZ535, DZ637 and DZ647, dropped their “Blockbuster” bombs on Düsseldorf (Germany). The major differences from the standard production B.Mk.IV Series II were the bulged bomb bay doors, the larger horn balances fitted to tail plane elevators and the late-type glazed observer/bomb aimer's nose (that reduced the fuselage length to 40ft-6in - 12.344m) without center frames and with the flat observation panel of the “dry air” sandwich type. The rear fuselage camera was removed from the lower surface and some 60 pounds (27.2kg) of ballast were fitted in the nose to improve the aircraft's stability. Some bombers were equipped with the paddle blade propellers. These bombers were usually operated with a fuel capacity of 497gal (2,259 liters) including two 50gal (227.3-liter) underwing slipper-type drop tanks.

① 後期型の中空合わせガラス式偵察／爆撃照準用平面ガラス
② 後期型の機首透明ガラス
③ 防氷スプレーパイプ・フェアリング
④ 航法灯
⑤ 尾部スペアソケット・フェアリング
⑥ 爆弾倉バルジ後部フェアリング
⑦ 4,000ポンド爆弾用にバルジ化された爆弾倉ドア
⑧ 曳航式アンテナ・フェアリング
⑨ 主翼取付アタッチメント・フェアリング
⑩ 爆弾搭載確認ハッチ
⑪ 燃料パイプ・フェアリング
⑫ 4,000ポンド爆弾用内蔵ウィンチハンドル・アクセスパネル
⑬ 爆弾倉バルジ前方フェアリング
⑭ 地上用敵味方識別灯（前方が赤、中央が緑、後方がアンバー）
⑮ パネルで塞がれたカメラ窓（胴体下面にカメラ窓はない）
⑯ 増積された昇降舵ホーンバランス

① “Dry air” sandwich type flat observation/bomb aiming Perspex panel (late-type)
② Perspex nose glazing (late-type)
③ Deicing spray pipe fairing
④ Tail navigation light
⑤ Spare tail-socket fairing
⑥ Rear bulged bomb bay doors fairing
⑦ Bulged bomb bay door for 4,000-lb bomb
⑧ Trailing antenna fairlead fairing
⑨ Wing fix attachment fairing
⑩ Bomb loading check hatch
⑪ Fuel pipe fairing
⑫ Access hatch for 4,000-lb bomb winch handle
⑬ Front bulged bomb bay doors fairing
⑭ Downward identification code lights (forward: red, middle: green, aft: amber)
⑮ Cover panel (no vertical camera window on fuselage underside)
⑯ Enlarged elevator horn balance

Mosquito P.R.Mk.IV Left view
左側面図

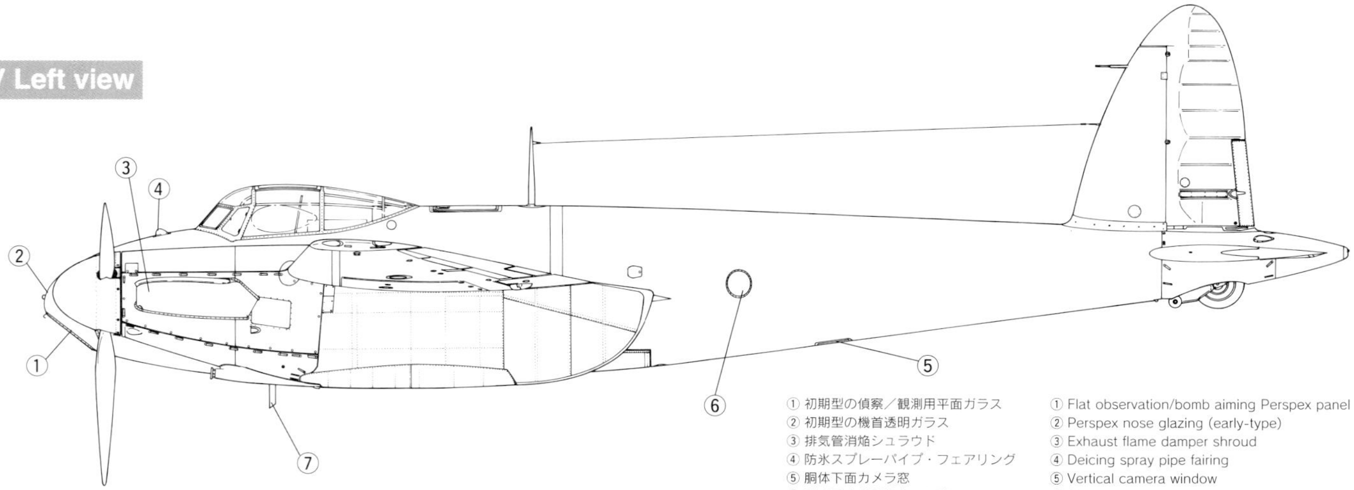

① 初期型の偵察／観測用平面ガラス
② 初期型の機首透明ガラス
③ 排気管消焔シュラウド
④ 防氷スプレーパイプ・フェアリング
⑤ 胴体下面カメラ窓
⑥ 斜め下方撮影用カメラ窓
⑦ 曳航式アンテナ・フェアリング

① Flat observation/bomb aiming Perspex panel (early-type)
② Perspex nose glazing (early-type)
③ Exhaust flame damper shroud
④ Deicing spray pipe fairing
⑤ Vertical camera window
⑥ Oblique camera window
⑦ Trailing antenna fairlead fairing

モスキートP.R.Mk.IV

P.R.Mk.IVは、より多くの偵察機型モスキートを必要としたRAFの要請により、1942年から1943年にかけて30機以上のB.Mk.IV シリーズIIを改修して作られた写真偵察機タイプで、1号機（DK284）の初飛行は1942年4月29日に行われた。P.R.Mk.IVはP.R.Mk.Iと同じく胴体内に4基の偵察カメラを装備、爆弾倉内部に大型の燃料タンクを配し、航続距離を増加されている。P.R.Mk.IVはドイツ領内において数々の写真偵察ソーティを行っているが、そのなかでも特筆すべきなのは、V-2号ロケット兵器開発センターの撮影に成功したことであろう。本機の最大燃料搭載量は50ガロン（227.3リッター）入落下タンク２個使用時で660ガロン（3,000リッター）。

Mosquito P.R.Mk.IV

To address an RAF need to operate more P.R. Mosquitos, over thirty B.Mk.IV Series II airframes were converted to P.R.Mk.IV day and night photoreconnaissance variants from April 1942 to March 1943. The first No.1 PRU's personnel-modified bomber (DK284) made its maiden operational mission on April 29, 1942. Instead of the bomb load, the P.R.Mk.IV was equipped with the same number of cameras (four) installed on the P.R.Mk.I. In the course of numerous sorties over German territory, P.R.Mk.IVs successfully photographed V-2 research and development centers. This model could carry a maximum fuel load of 660gal (3,000 liters) including two 50gal (227.3-liter) drop tanks.

デ・ハヴィラント・モスキート　諸元

データはすべてデ・ハヴィラントD.H.98のオリジナル図面およびRAFの出版物より引用
ALL DATA TAKEN FROM ORIGINAL DE HAVILAND D.H.98 DRAWINGS AND RAF AIR PUBLICATIONS

MODEL 型　式	全幅(m) Wingspan(m)	全長(m) Length Overall(m)	全高(m) Height(m)	主翼面積(m²) Wing Area(m²)	自重(kg) Weight(kg)	全備重量(kg) Loaded Weight(kg)	出力(hp) Rolls-Royce Merlin Engine(hp)	プロペラ形式 Propeller	燃料搭載量(ℓ) Fuel Capacity (ℓ)	最大速度(km/h) Max.Speed (km/h)	上昇速度(m/min) Initial Climb (m/min)	巡航速度(km/h) Cruising Speed (km/h)	実用上昇高度(m) Service Ceiling (m)	航続距離(km) Range(km)	標準兵装 Standard Armament
P.R.MK.I	16.510	12.433	5.309	41.806	5,817	8,922	21(1,480)	D.H.Hydromatic	3,317	615	869	410	10,700	3,510	—
F. / N.F.Mk.II	16.510	12.548	5.309	40.469	6,092	8,413	21(1,480)or 22(1,480)or 23(1,480)	D.H.Hydromatic	2,487	595	914	410	11,000	2,750	20mm cannon x 4+ .303 in MG x 4
B.Mk.IV Series I	16.510	12.433	5.309	41.806	5,940	9,467	21(1,480)	D.H.Hydromatic	3,317	612	830	426	8,800	3,700	500-lb bomb x 2 or 250-lb bomb x 4
B.Mk.IV Series II	16.510	12.433	5.309	40.469	6,078	10,238	21(1,480) or 23(1,480)	D.H.Hydromatic	2,450	620	762	426	8,200	3,285	500-lb bomb x 4 or 1,000-lb bomb x 2 or 4,000-lb bomb x 1
F.B.Mk.VI Series I	16.510	12.548	5.309	40.469	6,476	10,005	21(1,480) or 23(1,480)	D.H.Hydromatic	3,255	608	869	410	10,100	2,985	20mm cannon x 4 + .303 in MG x 2 + 250-lb bomb x 4
F.B.Mk.VI Series II	16.510	12.548	5.309	40.469	6,506	10,124	25(1,640)	D.H.Hydromatic	3,255	611	655	402	11,000	2,750	20mm cannon x 4 + .303 in MG x 4 + 500-lb bomb x 2 + 60-lb RP x 8 or 500-lb bomb x 4
B.Mk.IX	16.510	12.497	5.309	40.469	6,642	11,227	72(1,680)	D.H.Hydromatic	3,169	644	835	402	11,000	2,390	500-lb bomb x 6 or 4,000-lb bomb x 1
N.F.Mk.XII	16.510	12.675	5.309	40.469	6,212	8,936	21(1,480) or 23(1,480)	D.H.Hydromatic	2,487	595	914	410	11,000	2,745	20mm cannon x 4
B.Mk.XVI	16.510	12.497	5.309	40.469	6,759	11,526	72*/73**(1,680) or 76*/77**(1,710)	D.H.Hydromatic	2,714	657	853	370	11,300	2,205	500-lb bomb x 6 or 4,000-lb bomb x 1
P.R.Mk.XVI	16.510	12.497	5.309	40.469	6,638	10,138	72*/73**(1,680) or 76*/77**(1,710)	D.H.Hydromatic	3,910	688	884	394	10,750	3,945	—
N.F.Mk.XVII	16.510	12.675	5.309	40.469	5,998	8,709	21(1,480) or 23(1,480)	D.H.Hydromatic	2,487	595	914	402	11,000	2,745	20mm cannon x 4
N.F.Mk.30	16.510	12.598	5.309	40.469	6,985	9,849	72(1,680)or 76(1,710)or 113(1,690)	D.H.Hydromatic	3,225	655	869	402	11,900	2,100	20mm cannon x 4
P.R.Mk.34	16.510	12.497	5.309	40.469	6,432	11,567	76*/77**(1,710) or 113*/114**(1,690)	D.H.Hydromatic	5,769	684	610	483	13,100	5,375	—
B.Mk.35	16.510	12.497	5.309	40.469	6,638	11,431	113*/114*(1,690)	D.H.Hydromatic	2,714	668	610	444	12,800	2,235	500-lb bomb x 6 or 4,000-lb bomb x 1

*右エンジンの値　**与圧キャビン用過給器付左エンジンの値
*Starboard enjine　**Port engine driving pressurized cabin supercharger

Mosquito B.Mk.IX Left view

左側面図

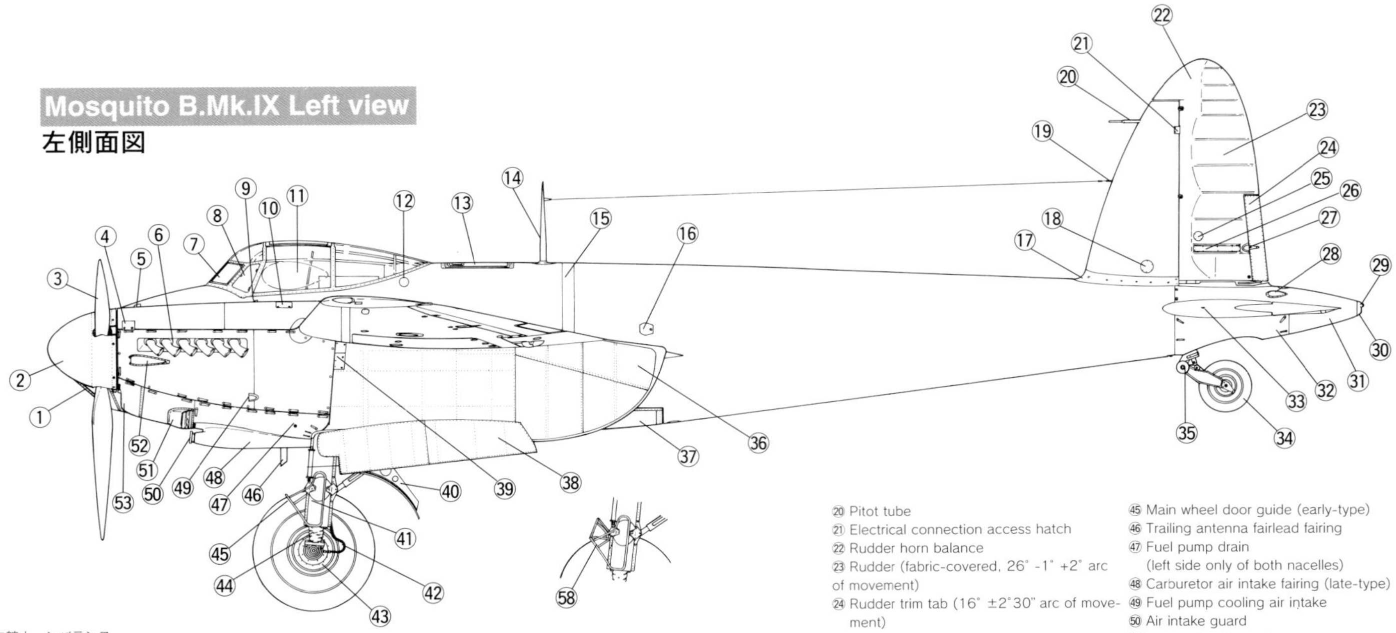

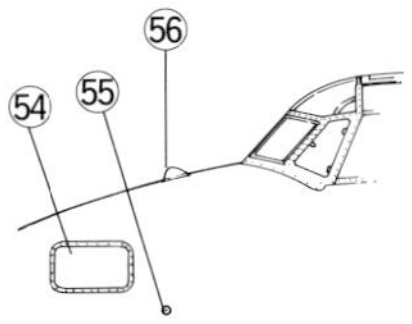

機首左側面
Nose Left side

① 後期型の中空合わせガラス式偵察／爆撃照準用平面ガラス
② スピナー
③ デ・ハヴィランド・油圧式3翅プロペラ（ニードルブレードタイプの初期型）
④ 冷却液注入ハッチ（左右ナセル左側のみ）
⑤ 冷却液タンク・ベント
⑥ 推力式単排気管
⑦ V字タイプの初期型前部風防
⑧ 内側開閉式の前部風防側面窓
⑨ 中間冷却器タンク・ベント
⑩ 中間冷却器タンク冷却液注入ハッチ
⑪ 側面窓のブリスター
⑫ 燃料注入ハッチ
⑬ 救命ボート収納部パネル（投棄式）
⑭ アンテナ支柱
⑮ 胴体補強用帯金
⑯ 地上スターターソケットハッチ
⑰ 垂直尾翼基部フェアリング
⑱ 垂直尾翼基部点検ハッチ
⑲ アンテナ空中線接続部
⑳ ピトー管
㉑ 外部電源接続部パネル
㉒ 方向舵ホーンバランス
㉓ 方向舵（羽布張り、作動角26°-1°+2°）
㉔ 方向舵トリムタブ（作動角16°±2°30'）
㉕ 方向舵操舵装置点検ハッチ
㉖ 方向舵トリムタブ操作ロッド・アクセスパネル
㉗ 方向舵トリムタブ操作ロッド・フェアリング
㉘ 昇降舵操舵リンケージ・フェアリング
㉙ 航法灯
㉚ 編隊灯
㉛ 初期型のテイルコーン・フェアリング（取り外し式）
㉜ 引込式尾脚アクセスパネル
㉝ アンテナ空中線接続部(アンテナ空中線はなし)
㉞ 後期型のシミーダンパー付"マーストランド"タイヤの引込式尾脚
㉟ 後期型の尾輪ストラット
㊱ ナセル後部フェアリング
㊲ 爆弾倉ドア
㊳ 初期型の主脚収納ドア
㊴ ジャッキ用アクセスハッチ(両ナセル外側のみ)
㊵ マッドガード
㊶ ショック・アブゾーバー・ストラット
㊷ ブレーキパイプ
㊸ 後期型の主車輪
㊹ 衝撃吸収用ラバーパッド
㊺ 初期型の主脚収容ドア・ガイド
㊻ 曳航式アンテナ用フェアリング
㊼ 燃料ポンプ・ドレイン（両ナセル左側のみ）
㊽ 後期型の気化器空気取入口フェアリング
㊾ 燃料ポンプ冷却空気取入口
㊿ 空気取入れ口保護用防氷カバー
51 中間冷却器用空気排出口
52 磁石発電機冷却用空気取入口
53 中間冷却器用空気取入口
54 機首側面窓
55 ピトー管静圧ヘッド
56 防氷スプレーパイプ・フェアリング
57 エンジン始動用カートリッジスターター用点検ハッチ（両ナセル右側のみ）
58 後期型の主脚収容ドア・ガイド
59 後期型の主脚収容ドア
60 主脚収容ドア開閉ハッチ・アクセスパネル
61 4,000ポンド爆弾用にバルジ化された爆弾倉ドア
62 爆弾倉バルジ後方フェアリング

① "Dry air" sandwich type flat observation /bomb aiming Perspex panel (late-type)
② Spinner
③ De Havilland Hydromatic needle blade propeller (three-blade, early-type)
④ Coolant fill hatch (left side only of both nacelles)
⑤ Coolant header tank vent
⑥ Stub exhaust pipes
⑦ Perspex V-windscreen (early-type)
⑧ Direct vision panel (opens inwards)
⑨ Intercooler header tank vent
⑩ Intercooler header tank fill hatch
⑪ Side panel blister
⑫ Fuel fill hatch
⑬ Dinghy access panel (jettisonable)
⑭ Antenna mast
⑮ Reinforcing strip
⑯ Ground starter socket hatch
⑰ Vertical stabilizer root fairing
⑱ Root fitting inspection hatch
⑲ Antenna wire mount
⑳ Pitot tube
㉑ Electrical connection access hatch
㉒ Rudder horn balance
㉓ Rudder (fabric-covered, 26° -1° +2° arc of movement)
㉔ Rudder trim tab (16° ±2°30" arc of movement)
㉕ Rudder controls inspection hatch
㉖ Rudder trim tab control rod access panel
㉗ Rudder trim tab control rod fairing
㉘ Elevator operating linkage fairing
㉙ Tail navigation light
㉚ Spare tail-socket fairing
㉛ Detachable tail cone fairing (early-type)
㉜ Retracting gear access panel
㉝ Antenna wire mount (antenna wire not fitted)
㉞ Rearward-retracting tail wheel with Marstrand two-track anti-shimmy tire (late-type)
㉟ Tail wheel strut construction (late-type)
㊱ Nacelle tail fairing
㊲ Bomb bay door
㊳ Main wheel door (early-type)
㊴ Access hatch for jacking (outboard side only of both nacelles)
㊵ Mudguard
㊶ Shock-absorber strut
㊷ Brake pipe
㊸ Plan-type main wheel (late-type)
㊹ Rubber shock pads
㊺ Main wheel door guide (early-type)
㊻ Trailing antenna fairlead fairing
㊼ Fuel pump drain (left side only of both nacelles)
㊽ Carburetor air intake fairing (late-type)
㊾ Fuel pump cooling air intake
㊿ Air intake guard
51 Intercooler duct exhaust
52 Magneto cooling air intake
53 Intercooler radiator air intake
54 Nose compartment side window
55 Static head
56 Deicing spray pipe fairing
57 Priming pump inspection hatch (right side only of both nacelles)
58 Main wheel door guide (late-type)
59 Main wheel door (late-type)
60 Door locking latch access hatch
61 Bulged bomb bay door for 4,000-lb bomb
62 Rear bulged bomb bay doors fairing

モスキートB.Mk.IX

B.Mk.IXは中間冷却器付ロールスロイス・マーリン72、2段2速過給器エンジンを装備した、モスキート最初の高高度爆撃機である。与圧キャビンこそなかったものの、さまざまな電子機器、たとえば"オーボエ"（全天候爆撃／航法システム）、Gee-H（航法／計器爆撃用ビームレーダー）、そしてH_2S（10cm波使用の地形表示スキャニング・レーダー）を装備した一部の機体は、高度30,000フィート（約9,000m）からの爆撃を達成することが可能な性能をもっていた。量産機の初飛行は1943年3月24日。部隊配備は同年4月21日から開始され、No.109Sqnが最初の機体を受領。同年11月までに残りの54機が完成している。初期の爆撃機型モスキートB.Mk.IVシリーズIIとくらべてもっとも異なる点は延長されたエンジン・ナセルである。エンジン・ナセルの前方下面には、チン・タイプの中間冷却器用空気取入口が設けられ、気化器空気取入口のフェアリングも大型化されている。エンジン・ナセルが延長された結果、全長は41フィート（12.497m）になった。エンジンの排気管は推力式単排気管で、通常、排気管シュラウドは装着されなかった。機首は後期型の偵察／爆撃照準用機首透明ガラスで、また、本機は"ユニバーサルウイング"を採用した最初のタイプである。"ユニバーサルウイング"は爆弾か落下タンクのどちらか選択でき、500ポンド爆弾（226.8kg）2発か50／100ガロン（227.3／454.6リッター）入落下タンク2個のどちらかを懸吊することが可能であった。通常、翼端後方にはレジンランプと航法灯が装備されていた。B.Mk.IXの主脚、尾脚はともに後期タイプ。アンテナ支柱と左水平安定板を結ぶアンテナ空中線は取り外されている。ほとんどの機体の主脚収容ドア、主脚収容ドア・ガイド、テイルコーン、そしてプロペラは、生産上の関係で後期型のものが採用されている。燃料搭載量は通常時で497ガロン（2,259リッター）、100ガロン（454.6リッター）入落下タンク2個使用時で最大697ガロン（3,169リッター）であった。B.Mk.IXのうちごく一部の機体が、1943年から1944年にかけて4,000ポンド爆弾用バルジドアをレトロフィットされた。このタイプの燃料搭載量は約2,280リッター。4,000ポンド爆弾搭載型B.Mk.IXが最初に記録したドイツへの昼間ソーティは、1944年11月29日、ドゥイスブルクに対して行われた。

Mosquito B.Mk.IX

This model was the first high-altitude bomber Mosquito, powered by the Rolls-Royce Merlin 72 intercooled-engine with a two-speed, two-stage supercharger. Although lacking a pressurized cabin, some Mosquito crews were able to carry out attacks from 30,000 feet (about 9,000 meters) using various electronic devices such as Oboe (blind bombing system), Gee-H (beam radar for navigation or blind bombing) and H_2S (airborne/ground scanning radar operating on 10cm waveband). The first production B.Mk.IX made its maiden flight on March 24, 1943. Deliveries started on April 21, 1943, when No.109 Squadron received the first aircraft, and ended in November 1943 after 54 examples had been built. The major external differences from the earlier B.Mk.IV Series II model included the modified and lengthened engine nacelles having a chin-type air intake fairing added to the front underside for the supercharger intercooler and an enlarged carburetor air intake fairing. With the new engine nacelles the aircraft had an overall length of 41ft-0in (12.497m). The engines had six stub ejector exhausts usually fitted without exhaust shrouds. The late-type glazed observer/bomb aimer's nose and the "universal wing" were fitted for the first time. This wing allowed either two 500-lb (226.8kg) bombs or two 50-100gal (227.3 - 454.6-liter) fuel drop tanks mounted on universal underwing carriers to be fitted. Each standard wingtip had a rearward resin lamp and a navigation light. B.Mk.IXs had both late-type main wheels and tail wheel strut construction. The antenna wire fitted between the antenna mast and the left horizontal stabilizer was removed. In the course of production, the type was also fitted with late-type main wheel doors, late-type main wheel door guides, a late-type tail cone fairing and late-type de Havilland Hydromatic paddle blade propellers. The normal fuel capacity was 497gal (2,259 liters), while the maximum fuel load was 697gal (3,169 liters) including two 100gal (454.6-liter) wing drop tanks. Only a few B.Mk.IXs were modified during 1943-44 with the retrofitting of the bulged bomb bay doors to carry a 4,000-lb bomb with a fuel load of about 2,280 liters. The first daylight sortie over Germany with B.Mk.IXs carrying 4,000-lb bombs was mounted on November 29, 1944, on Duisburg.

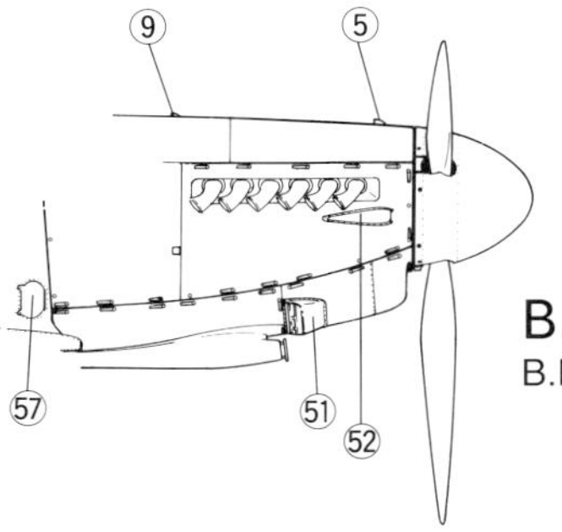

B.Mk.IXのナセル右側面
B.Mk.IX Right side Nacelle

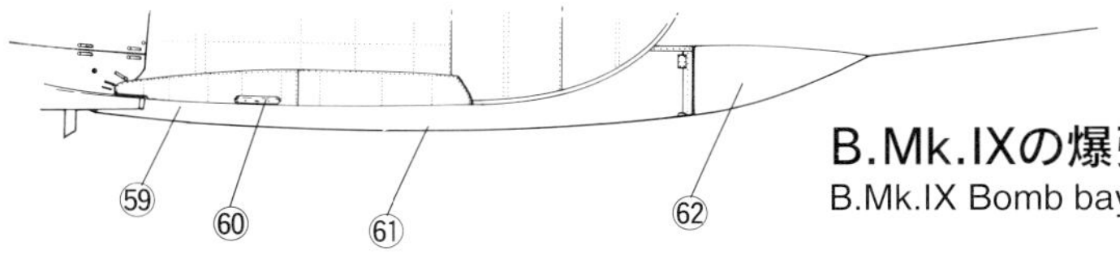

B.Mk.IXの爆弾倉左側面
B.Mk.IX Bomb bay left side

モスキートB.Mk.XVI

B.Mk.XVIは2段過給器付エンジンを装備したB.Mk.IXから発展し、与圧キャビンをそなえた高高度爆撃機型モスキートの量産タイプである。B.Mk.IVシリーズII(DZ540)をもとに開発され、ロールスロイス・マーリン72エンジン(1,680hp)を装備したプロトタイプは、1943年6月に初飛行した。B.Mk.XVIの実用作戦高度は約35,000フィート(11,500m)であった。"オーボエ"、"アルバム・リーフ"、"フィッシュポンド"、"Gee-H"、"H_2S"や敵レーダー電波を感知する"ブザー"、"モニカ"など各種電子機器を装備。"H_2S"レーダーアンテナを機首レドーム、または胴体下面レドーム内に装備する機体もあった。"ユニバーサルウイング"を装備した量産機のエンジンは、1,680hpのマーリン72(右ナセル)とマーリン73(左ナセル)か、1,710hpのマーリン76(右ナセル)とマーリン77(左ナセル)の組みあわせが用いられている。プロペラは後期型のパドルブレードタイプ。左主翼エンジン・ナセルのマーリン73と77は、与圧キャビン用過給器付であった。B.Mk.XVIを最初に受領したのは1943年12月、No.109Sqnで、量産機の配備は1945年12月まで続けられた。最終的に402機が生産され、No.105、109、128、139、163、571、608、692SqnがB.Mk.XVIを使用している。B.Mk.XVIの原型となったB.Mk.IXとの大きな相違点は、4,000ポンド爆弾用のバルジタイプ爆弾倉ドアである。しかしこのバルジタイプ爆弾倉ドアは最初からB.Mk.XVIに装備されていたわけではなく、生産第12号機までは通常の爆弾倉ドアで、搭載爆弾も胴体内に500ポンド爆弾4発、主翼下面に500ポンド爆弾2発であった。4,000ポンド爆弾用のバルジタイプ爆弾倉ドアは生産第13号機から採用されている。4,000ポンド爆弾用のバルジタイプ爆弾倉ドアをつけたモスキートは1944年4月から実戦で使用され、のちに4,000ポンド爆弾ラックにかわって、アブロ社で開発された500ポンド爆弾を6発懸吊するタイプに改修されている。B.Mk.XVIのキャノピー側面には中空合わせガラス式ブリスターが採用され、多くの機体のテイルコーン・フェアリングは後期型のものであった。また、4,000ポンド爆弾用に大型化された爆弾倉後部フェアリングにより、赤色の地上用敵味方識別灯は、緑とアンバーの地上用敵味方識別灯のそばに移動された。曳航式アンテナ・フェアリングと胴体上面のアンテナ支柱は撤去され、かわりにウィップアンテナが採用されている。また、後期生産機の一部には冷却用ルーバー付気化器空気取入口フェアリングがみられる。B.Mk.XVIの最大燃料搭載量は機体が597ガロン(2,714リッター)、しかし通常の作戦時は、497ガロン(2,259リッター)で運用された。

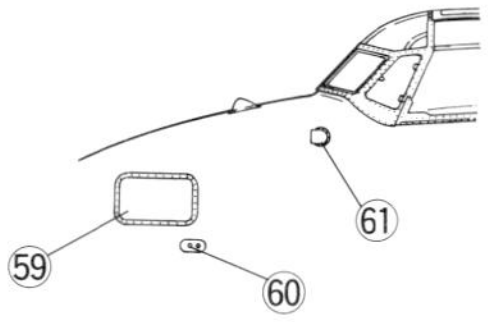

B.Mk.XVIの機首左側面
B.Mk.XVI Nose Left side

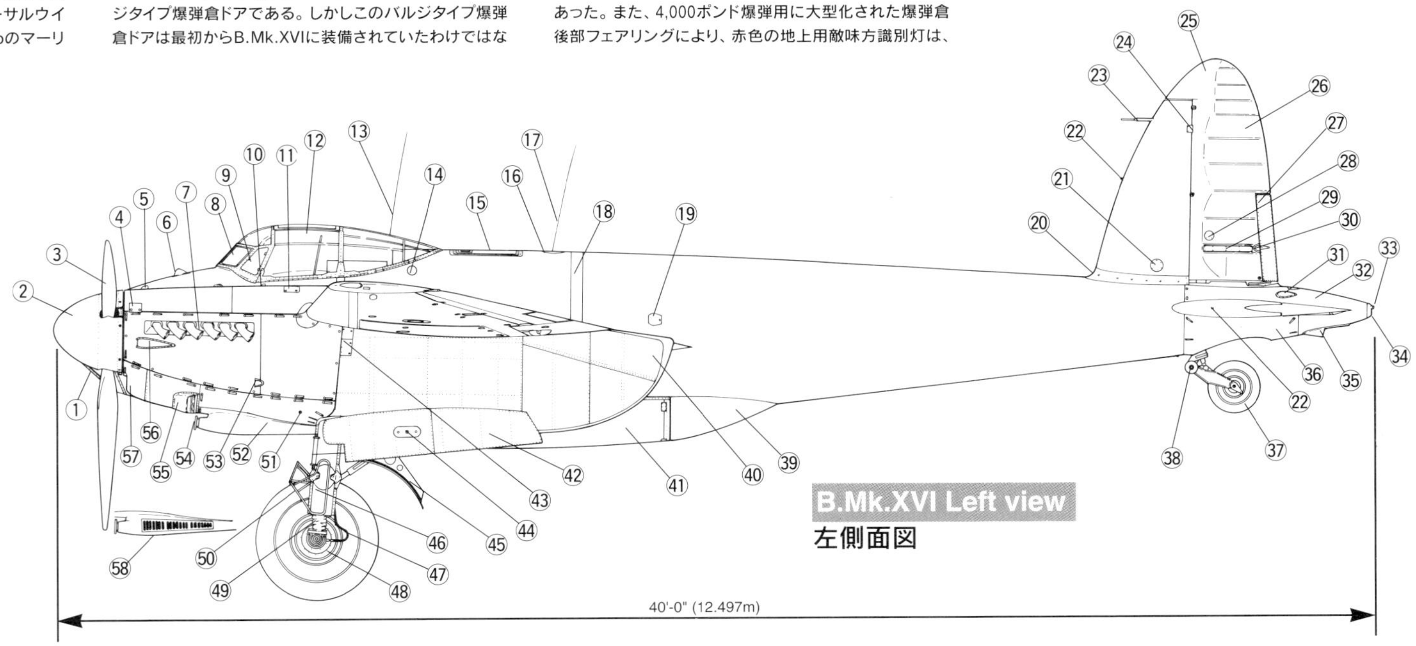

B.Mk.XVI Left view
左側面図

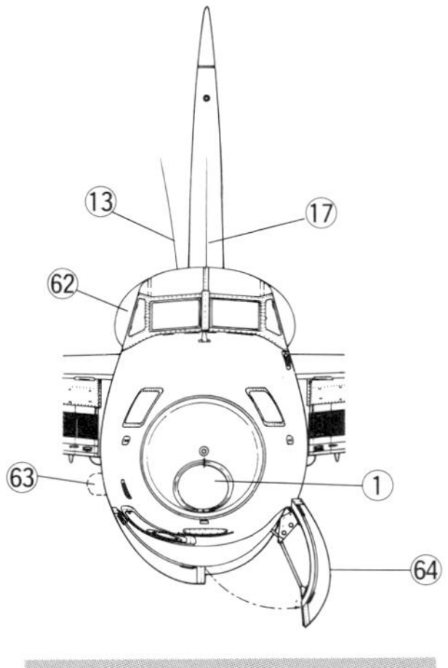

B.Mk.XVI Front view
正面図

① 後期型の中空合わせガラス式偵察／爆撃照準用平面ガラス
② スピナー
③ デ・ハヴィランド・油圧式3翅プロペラ(パドルブレードタイプの後期型、直径12 ft-6 in／3.810m)
④ 冷却液注入ハッチ(左ナセル左側のみ)
⑤ 冷却液タンク・ベント
⑥ 防氷スプレーパイプ・フェアリング
⑦ 推力式単排気管
⑧ 後期型の中空合わせガラス式V字タイプ前部風防
⑨ 内側開閉式の前部風防側面窓
⑩ 中間冷却器タンク・ベント
⑪ 中間冷却器タンク冷却液注入ハッチ
⑫ 与圧コクピット・キャノピー
⑬ ウィップアンテナ(後部キャノピー右側に装備)
⑭ 燃料注入ハッチ
⑮ 救命ボート収納部パネル(投棄式)
⑯ ベークライト製アンテナ保護板
⑰ ウィップアンテナ
⑱ 胴体補強用帯金
⑲ 地上スターターソケットハッチ
⑳ 垂直尾翼基部フェアリング
㉑ 垂直尾翼基部点検ハッチ
㉒ アンテナ空中線接続部(アンテナ空中線はなし)
㉓ ピトー管
㉔ 外部電源接続部パネル
㉕ 方向舵ホーンバランス
㉖ 方向舵(羽布張り、作動角26°±1°)
㉗ 方向舵トリムタブ(作動角16° ±2°30')

① "Dry air" sandwich type flat observation/bomb aiming Perspex panel (late-type)
② Spinner
③ De Havilland Hydromatic paddle blade propeller (three-blade, late-type, 12ft-6in - 3.810m diameter)
④ Coolant fill hatch (left side only of both nacelles)
⑤ Coolant header tank vent
⑥ Deicing spray pipe fairing
⑦ Stub exhaust pipes
⑧ "Dry air" sandwich type Perspex V-windscreen (late-type)
⑨ Direct vision panel (opens inwards)
⑩ Intercooler header tank vent
⑪ Intercooler header tank fill hatch
⑫ Pressurized cockpit canopy
⑬ Whip antenna (fitted on the right side of the rear canopy)
⑭ Fuel fill hatch
⑮ Dinghy access panel (jettisonable)
⑯ Antenna bakelite securing plate
⑰ Whip antenna
⑱ Reinforcing strip
⑲ Ground starter socket hatch
⑳ Vertical stabilizer root fairing
㉑ Root fitting inspection hatch
㉒ Antenna wire mount(antenna wire not fitted)
㉓ Pitot tube

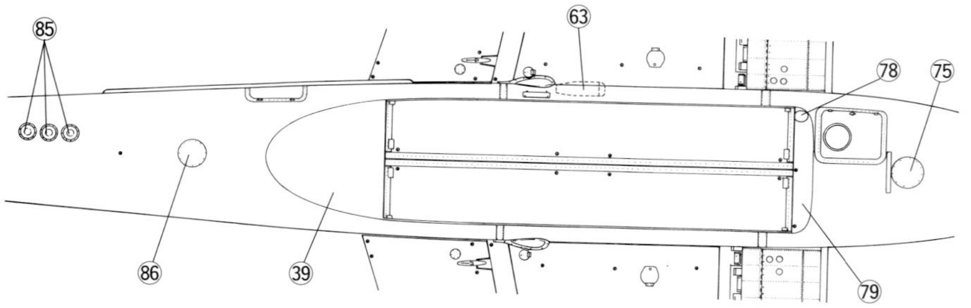

B.Mk.XVI Bottom view
下面図

28 方向舵操舵装置点検ハッチ
29 方向舵トリムタブ操作ロッド・アクセスパネル
30 方向舵トリムタブ操作ロッド・フェアリング
31 昇降舵操舵リンケージ・フェアリング
32 後期型のテイルコーン・フェアリング(取り外し式)
33 航法灯
34 尾部スペアソケット・フェアリング
35 後方警戒レーダー用フェアリング
36 引込式尾脚アクセスパネル
37 後期型のシミーダンパー付"マーストランド"タイヤの引込式尾脚
38 尾輪ストラット
39 爆弾倉バルジ後方フェアリング
40 ナセル後部フェアリング
41 4,000ポンド爆弾用にバルジ化された爆弾倉ドア
42 主脚収容ドア
43 ジャッキ用アクセスハッチ(両ナセル外側のみ)
44 主脚収容ドア開閉ハッチ・アクセスパネル
45 マッドガード
46 ショック・アブゾーバー・ストラット

24 Electrical connection access hatch
25 Rudder horn balance
26 Rudder (fabric-covered, 26°±1° arc of movement)
27 Rudder trim tab(16°±2°30" arc of movement)
28 Rudder controls inspection hatch
29 Rudder trim tab control rod access panel
30 Rudder trim tab control rod fairing
31 Elevator operating linkage fairing
32 Detachable tail cone fairing (late-type)
33 Tail navigation light
34 Spare tail-socket fairing
35 Tail warning radar mount fairing
36 Retracting gear access panel
37 Rearward-retracting tail wheel with Marstrand two-track anti-shimmy tire
38 Tail wheel strut construction
39 Rear bulged bomb bay doors fairing
40 Nacelle tail fairing
41 Bulged bomb bay door for 4,000-lb bomb

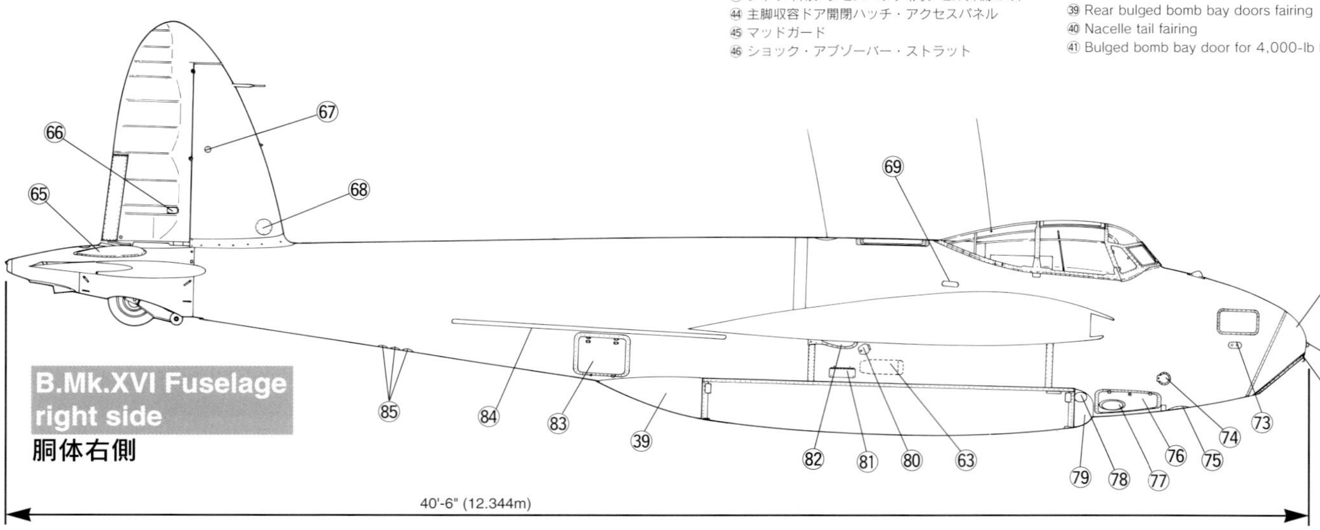

B.Mk.XVI Fuselage right side
胴体右側

47 ブレーキパイプ
48 主車輪ホイール
49 衝撃吸収用ラバーパッド
50 後期型の主脚収容ドア・ガイド
51 燃料ポンプ・ドレイン(両ナセル左側のみ)
52 気化器空気取入口フェアリング
53 燃料ポンプ冷却用空気取入口
54 空気取入れ口保護用防氷カバー
55 中間冷却器用空気排出口
56 磁石発電機冷却用空気取入口
57 中間冷却器用空気取入口
58 冷却用ルーバー(後期生産分)
59 機首側面窓
60 ピトー管静圧ヘッド
61 冷却用空気取入口
62 後期型の中空2重ガラス式ブリスター
63 燃料冷却器(B.Mk.XVIとP.R.Mk.XVIの一部にみられる)
64 爆弾倉ドア開位置
65 昇降舵内側マスバランスのフェアリング
66 左側と位置が異なる方向舵操舵装置点検ハッチ
67 方向舵中央ヒンジ点検ハッチ
68 左側と位置が異なる垂直尾翼基部点検ハッチ
69 主翼取付部フェアリング
70 後期型の機首透明ガラス
71 航法灯
72 防氷スプレーパイプ
73 ピトー管静圧ヘッド
74 アクセスハッチ
75 機首下面アクセスハッチ
76 搭乗員乗降ハッチ
77 下方確認窓
78 曳航式アンテナ・フェアリング・カバーパネル
79 爆弾倉バルジ前方フェアリング
80 点検ハッチ
81 燃料廃棄口ハッチ
82 燃料ポンプ・フェアリング
83 乗降ハッチ
84 胴体補強用ストレーキ
85 地上用敵味方識別灯(前方が赤、中央が緑、後方がアンバー)
86 カバーパネル

42 Main wheel door (late-type)
43 Access hatch for jacking (outboard side only of both nacelles)
44 Door locking latch access hatch
45 Mudguard
46 Shock-absorber strut
47 Brake pipe
48 Plan-type main wheel
49 Rubber shock pads
50 Main wheel door guide (late-type)
51 Fuel pump drain (left side only of both nacelles)
52 Carburetor air intake fairing
53 Fuel pump cooling air intake
54 Air intake guard
55 Intercooler duct exhaust
56 Magneto cooling air intake
57 Intercooler radiator air intake
58 Cooling louver plate (later production aircraft)
59 Nose compartment side window
60 Static head
61 Cooling air intake
62 "Dry air" sandwich type blister (latetype)
63 Fuel cooler (non-standard equipment, seen on a number of B.Mk.XVIs and P.R.Mk.XVIs)
64 Bulged bomb bay door opening position
65 Elevator static balance fairing
66 Rudder controls inspection hatch (different position on this side)
67 Rudder center hinge inspection hatch
68 Root fitting inspection hatch (different position on this side)
69 Wing fix attachment fairing
70 Perspex nose glazing (late-type)
71 Forward navigation light
72 Deicing spray pipe
73 Static head (different position on this side)
74 Access hatch
75 Under floor access hatch
76 Crew entry door
77 Drift sight window
78 Trailing antenna fairlead fairing mount covering panel
79 Front bulged bomb bay doors fairing
80 Inspection hatch
81 Fuel gallery hatch (right side only)
82 Fuel pipe fairing
83 Entry/access door
84 Reinforcing strake
85 Downward identification code lights (forward: red, middle: green, aft: amber)
86 Cover panel

Mosquito B.Mk.XVI

This model was the standard production bomber, a pressurized-cabin version of the B.Mk.IX, equipped with two-stage supercharger engines. The prototype, a modified B.Mk.IV Series II serialled DZ540, made its maiden flight in July 1943 powered by 1,680hp Rolls-Royce Merlin 72 engines. The B.Mk.XVI was able to reach an operational ceiling of about 35,000 feet (11,500 meters), equipped with various electronic detection/counter-measures devices such as Oboe, Album Leaf, Fishpond, Gee-H, H_2S and Boozer or Monica, the latter two being radar warning devices giving notice of detection by enemy radar. Some aircraft were modified with the H_2S radar antenna fitted in a nose radome or a ventral radome. The production aircraft had the "universal wing" and were powered by either the 1,680hp Merlin 72 (right side) and 73 (left side) or the 1,710hp Merlin 76 (right side) and 77 (left side) engines, both driving paddle blade propellers. Both Merlin 73 and 77, fitted to the left wing engine nacelle, drove a Marshalls cabin supercharger. No.109 Squadron was the first unit to receive this model in December 1943, and production aircraft deliveries concluded in December 1945, totaling 402 aircraft. No.105, 109, 128, 139, 163, 571, 608 and 692 Squadrons of the Light Night Striking Force operated B.Mk.XVIs. The major apparent differences from the original B.Mk.IX were the bulged bomb bay doors allowing the aircraft to carry a 4,000-lb bomb. However, this modification was introduced from the 13th aircraft with the first twelve aircraft being fitted with ordinary bomb bay doors and carrying the standard load of four 500-lb bombs and two 500-lb bombs on the underwing racks. Mosquitoes with bulged bomb bay doors started war operations in April 1944, and later they were also capable of carrying six 500-lb bombs, in place of the 4,000-lb bomb, with the fitting of a special carrier developed by the Avro Company. These had a modified canopy frame structure fitted with "dry air" sandwich blisters on side panels and most were equipped with late-type tail cone fairings. Because of the fitting of the rear bulged bomb bay door fairing, the red downward identification code light was moved from the left underside of the mid-fuselage to the right underside of the rear fuselage, near the green and amber downward identification code lights. The trailing antenna fairlead fairing and the antenna mast were removed and different types of whip antenna fittings were used. Some later production aircraft had the carburetor air intake fairings fitted with cooling louver plates. The B.Mk.XVI had a fuel capacity of 597gal (2,714 liters), but usually it carried 497gal (2,259 liters) for operational missions.

P.R.Mk.XVI Left view
左側面図

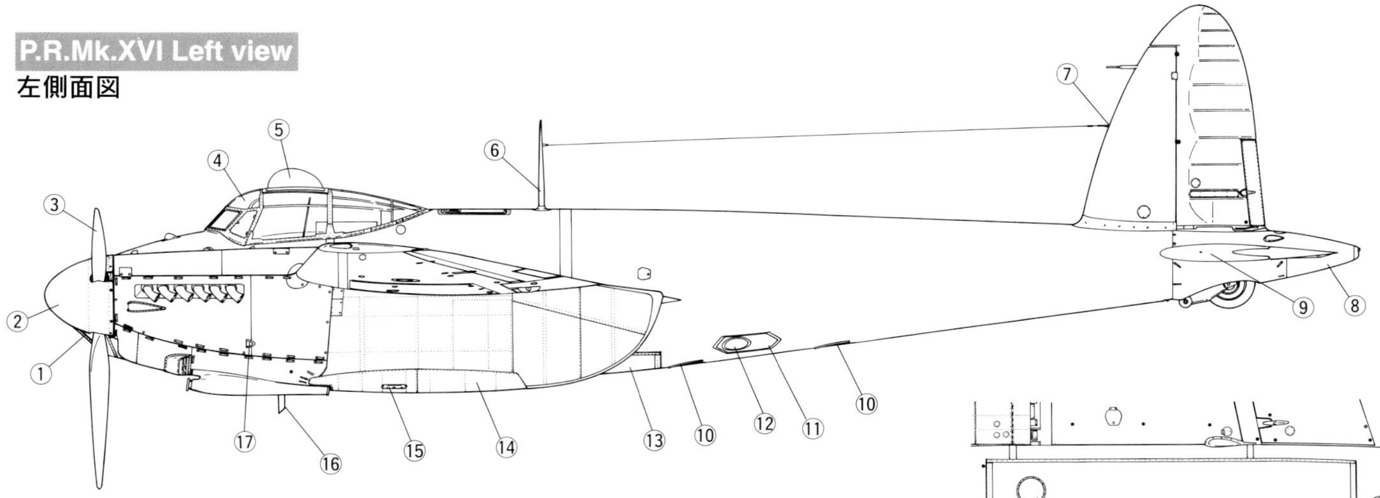

P.R.Mk.XVI Bottom view
下面図

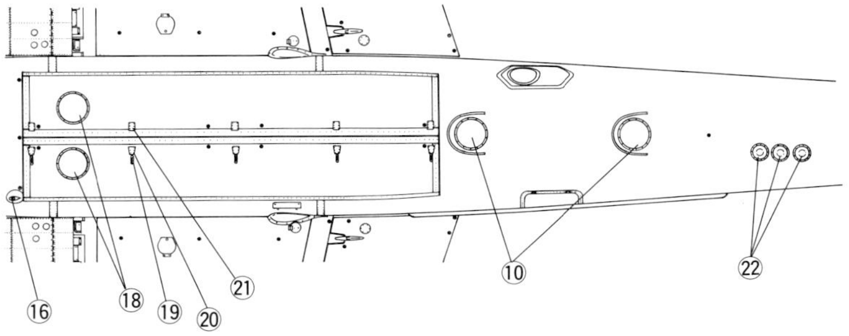

モスキートP.R.Mk.XVI

P.R.Mk.XVIは第二次大戦中の写真偵察機型モスキートのなかで、もっとも主要なタイプである。配備は1943年12月から開始され、No.140Sqnが最初に受領している。P.R.Mk.XVIは4,000ポンド爆弾用大型爆弾倉を装備した爆撃機型のB.XVIによく似ているが、爆弾倉は通常の胴体コンポーネント・ドアが用いられ、燃料タンク3個が増設されている。偵察カメラは4基を装備。そのアレンジはP.R.Mk.I、P.R.Mk.IVと異なり、胴体後方に垂直方向撮影用のカメラ窓が新設されていた。また、ほとんどの機体は与圧コクピットの航法士頂部非常用脱出ハッチに、中空合わせガラス式ブリスター（天測ドーム）を採用し、後方視界を改善していた。エンジンはロールスロイス・マーリン72／73か、マーリン76／77の組みあわせが用いられ、後期生産機の一部には右主翼前縁に着陸／滑走灯が取り付けられた。総生産機数は432機。P.R.Mk.XVIの最終生産機は1945年5月に配備されている。また、1944年に100機以上のP.R.Mk.XVIがアメリカ第8空軍に配備され、これらのうちほとんどの機体は、主に天候観測機と写真偵察機として使用されている。H_2Xを装備する機体もあった。H_2Xはアメリカで開発された3cm波を用いる地形表示レーダーである。また、P.R.Mk.XVIはレーダー妨害のための"ウインドウ"（チャフ）投下機としても使用され、さらにフォト・フラッシュ爆弾を使用した夜間写真偵察も行っている。これらの機体は胴体コンパートメント内前後の燃料タンクを撤去し、コンパートメント・ドアも油圧ジャッキで開閉できるようになっていた。本機の最大燃料搭載量は860ガロン（3,910リッター）、通常は539ガロン（2,450リッター）であった。

Mosquito P.R.Mk.XVI

The P.R.Mk.XVI was the most important photoreconnaissance Mosquito operated in the course of WWII. Deliveries began in December 1943 and No.140 Squadron was the first unit equipped with the model. It was quite similar to the bomber variant B.Mk.XVI except for the fitting of the ordinary ventral compartment doors and the installation of three extra fuel tanks. The model had four vertical cameras fitted, instead of three as in the P.R.Mk.I and P.R.Mk.IV, having an additional rear vertical camera that altered the arrangement of the cameras installed in the rear underside of the fuselage. Most of the aircraft had the "dry air" sandwich panel of the pressurized cockpit canopy emergency roof fitted with a navigator's top blister (astrodome) improving rearward visibility. The P.R.Mk.XVIs were powered by either the Rolls-Royce Merlin 72/73 or 76/77 engines. Some late production aircraft had landing/taxing lights added to the right wing leading edge. A total of 432 aircraft were completed when the last P.R.Mk.XVI was delivered in May 1945. In the course of 1944, more than 100 P.R.Mk.XVIs were delivered to the USAAF 8th Air Force, and these Mosquitos were mainly operated as weather reconnaissance and photographic aircraft. A humber were modified and fitted with various electronic devices, some carrying the H_2X radar, the American development of the British airborne/ground scanning radar operating on the 3cm band. P.R.Mk.XVIs were also operated to dispense radar-jamming "chaff" (the American code-name for the British "window") and for night photography using photoflash bombs. Depending on the mission these operations required the removal of front/rear fuel tanks located in the bomb bay and the installation of bomb bay doors fitted with hydraulic jacks. The maximum fuel capacity was 860gal (3,910 liters), and the usual fuel load was 539gal (2,450 liters).

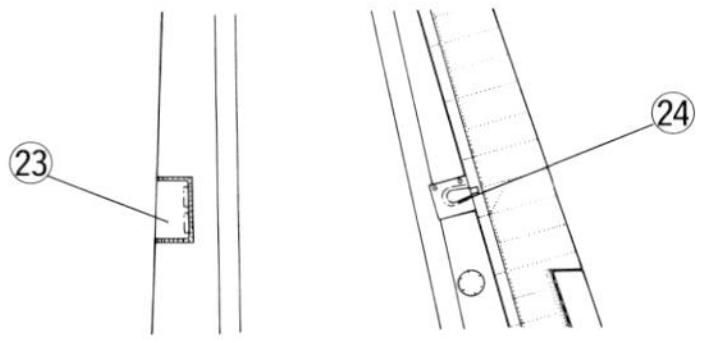

P.R.Mk.XVIの右主翼上面
P.R.Mk.XVI Right wing upper side

① 後期型の中空合わせガラス式偵察／観測用平面ガラス
② スピナー
③ デ・ハヴィランド・油圧式3翅プロペラ（パドルブレードタイプの後期型）
④ 与圧コクピット・キャノピー
⑤ 航法士席頂部ブリスター
⑥ アンテナ支柱
⑦ アンテナ空中線接続部
⑧ 初期型のテイルコーン・フェアリング（取り外し式）
⑨ アンテナ空中線接続部（アンテナ空中線はない）
⑩ 胴体後方下面カメラ窓
⑪ カメラ窓フェアリング・パネル
⑫ 斜め下方撮影用カメラ窓
⑬ 胴体下面コンパートメント・ドア
⑭ 後期型の主脚収容ドア
⑮ ドアロック用ラッチ・アクセスパネル
⑯ 曳航式アンテナ・フェアリング
⑰ 燃料ポンプ冷却空気取入口
⑱ 胴体前方下面カメラ窓
⑲ ネジ式ドアロック
⑳ ドア開閉フック・フェアリング
㉑ ドア開閉ラッチ・フェアリング
㉒ 地上用敵味方識別灯（前方が赤、中央が緑、後方がアンバー）
㉓ 着陸／滑走灯（後期生産機の右主翼のみ）
㉔ 補助翼中央ヒンジ・フェアリング

① "Dry air" sandwich type flat observation aiming Perspex panel (late-type)
② Spinner
③ De Havilland Hydromatic paddle blade propeller (three-blade, late-type)
④ Pressurized cockpit canopy
⑤ Navigator's top blister
⑥ Antenna mast
⑦ Antenna wire mount
⑧ Detachable tail cone fairing (early-type)
⑨ Antenna wire mount (antenna wire not fitted)
⑩ Rear vertical camera window
⑪ Camera fairing panel
⑫ Oblique camera window
⑬ Ventral compartment door (fitted with door locks)
⑭ Main wheel door (late-type)
⑮ Door locking latch access hatch
⑯ Trailing antenna fairlead fairing
⑰ Fuel pump cooling air intake
⑱ Front vertical camera window
⑲ Screwdriver opening for door lock
⑳ Lock hook screw operated mount fairing
㉑ Lock catch mount fairing
㉒ Downward identification code lights (forward: red, middle: green, aft: amber)
㉓ Landing/taxi lights (right wing only, late production aircraft)
㉔ Aileron center hinge fairing

P.R.Mk.34 Left view
左側面図

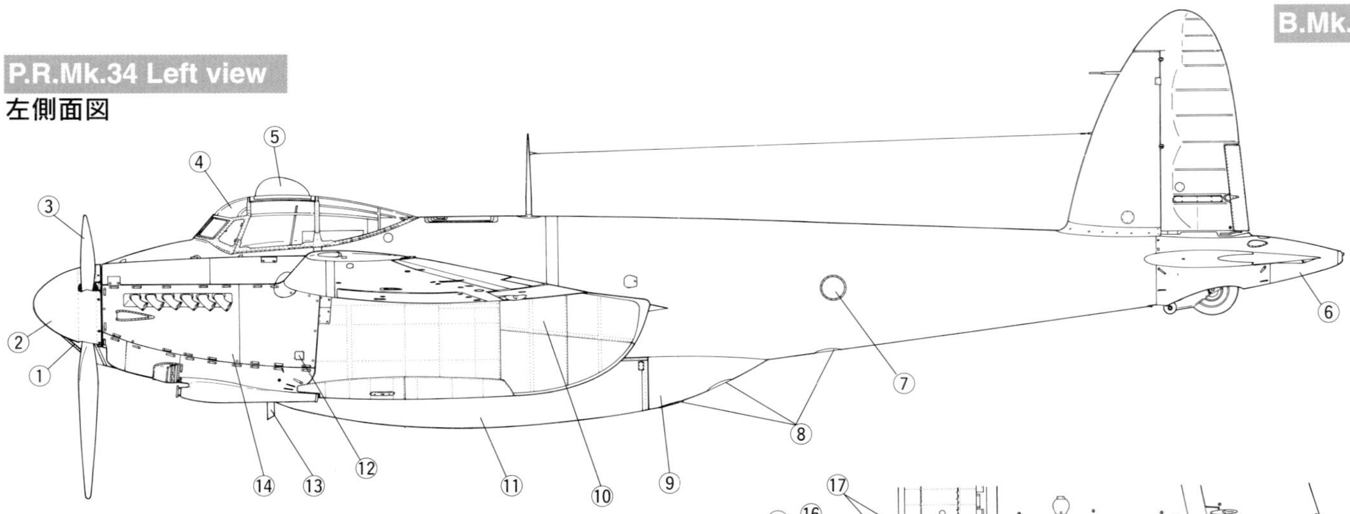

B.Mk.34 Front view
正面図

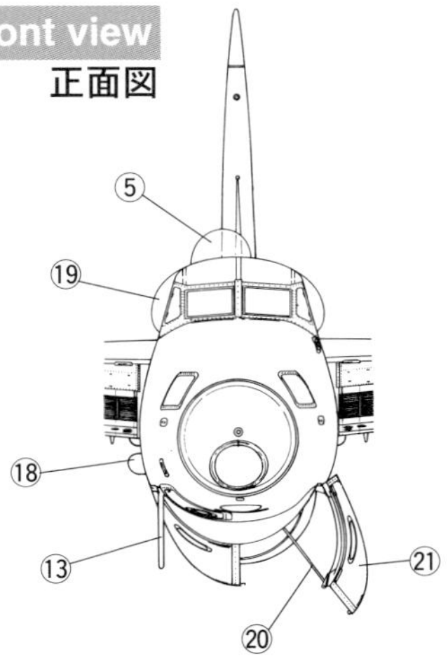

P.R.Mk.34 Bottom view
下面図

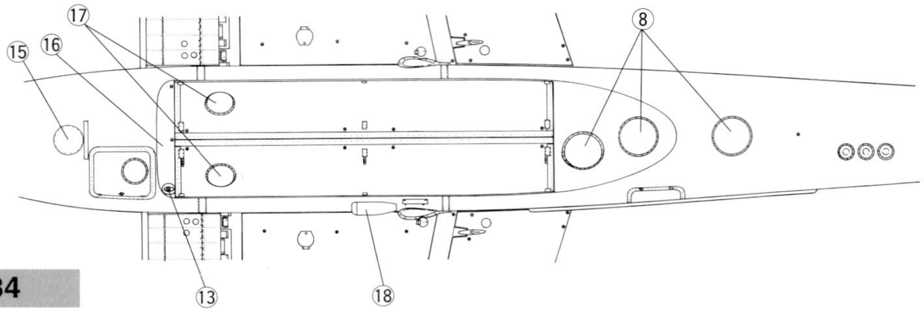

モスキートP.R.Mk.34

P.R.Mk.34は極東での作戦用に開発された、写真偵察機型最後のバリエーションである。初期のP.R.Mk.XVI同様、エンジンは2段2速過給器付ロールスロイス・マーリン76/77、のちにマーリン113（右側）、114（左側）の組みあわせとなった（高高度作戦用与圧キャビンの過給器は、左エンジンに装備）。1号機の初飛行は1944年12月4日。181機を生産する予定であったが、終戦により契約が解除されたため、完成したのは50機あまりだった。第二次大戦中、すべてのP.R.Mk.34はNo.684Sqnに配備され、マラヤ、シンガポール、オランダ領東インドでの偵察任務に用いられた。P.R.Mk.XVIともっとも異なる部分は、4,000ポンド爆弾用バルジよりさらに大型化した胴体下面コンポーネント・ドアと、B.Mk.IVスペシャルと同様に増積された昇降舵のホーンバランスである。搭載するカメラも数が増やされ、垂直撮影用カメラ5基とF24斜め下方撮影用カメラが搭載された。さらにF24斜め下方撮影用カメラのかわりに、K17測量カメラの搭載も可能だった。また、カメラ窓のアレンジも変更されている。P.R.Mk.34は胴体下面コンパートメント・ドアを大型化したことにより燃料搭載量が増大し、胴体内部タンクは通常で1,269ガロン（5,769リッター）、200ガロン（909.2リッター）入落下タンク2個を併用して約5,375kmの航続距離を得ることができた。また、主翼下面の着陸灯システム、操縦席と燃料タンクの防弾設備を取り外した機体の上昇限度は3,000フィート（約910m）向上した。キャノピーの緊急脱出用ハッチは航法士頂部のブリスター部分に設けられ、後期生産分の機体では冷却用ルーバー気化器空気取入口フェアリングが採用されている。

Mosquito P.R.Mk.34

The P.R.Mk.34 was the last Mosquito photoreconnaissance variant built, being developed especially for operational service in the Far East. Similar to the earlier P.R.Mk.XVI, this model was initially equipped with the Rolls-Royce Merlin 76/77 two-stage supercharger engines, and later with Merlin 113 (right side) and 114 (left side) engines, with the left engine driving a cabin supercharger for operations at high altitudes. The first production aircraft made its maiden flight on December 4, 1944, and about 50 aircraft, among the 181 originally ordered, were built and delivered in the course of 1945 before the contract was canceled at the end of the war. All P.R.Mk.34s involved in war missions were flown by No.684 Squadron, seeing operations over Malaya, Singapore and the Dutch East Indies. The major changes from the P.R.Mk.XVI were the bulged ventral compartment doors, larger than the bulged bomb bay doors fitted to bomber variants carrying a 4,000lb bomb, larger horn balances fitted to the tail plane elevators, similar to those operated by the B.Mk.IV Special, the addition of a fuselage underside camera (the aircraft was equipped with five vertical cameras and one F24 oblique camera that could be replaced by a K17 camera for survey work) and a new arrangement and layout of the camera windows. With the introduction of the deepened bomb bay doors the P.R.Mk.34 was capable of carrying additional fuel tanks in the belly, having a normal fuel capacity of about 1,269gal (5,769 liters), including two 200gal (909.2-liter) underwing drop tanks, and a range of about 5,375 kilometers. Underwing landing lights, all armored fittings and fuel tank protection were removed to give a further 3,000 feet (about 910 meters) increase in the aircraft's ceiling. The canopy emergency roof exit was fitted with the navigator's top blister. Late production aircraft had the carburetor air intake fairings fitted with cooling louver plates.

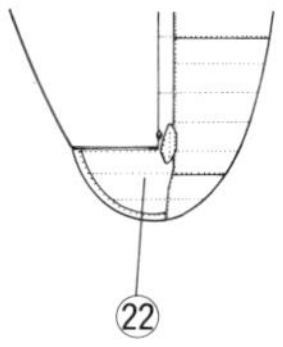

P.R.Mk.34の昇降舵ホーンバランス
P.R.Mk.34 Elevator horn balance

① 後期型の中空合わせガラス式偵察／観測用平面ガラス
② スピナー
③ デ・ハヴィランド・油圧式3翅プロペラ（パドルブレードタイプの後期型）
④ 与圧コクピット・キャノピー
⑤ 航法士席頂部ブリスター
⑥ 初期型のテイルコーン・フェアリング
⑦ 斜め下方撮影用カメラ窓
⑧ 通常の偵察タイプより大型のカメラ窓
⑨ 胴体コンパートメント後方バルジ
⑩ パネルを再設計したナセル後部フェアリング
⑪ バルジタイプの胴体コンパートメント・ドア
⑫ アクセスハッチ
⑬ 曳航式アンテナ・フェアリング
⑭ 燃料ポンプ冷却空気取入口がない
⑮ 機首下面アクセスハッチ
⑯ 胴体コンパートメント前方バルジ
⑰ 垂直撮影用前方カメラ窓
⑱ 燃料冷却器
⑲ 後期型の中空2重ガラス式ブリスター
⑳ 胴体コンパートメント・ドア前方ステイ
㉑ 胴体コンパートメント・ドア前方開位置
㉒ 昇降舵ホーンバランスを増積

① "Dry air" sandwich type flat observation aiming Perspex panel (late-type)
② Spinner
③ De Havilland Hydromatic paddle blade propeller (three-blade, late-type)
④ Pressurized cockpit canopy
⑤ Navigator's top blister
⑥ Detachable tail cone fairing (early-type)
⑦ Oblique camera window
⑧ Camera window (larger than type fitted to former P.R. models)
⑨ Rear bulged ventral compartment doors fairing
⑩ Nacelle tail fairing (redesigned panels)
⑪ Bulged ventral compartment door (fitted with door locks)
⑫ Fire access hatch
⑬ Trailing antenna fairlead fairing
⑭ Fuel pump cooling air intake removed
⑮ Under floor access hatch
⑯ Front bulged ventral compartment door fairing
⑰ Camera window
⑱ Fuel cooler
⑲ "Dry air" sandwich type blister (late-type)
⑳ Front bulged ventral compartment door stay
㉑ Bulged ventral compartment door open position
㉒ Enlarged elevator horn balance

B.Mk.35 Left view
左側面図

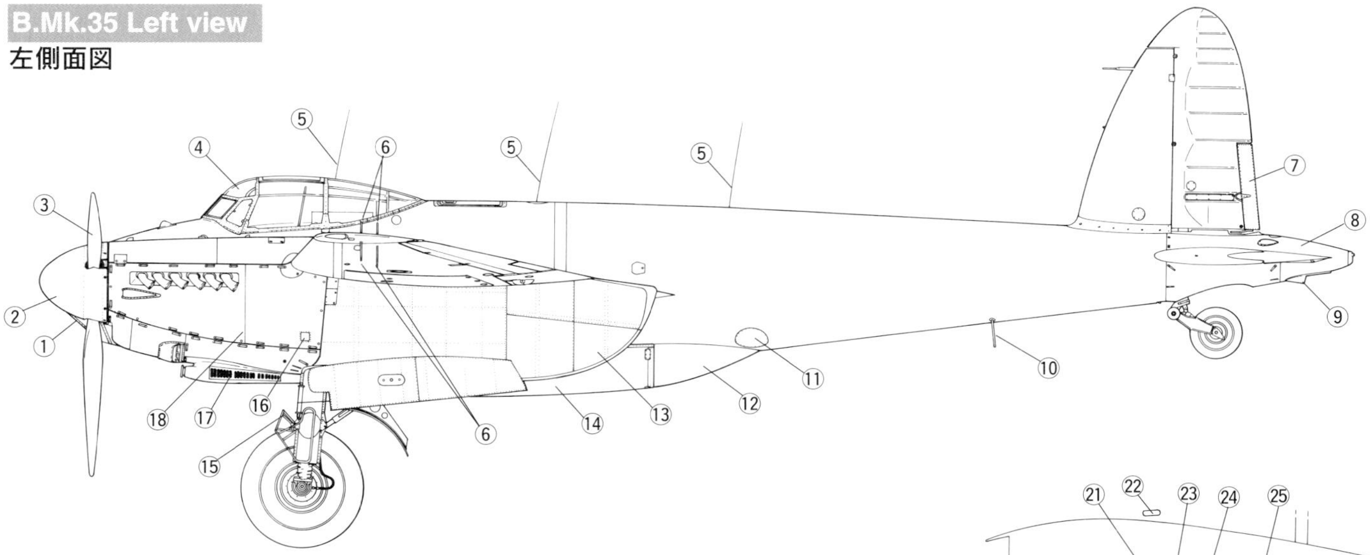

B.Mk.35の爆弾倉左側面
B.Mk.35 Bomb bay left side

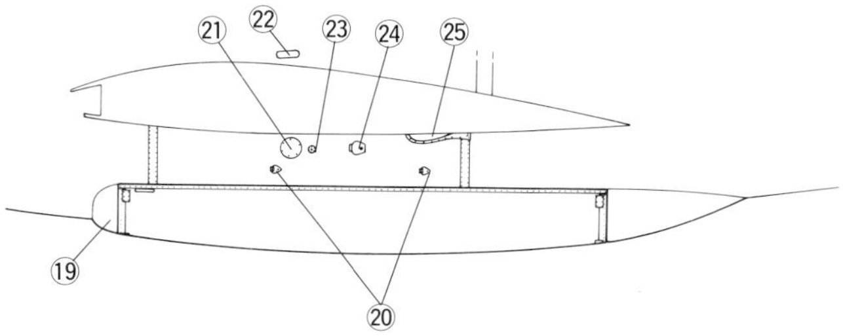

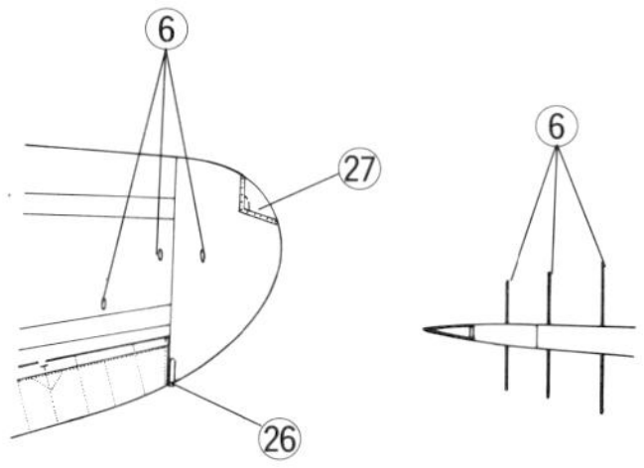

B.Mk.35の右主翼端
B.Mk.35 Right wing

モスキートB.Mk.35

B.Mk.35はB.MkXVIとたいへんによく似た、爆撃機型モスキート最後の量産タイプである。原型機の初飛行は1945年3月12日。同月中にボマーコマンドのNo.109とNo.139Sqnへ配備が開始された。B.Mk.35は、実戦には間に合わなかったが、1948年春までにハットフィールドで計274機が生産されている。バルジタイプの爆弾倉を装備したB.Mk.35は、与圧キャビンをもち、エンジンは2段2速過給器付1,690hpのロールスロイス・マーリン113（右側）と、与圧キャビン用過給器を備えたマーリン114（左側）であった。主翼端には4本から6本のGee-H爆撃／後方システム用ダイポールアンテナが装備され、胴体前方下面の曳航式アンテナは廃止。胴体上面のアンテナ支柱も撤去され、かわりにキャノピー上部と胴体上面にウィップアンテナが装備された。後期生産機の気化器空気取入口フェアリングは冷却用ルーバー付。通常作戦時には、4,000ポンド爆弾装備状態で50ガロン入落下タンク2個、500ポンド爆弾4発装備状態で100ガロン入落下タンク2個を使用する。戦後、B.Mk.35は旧西ドイツ地域に駐留したBAFO（占領国イギリス空軍）のNo.14、69、98Sqnに配備され、また、105機のB.Mk.35が標的曳航機型T.T.Mk.35に、9機が写真偵察機型 P.R.Mk.35に改修されている。マーリン113A／114Aを装備したB.Mk.35の多くは、後期型の長いスピナーを装着し、全長が41フィート6インチ（12.649m）に延長された。

Mosquito B.Mk.35

This model, very similar to the B.Mk.XVI, was the last Mosquito bomber variant manufactured. The prototype made its first flight on March 12, 1945, and deliveries commenced in March 1945 to No.109 and 139 Squadrons of Bomber Command. By May 31, 1945, 57 B.Mk.35s had been delivered from Hatfield, but these were too late to see war operations. In all, production totaled 274 aircraft, with deliveries ending in spring of 1948. The B.Mk.35, fitted with bulged bomb bay doors and a pressurized cabin, was powered by the 1,690hp Rolls-Royce Merlin 113 (right side) and 114 (left side) two-stage supercharger engines with the Merlin 114 driving a cabin supercharger. This bomber was fitted with four to six wingtip-mounted dipole antennas for the Gee-H bombing /navigational device. The trailing antenna fairlead fairing was removed from the front fuselage underside and the antenna mast was replaced by whip antennas located on the roof of the cockpit canopy and the fuselage upper side. The carburetor air intake fairing of late production aircraft was fitted with cooling louver plates. Usually the operational configuration provided for two 50gal wing drop tanks with a bomb load of 4,000-lb, or two 100gal wing drop tanks with a bomb load of four 500-lb bombs. In the post-war years, this model served with Nos.14, 69 and 98 Squadrons of the BAFO (British Air Force of Occupation) in former West Germany. About 105 airframes were transformed into target tug aircraft as T.T.Mk.35 and nine into reconnaissance aircraft as P.R.Mk.35. A number of aircraft, also equipped with the Merlin 113A/114A engines, were fitted with late-type longer propeller spinners which extended the overall length to 41ft-6in (12.649m).

① 後期型の中空合わせガラス式偵察／爆撃照準用平面ガラス
② スピナー
③ デ・ハヴィランド・油圧式3翅プロペラ（パドルブレードタイプの後期型）
④ 与圧コクピット・キャノピー
⑤ ウィップアンテナ
⑥ ダイポールアンテナ
⑦ 方向舵トリムタブ（作動角16°±1°15"）
⑧ 後期型のテイルコーン・フェアリング（取り外し式）
⑨ 後方警戒レーダー用フェアリング
⑩ アンテナ
⑪ カバーパネル
⑫ 爆弾倉バルジ後方フェアリング
⑬ パネルを再設計したナセル後部フェアリング
⑭ 4,000ポンド爆弾用にバルジ化された爆弾倉ドア
⑮ 後期型の主脚収容ドア・ガイド
⑯ アクセスハッチ
⑰ 後期生産期の冷却用ルーバー付気化器空気取入口フェアフェアリング
⑱ 燃料ポンプ冷却空気取入口がない
⑲ 爆弾倉バルジ前方フェアリング
⑳ 4,000ポンド爆弾用内蔵ウィンチハンドル・アクセスハッチ
㉑ 点検ハッチ
㉒ 主翼取付部フェアリング
㉓ アクセスハッチ
㉔ 爆弾搭載確認ハッチ
㉕ 燃料パイプ・フェアリング
㉖ レジンランプ
㉗ 右主翼端航法灯（緑）

① "Dry air" sandwich type flat observation/bomb aiming Perspex panel (late-type)
② Spinner
③ De Havilland Hydromatic paddle blade propeller (three-blade, late-type)
④ Pressurized cockpit canopy
⑤ Whip antenna
⑥ Dipole antenna
⑦ Rudder trim tab (16°±1°15" arc of movement)
⑧ Detachable tail cone fairing (late-type)
⑨ Tail warning radar mount fairing
⑩ Antenna
⑪ Cover panel
⑫ Rear bulged bomb bay doors fairing
⑬ Nacelle tail fairing (redesigned panel)
⑭ Bulged bomb bay door for 4,000-lb bomb
⑮ Main wheel door guide (late-type)
⑯ Fire access hatch
⑰ Cooling louver plate (late production aircraft)
⑱ Fuel pump cooling air intake removed
⑲ Front bulged bomb bay doors fairing
⑳ Access hatch for 4,000-lb bomb winch handle
㉑ Inspection hatch
㉒ Wing fix attachment fairing
㉓ Access hatch
㉔ Bomb loading check hatch
㉕ Fuel pipe fairing
㉖ Resin lamp
㉗ Right navigation light (green)

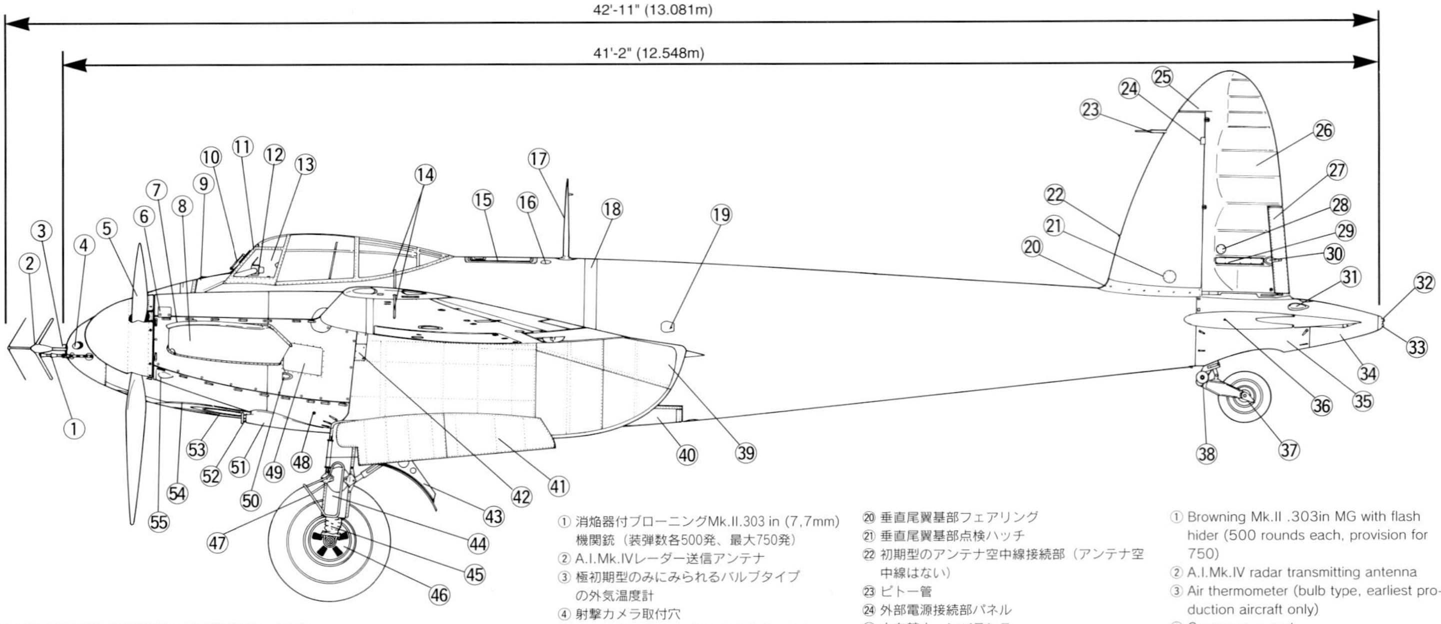

F. / N.F.Mk.II Left view
左側面図

① 消焔器付ブローニングMk.II.303 in (7,7mm)機関銃（装弾数各500発、最大750発）
② A.I.Mk.IVレーダー送信アンテナ
③ 極初期型のみにみられるバルブタイプの外気温度計
④ 射撃カメラ取付穴
⑤ デ・ハヴィランド・油圧式3翅プロペラ（ニードルブレードタイプの初期型）
⑥ 冷却液注入ハッチ
⑦ 排気マニフォールド・フランジ冷却空気ダクト
⑧ 排気管消焔シュラウド
⑨ 防氷スプレーパイプ
⑩ 前部風防ワイパー
⑪ 1枚板耐弾ガラスの前部風防
⑫ 光像式射撃照準器
⑬ スライド式側面窓
⑭ A.I.Mk.IVレーダー方位受信アンテナ
⑮ 救命ボート収納部パネル（投棄式）
⑯ 燃料注入ハッチ
⑰ アンテナ支柱（アンテナ空中線はない）
⑱ 胴体補強用帯金
⑲ 地上スターターソケットハッチ
⑳ 垂直尾翼基部フェアリング
㉑ 垂直尾翼基部点検ハッチ
㉒ 初期型のアンテナ空中線接続部（アンテナ空中線はない）
㉓ ピトー管
㉔ 外部電源接続部パネル
㉕ 方向舵ホーンバランス
㉖ 方向舵（羽布張り、作動角26°-1°+2°）
㉗ 方向舵トリムタブ（作動角16°±2°30"）
㉘ 方向舵操舵装置点検ハッチ
㉙ 方向舵トリムタブ操作ロッド・アクセスパネル
㉚ 方向舵トリムタブ操作ロッド・フェアリング
㉛ 昇降舵操舵リンケージ・フェアリング
㉜ 航法灯
㉝ 編隊灯
㉞ 初期型のテイルコーン・フェアリング（取り外し式）
㉟ 引込式尾脚アクセスパネル
㊱ アンテナ空中線接続部(アンテナ空中線はない)
㊲ 初期型タイヤ付き引込式尾輪
㊳ 初期型の尾輪ストラット
㊴ 後期型のナセル後部フェアリング
㊵ 胴体下面コンパートメント・ドア
㊶ 初期型の主脚収容ドア
㊷ ジャッキ用アクセスハッチ(両ナセル外側のみ)
㊸ マッドガード
㊹ ショック・アブゾーバー・ストラット
㊺ 衝撃吸収用ラバーパッド
㊻ ドラム左側が鋳造スロット、右側がブレーキドラムの初期型主車輪ホイール
㊼ 初期型の主脚収容ドア・ガイド
㊽ 燃料ポンプ・ドレイン（両ナセル左側のみ）
㊾ 排気用耐熱パネル
㊿ 燃料ポンプ冷却空気取入口
(51) 初期型の気化器空気取入口フェアリング
(52) 空気取入れ口保護用防氷カバー
(53) イスパノMk.I 20mm機関砲（装弾数150発、最大175発）
(54) 機関砲フェアリング
(55) 冷却液パイプのバルジ（両ナセル左側のみ）
(56) 航法灯（左翼が赤、右翼が緑）
(57) 編隊灯
(58) 補助翼中央ヒンジ・フェアリング
(59) 補助翼サーボ・アクセスハッチ
(60) ベントパイプ・アクセスハッチ
(61) 燃料注入ハッチ
(62) A.I.Mk.IVレーダー受信アンテナ（右翼のみ）

① Browning Mk.II .303in MG with flash hider (500 rounds each, provision for 750)
② A.I.Mk.IV radar transmitting antenna
③ Air thermometer (bulb type, earliest production aircraft only)
④ Camera gun port
⑤ De Havilland Hydromatic needle blade propeller (three-blade, early-type)
⑥ Coolant fill hatch
⑦ Cooling spouts air duct fairing for exhaust manifold flanges
⑧ Exhaust flame damper shroud
⑨ Deicing spray pipe
⑩ Windscreen wiper
⑪ Flat bulletproof windscreen
⑫ Reflecting gunsight
⑬ Sliding side panel
⑭ A.I.Mk.IV radar receiving azimuth antenna
⑮ Dinghy access panel (jettisonable)
⑯ Fuel fill hatch
⑰ Antenna mast (antenna wire not fitted)
⑱ Reinforcing strip
⑲ Ground starter socket hatch
⑳ Vertical stabilizer root fitting
㉑ Root fitting inspection hatch
㉒ Antenna wire mount (early position,antenna wire not fitted)
㉓ Pitot tube
㉔ Electrical connection access hatch
㉕ Rudder horn balance
㉖ Rudder(fabric-covered, 26°-1°+2° arc of movement)
㉗ Rudder trim tab (16° ±2°30"arc of movement)
㉘ Rudder controls inspection hatch
㉙ Rudder trim tab control rod access panel
㉚ Rudder trim tab control rod fairing
㉛ Elevator operating linkage fairing
㉜ Tail navigation light
㉝ Tail formation light
㉞ Detachable tail cone fairing (early-type)
㉟ Retracting gear access panel
㊱ Antenna wire mount (antenna wire not fitted)
㊲ Rearward-retracting tail wheel with one-track tire (early-type)
㊳ Tail wheel strut construction (early-type)
㊴ Nacelle tail fairing (late-type)
㊵ Ventral compartment door(fitted with door locks)
㊶ Main wheel door (early-type)
㊷ Access hatch for jacking (outboard side only of both nacelles)
㊸ Mudguard
㊹ Shock-absorber strut
㊺ Rubber shock pads
㊻ Main wheel with slots cast on the left side and single brake drums on the right side (early-type)
㊼ Main wheel door guide (early-type)
㊽ Fuel pump drain (left side only of both nacelles)
㊾ Heat resistant panel
㊿ Fuel pump cooling air intake
(51) Carburetor air intake fairing (early-type)
(52) Air intake guard
(53) Hispano Mk.I 20mm cannon (150 rounds each, provision for 175)
(54) Cannon under fairing
(55) Coolant pipe bulge (left side only of both nacelles)
(56) Navigation light (left: red, right: green)
(57) Formation light
(58) Aileron center hinge fairing
(59) Aileron servo access hatch
(60) Vent pipe access hatch
(61) Fuel fill hatch
(62) A.I.Mk.IV radar receiving elevation antenna (right wing only)

モスキートF. / N.F.Mk.II

F. / N.F.Mk.IIはヨーロッパ、中東、極東のさまざまな戦域で昼間戦闘機、夜間戦闘機、イントルーダー、レンジャー機として活躍した戦闘機型モスキート最初のバリエーションである。原型機（W4052）の初飛行は1941年3月15日。1943年3月までに392機が生産されている。量産第1号機は同年12月に完成し、1942年4月3日にNo.157Sqnの所属機が実戦デビューをはたした。写真偵察機型および爆撃機型とのいちばんの違いは、ソリッドノーズ化された機首である。F./ N.F.Mk.IIは機首にブローニング .303 in(7.7mm)機関銃4挺と射撃カメラを、爆弾倉のスペースにイスパノMk.I 20mm機関砲を装備。機関銃の先端にはのちに消焔器が追加されている。また、搭乗員昇降ハッチは機首右側面に移動し、キャノピーの前部風防は1枚板のワイパー付き耐弾ガラスに変更された。N.F.Mk.IIの電子機器には、はじめにA.I.Mk.IVレーダーを搭載され、のちにA.I.Mk.Vに更新されている。機首には"矢形"のレーダー送信アンテナが取り付けられ、機体全長は42フィート11インチ(13.081m)になった。レーダー波の受信用としては両主翼端の上下に2本ずつ、計8本のダイポールアンテナが立てられている。また、右主翼上下面にもアンテナを装備。主翼桁は戦闘中のはげしい運動に耐えられるよう補強された。初期生産機の排気管は消焔シュラウド付サキソフォーン形であったが、後期生産分の昼間戦闘専用機は、シュラウドのない5本の推力式単排気装をつけられている。初期生産機の主翼は、翼端前後に航法灯と編隊灯のあるタイプだったが、1943年初頭に作られた後期生産分から、翼端前方に航法灯、後方にレジンランプを装備した"ユニバーサルウイング"が採用された。敵味方識別灯は緑とアンバーの2灯が胴体後部右下面に、赤の敵味方識別灯は胴体下面コンポーネント・ドア直後左側の位置に、それぞれ配されている。F./ N.F.Mk.IIは基本的にロールスロイス・マーリン21、または23、1段2速過給器付エンジン(1,480hp)を装備したが、多くの機体はマーリン22エンジン(1,480hp)に換装されている。初期に生産されたF./N.F.Mk.IIは主車輪、主脚収容ドア・ガイド、尾脚にモスキート初期型のものを使用していたが、生産段階で胴体右側面の補強用ストレーキが取り付けられるようになると同時に、左側にスロットのない後期型の主車輪、シミーダンパー付"マーストランド"タイヤの後期型引込式尾脚が採用された。キャノピーは航法士席右側面がバルジ形に膨らみ、垂直安定板前縁のアンテナ空中線取付位置はモスキート初期型にくらべて上方に移動したが、空中線は取り付けられていない。本機の最大燃料搭載量は547ガロン（2,487リッター）であったが、通常、作戦時には403ガロン（1,832リッター）で行動した。なおNo.100グループに所属するN.F.Mk.II数機が、1943年10月から実験的に後方警戒レーダーを装備している。

F. / N.F.Mk.IIの右主翼
F. / N.F.Mk.II Right wing

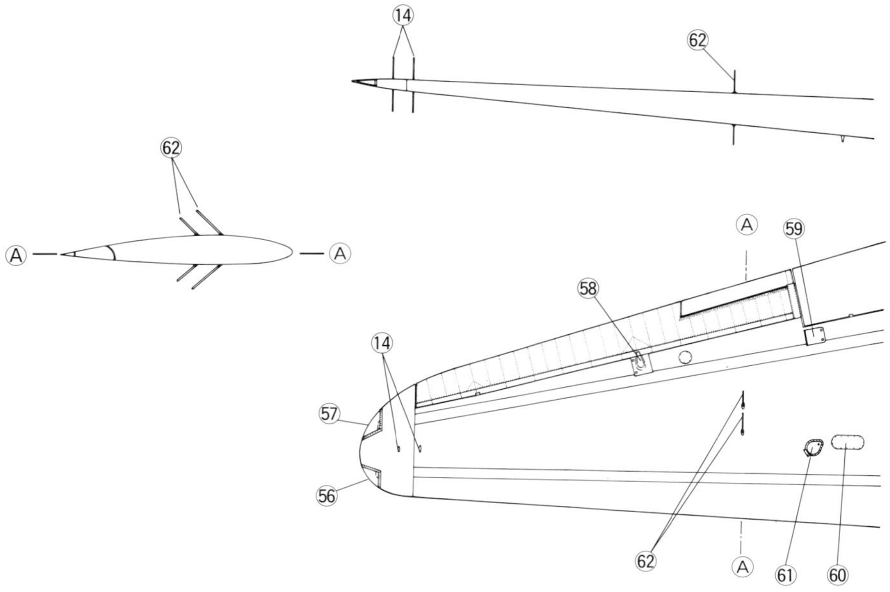

Mosquito F. / N.F.Mk.II

This model was the first Mosquito fighter variant built and operated as day and night fighter, intruder and ranger aircraft, serving in different war theaters over Europe, the Middle East and the Far East. The prototype, serialled W4052, made its maiden flight on May 15, 1941, and the first production aircraft were completed in the December of 1941, debuting in operational service with No.157 Squadron on April 3, 1942. A total of 392 aircraft were manufactured, with the last deliveries carried out in May 1943. The major external differences from the photo-reconnaissance and bomber variants were the solid nose housing four Browning .303in machine guns together with a gun camera and four Hispano 20mm cannons installed in the belly using part of the space usually taken up by the bomb fittings. The machine guns were later fitted with flash-hiders. The crew entry door moved to the right side of the front fuselage, and the canopy was fitted with a flat bullet-proof windscreen with a wiper arm. The N.F.Mk.II was equipped with A.I.Mk.IV radar equipment, later replaced by the A.I.Mk.V, with a "bow and arrow" transmitting antenna fitted on the nose, extending the overall fuselage length to 42ft-11in (13.081m), with two receiving dipole antennas extending above and below each wingtip and two further elevation antennas above and below the right wing only. The wing spars were reinforced to allow the aircraft to take higher stress loads during air combat maneuvers. Early production aircraft had "saxophone" exhausts covered by flame damper shrouds, while late production aircraft were fitted with five stub exhausts without shrouds only for daylight operations. Initially the wings were fitted with a pair of wingtip-mounted navigation and formation lights, but from early 1943 late production Mosquitoes of this variant mounted the new "universal wing" fitted with a rearward resin lamp and one navigation light on the wingtips. The aircraft had two green and amber downward identification code lights on the right underside of the rear fuselage, and a third red downward identification code light mounted on the left underside of the fuselage just behind the ventral compartment door. Mk.IIs were mainly powered by either the 1,480hp Rolls-Royce Merlin 21 or 23 one-stage supercharger engines, while a number of aircraft were re-engined with the 1,480hp Merlin 22s. Initially the aircraft had the early-type main wheels with slots on the left side, the early-type main wheel door guides and the early-type tail wheel strut construction with one-track tire. In the course of production a standard reinforcing strake was fitted as a modification on the right side of the rear fuselage of the aircraft and it was also equipped with the late-type main wheels without the slots located on the left side, and Marstrand two-track anti-shimmy tires. An observer's bulged window panel was fitted to the right side of the cockpit canopy and the antenna wire mount fitted to the leading edge of the vertical stabilizer was moved to a new position higher than the earlier location even though the aircraft was not fitted with antenna wires. This model carried a maximum fuel capacity of 547gal (2,487 liters) but usually it carried 403gal (1,832 liters) for operational sorties. Some of No.100 Group's N.F.Mk.IIs were equipped from October 1943 with rear-facing radar for experimental operations.

モスキートN.F.Mk.II(スペシャル)イントルーダー

No.23、25、85、151、157、264Sqn所属のN.F.Mk.IIを改修し、イントルーダーとした特別仕様機。敵地を襲撃するという任務の性格から、レーダーは捕獲される可能性を恐れて撤去し、Gee航法装置を装備。胴体の機関砲収容部後方に長距離用増加燃料タンクを追加し、20mm砲の装弾数も最大各175発から各225発に増大した。1942年12月20日、No.23Sqnはマルタ島へ移動し、地中海に配備された最初のモスキート部隊となった。No.23Sqnの主な任務はシシリーと南イタリアにあるドイツ軍飛行場の攻撃と、地上目標の破壊であった。なお、地中海方面で行動したモスキートの多くは、大型の気化器空気取入口フェアリングに、改修された空気取入口保護カバーを取り付けている。

Mosquito N.F.Mk.II (Special) Intruder

A number of N.F.Mk.IIs from No.23, 25, 85, 151, 157 and 264 Squadrons were modified as (Special) Intruder aircraft, having the Gee equipment installed in the place of airborne radar equipment, which was removed in order to prevent eventual capture by the enemy during air raids over hostile territory. The (Special) Intruders were equipped with long-range fuel tanks behind the cannon bay and the provision for 20mm cannons was increased from 175 to 225 rpg. On December 20, 1942, aircraft of No.23 Squadron were the first Mosquitos to move to Malta in the Mediterranean area, in order to attack mainly German airfields and various ground targets in Sicily and Southern Italy. Most Mk.IIs engaged in the Mediterranean theater were equipped with enlarged carburetor air intake fairings and revised intake guards.

N.F.Mk.II (Special) Intruder Nose left side
機首左側面

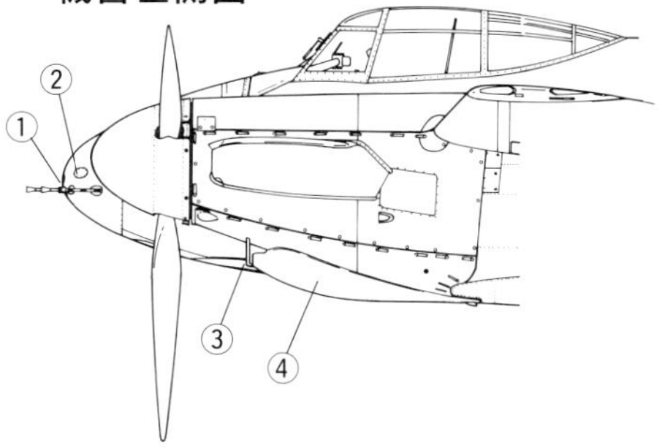

① レーダーを装備していないため、アンテナ・マウントをカバープレートで塞ぐ
② 射撃カメラ用開口部をカバープレートで塞ぐ
③ 空気取入口保護用防氷カバーを改修
④ 気化器空気取入口フェアリングを大型化

① A.I. radar antenna mount covering plate (radar equipment not installed)
② Camera port covering plate
③ Air intake guard (modified)
④ Carburetor air intake fairing (enlarged)

モスキート N.F.Mk.XII

A.I.Mk.VIIIレーダーを搭載した夜戦型。1943年1月から6月までのあいだにN.F.Mk.II後期生産分97機が、N.F.Mk.XIIへ改修され、最初のN.F.Mk.XIIは1943年2月28日、No.85スコードロンに配備された。N.F.Mk.XIIの機首はセンチメートル波を用いるA.I.Mk.VIIIレーダーをアンテナごと収める、新型の"シンブル（裁縫用の指ぬき）ノーズ"となり、これにともなって、機首先端の航法灯と .303 in 機関銃4挺は撤去され、射撃カメラは機首右側に移動。また、主翼両端のダイポールアンテナは従来どおり残されたが、右主翼上下面のアンテナは撤去されている。N.F.Mk.XIIのエンジンは左右ともマーリン21または23、プロペラは初期型のニードルブレードタイプが用いられていた。最大燃料搭載量は547ガロン(2,487リッター)、通常は403ガロン(1,832リッター)であった。

Mosquito N.F.Mk.XII

97 late production Mosquito N.F.Mk.IIs were converted into N.F.Mk.XIIs by Marshalls between January and June 1943 and the first operational aircraft was delivered on February 28, 1943 to No.85 Squadron. The main change was the fitting of the A.I.Mk.VIII centimetric radar that required a new "thimble nose" radome for the internal installation of the radar antenna. The four nose-mounted Browning .303in machine guns and the forward navigation light were removed, and the camera gun was moved to the right side of the front fuselage. The aircraft retained the two dipole antennas extending above and below each wingtip while the two elevation antennas above and below the right wing were removed. These Mosquitos were powered by either the Merlin 21 or 23 engines driving the early-type needle blade propellers. The top fuel capacity was 547gal (2,487 liters) and the normal load was 403gal (1,832 liters).

① 主翼取付部フェアリング
② ピトー管静圧ヘッド（右側のみ）
③ レーダー点検パネル
④ "シンブルノーズ"タイプのレドーム
⑤ 射撃カメラ・マウント・フェアリング
⑥ プレートタイプの外気温度計
⑦ イスパノMk.I 20mm機関砲（装弾数150発、最大175発）
⑧ 機関砲フェアリング
⑨ 搭乗員乗降ハッチ
⑩ 胴体下面コンパートメント・ドア（機関砲収容部）
⑪ 機関砲薬莢／リンク排出口

① Wing fix attachment fairing
② Static head (right side only)
③ Radar inspection panel
④ Thimble nose radome
⑤ Camera gun fairing mount
⑥ Air thermometer (plate type)
⑦ Hispano Mk.I 20mm cannon (150 rounds each, provision for 175)
⑧ Cannon under fairing
⑨ Crew entry door
⑩ Ventral compartment door (cannon bay)
⑪ Cannon shell case and link chutes ejection ports

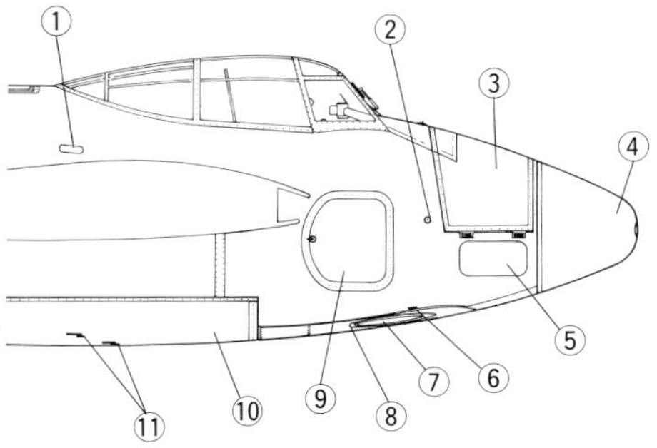

N.F.Mk.XII Nose right side
機首右側面

モスキートN.F.Mk.XVII

N.F.Mk.II 99機を改修し、アメリカ製のSCR720／729、またはイギリス製のA.I.Mk.X空中要撃レーダーを"ブルノーズ"とよばれる機首レドームに装備した夜戦型バリエーション。N.F.Mk.XIIを改修して空中要撃レーダーSCR720を装備した原型機(HK195／G)は、1943年3月に初飛行し、配備は1943年9月から1944年2月にかけて行われた。武装はN.F.Mk.XIIと同様に機首のブローニング .303 in 機関銃を撤去、イスパノ20mm機関砲4門はそのまま残されている。N.F.Mk.XVIIの主翼はすべて"ユニバーサルウイング"を採用。なお、一部の機体はA.I.Mk.X前方警戒レーダーと透明なテイルコーンの内部に"モニカ"後方警戒レーダーを装備していた。

Mosquito N.F.Mk.XVII

99 late production N.F.Mk.IIs were converted into N.F.Mk.XVIIs by the installation of either the American-built SCR720/729 airborne interception radar or the British-built A.I.Mk.X in a "bullnose" radome. The prototype, serialled HK195/G was a modified N.F.Mk.XII fitted with the SCR720 radar and it made its first flight in March 1943. The aircraft were delivered from September 1943 to February 1944. Likewise, the N.F.Mk.XII had the four nose-mounted Browning .303in machine guns removed, but it retained the four Hispano 20mm cannons. All the N.F.Mk.XVII had the "universal wing", and some aircraft were fitted with both the A.I.Mk.X forward-looking radar and Monica rearward-looking radar fitted in a Perspex tail cone.

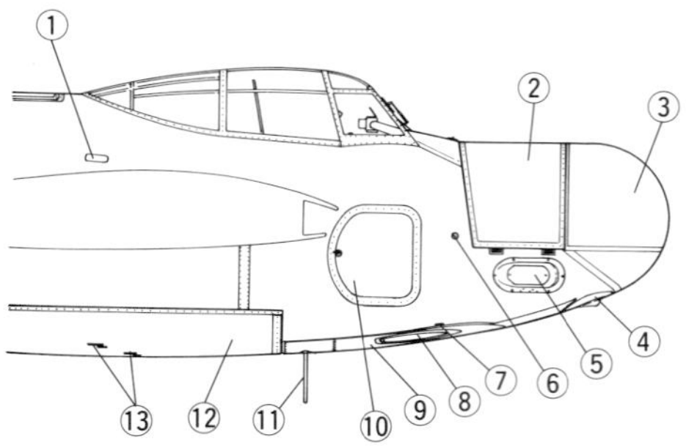

N.F.Mk.XVII Nose right side
機首右側面

① 主翼取付部フェアリング
② レーダー点検パネル
③ "ブルノーズ"タイプのレドーム
④ レーダースキャナ・マウント・フェアリング
⑤ 射撃カメラ・フェアリング
⑥ ピトー管静圧ヘッド（右側のみ）
⑦ プレートタイプの外気温度計
⑧ イスパノMk.I 20mm機関砲（装弾数150発、最大175発）
⑨ 機関砲フェアリング
⑩ 搭乗員乗降ハッチ
⑪ レーダーシステム用アンテナ
⑫ 胴体下面コンパートメント・ドア（機関砲収容部）
⑬ 機関砲薬莢／リンク排出口

① Wing fix attachment fairing
② Radar inspection panel
③ Bull nose radome
④ Radar scanner mount fairing
⑤ Camera gun fairing
⑥ Static head (right side only)
⑦ Air thermometer (plate type)
⑧ Hispano Mk.I 20mm cannon (150 rounds each, provision for175)
⑨ Cannon under fairing
⑩ Crew entry door
⑪ Antenna for radar system
⑫ Ventral compartment door
⑬ Cannon shell case and link chutes ejection ports

T.Mk.III Left view
左側面図

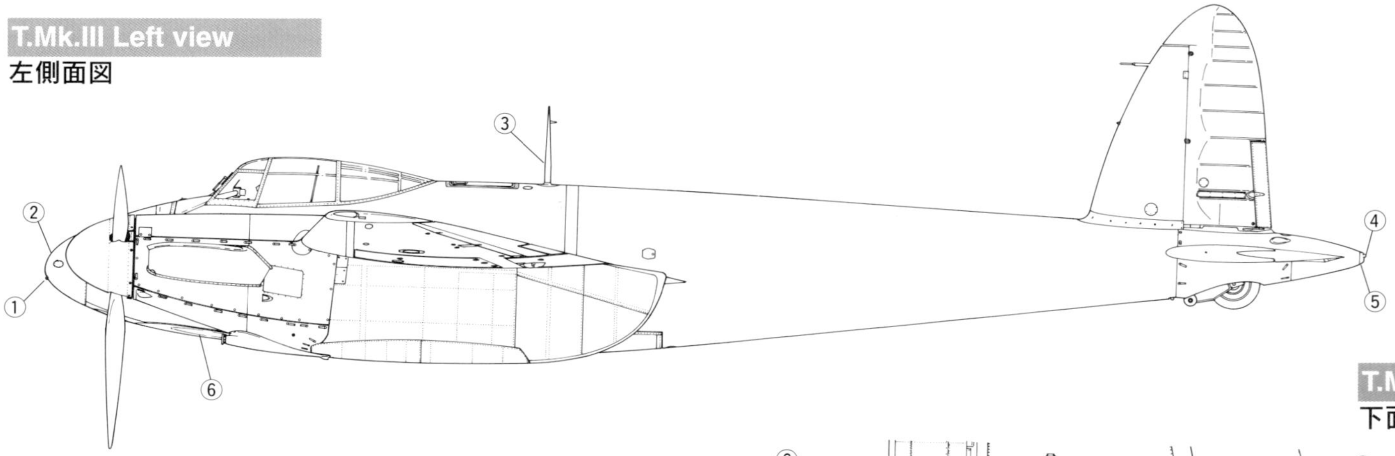

T.Mk.III Bottom view
下面図

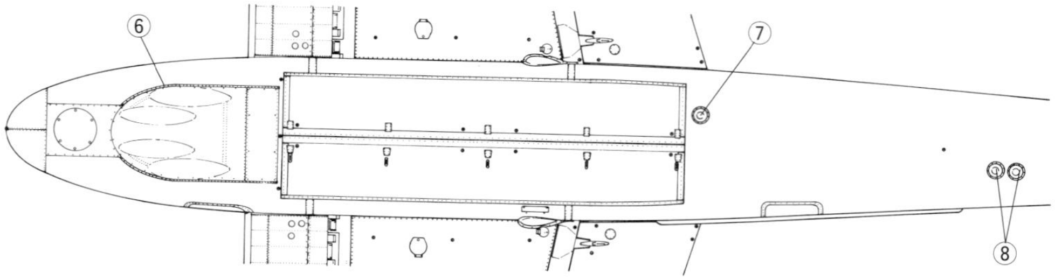

① 航法灯
② 射撃カメラ用開口部をカバープレートで塞ぐ
③ アンテナ柱（アンテナ空中線はない）
④ 航法灯
⑤ 尾部スペアソケット・フェアリング
⑥ 機関砲用開口部をカバープレートで塞ぐ
⑦ 地上用敵味方識別灯（赤）
⑧ 地上用敵味方識別灯（前方が緑、後方がアンバー）

① Forward navigation light
② Camera port covering plate
③ Antenna mast (antenna wire not fitted)
④ Tail navigation light
⑤ Spare tail-socket fairing
⑥ Cannon covering plate
⑦ Downward identification code light (red)
⑧ Downward identification code lights (forward: green, aft: amber)

モスキートT.Mk.III

F.／N.F.Mk.IIをベースにした練習機型モスキート。複式操縦装置を導入し、武装はない。エンジンは左右ともマーリン21、23または25を搭載。このうち4機（W4075、W4077、W4079、W4081）は複式操縦装置付きの戦闘機型MK.IIに改修されている。1942年5月から配備が開始されたT.Mk.IIIの量産機は、複式操縦装置を装備し、基本的に機首のブローニング .303 in 機関銃と胴体下面のイスパノ20mm機関砲を撤去。一部の機体のみは射撃訓練に用いられた。1948年までにおよそ348機を生産、このうち200機あまりが第二次大戦で配備についている。最大燃料搭載量50ガロン（227.3リッター）入落下タンク2個使用時で716ガロン（3,225リッター）。通常の燃料搭載量は453ガロン（2,059リッター）であった。

Mosquito T.Mk.III

This unarmed dual-control trainer variant was derived from the Mosquito F./N.F.Mk.II and was powered by either Merlin 21, 23 or 25 engines. The original four trainer aircraft were modified armed fighters (serials W4075, W4077, W4079, W4081) designated as Mk.II with the installation of dual controls. T.Mk.III production model deliveries started in May 1942, having been equipped with dual controls but lacking both the nose-mounted Browning .303in MGs and the belly-mounted Hispano 20mm cannons, with the exception of a few aircraft operated in the gunnery training role. A total of about 348 T.Mk.IIIs were built through late 1948, of which some 200 were delivered in the course of the war. The model had a fuel capacity of 716gal (3,255 liters) including two 50gal (227.3-liter) underwing drop tanks, and the usual fuel load was 453gal (2,059 liters).

① レーダースキャナ・マウント・フェアリング
② 後期生産機の"ブルノーズ"タイプ・レドーム
③ デ・ハヴィランド・油圧式3翅プロペラ（ニードルブレードタイプの初期型）
④ レーダー点検パネル
⑤ 方位受信レーダーアンテナ
⑥ 後期型のテイルコーン・フェアリング（取り外し式）
⑦ 航法灯
⑧ 尾部スペアソケット・フェアリング
⑨ 後方警戒レーダーのマウント・フェアリング
⑩ 後期型のシミーダンパー付"マーストランド"タイヤの引込式尾脚
⑪ 後期型の尾輪ストラット
⑫ 胴体下面コンポーネント・ドア
⑬ 後期型の主脚収容ドア
⑭ 主脚収容ドアロック用ラッチ・アクセスハッチ
⑮ レーダーシステム用アンテナ
⑯ イスパノMk.I 20mm機関砲（装弾数150発、最大175発）
⑰ 機関砲フェアリング
⑱ 初期生産機の"シンブルノーズ"タイプ・レドーム

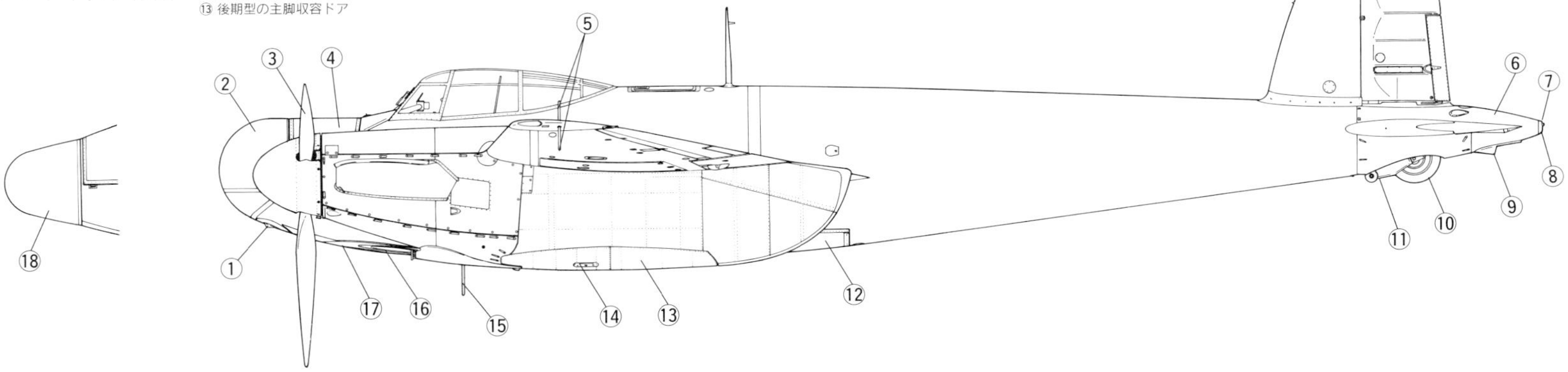

N.F.Mk.XIII Left view
左側面図

① Radar scanner mount fairing
② Bull nose radome (late production aircraft)
③ De Havilland Hydromatic needle blade propeller (three-blade, early-type)
④ Radar inspection panel
⑤ Radar receiving azimuth antenna
⑥ Detachable tail cone fairing (late-type)
⑦ Tail navigation light
⑧ Spare tail-socket fairing
⑨ Tail warning radar mount fairing
⑩ Rearward-retracting tail wheel with Marstrand two-track anti-shimmy tire (late-type)
⑪ Tail wheel strut construction (late-type)
⑫ Ventral compartment door
⑬ Main wheel door (late-type)
⑭ Door locking latch access hatch
⑮ Antenna for radar system
⑯ Hispano Mk.I 20mm cannon (150 rounds each, provision for 175)
⑰ Cannon under fairing
⑱ Thimble nose radome (early production aircraft)

モスキートN.F.Mk.XIII

N.F.Mk.XIIIはF.B.Mk.VIをベースに開発され、A.I.Mk.VIIIレーダーを"ブルノーズ"、または"ユニバーサルノーズ"と呼ばれるレドームに装備する夜間戦闘機型モスキート。また、"シンブルノーズ"を装備した機体も多く、この場合全長は41フィート11インチ（12.776m）になった。エンジンは左右とも1,480hpの1段2速過給器付マーリン21、または23を搭載。約270機が生産され、1943年10月から1944年6月にかけて配備された。N.F.Mk.XIIIを最初に装備した部隊は、No.488（RNZAF）Sqnであった。N.F.Mk.XIIIは、主翼下面に50英ガロン（227.3リッター）落下増槽2個、または爆弾2発を懸吊し、長距離輸送船団護衛戦闘任務とイントルーダー／レンジャー任務双方に用いられている。武装はイスパノ20mm機関砲。機首のブローニング .303 in 機関銃4挺は撤去され、射撃カメラも機首右側面に移動している。N.F.Mk.XIIIの主翼は全機とも"ユニバーサルウイング"を採用。テイルコーン・フェアリング、尾輪ストラット、主脚収容ドア、主脚と主脚収容ドア・ガイドはモスキート後期型のものである。最大燃料搭載量は716ガロン（3,225リッター）、通常は453ガロン（2,059リッター）。

Mosquito N.F.Mk.XIII

This night fighter variant was developed from the F.B.Mk.VI airframe equipped with the A.I.Mk.VIII radar fitted in a "bull nose" or "universal nose" radome. However a number of N.F.Mk.XIIIs were fitted with a longer-type "thimble nose" radome which extended the overall fuselage to 41ft-11in (12.776m). Powered by either the 1,480hp Merlin 21 or 23 one-stage supercharger engines, about 270 aircraft were built and delivered from October 1943 to June 1944. No.488 (RNZAF) Squadron was the first unit to equip with this model. The Mosquito N.F.Mk.XIII was operated both in the long-range convoy-escort fighter and intruder/ranger roles having the capability to carry either underwing 50gal (227.3-liter) drop tanks or a bomb load. The four Browning .303in machine guns were removed, retaining the four Hispano 20mm cannons, and the camera gun was fitted on the right side of the front fuselage. All were fitted with the "universal wing", late-type tail cone fairing, tail wheel strut construction, wheel doors, main wheels and wheel door guides. The max. fuel tankage was 716gal (3,255 liters), and the usual tankage was 453gal (2,059 liters).

モスキートN.F.Mk.XIX

N.F.Mk.XIXはN.F.Mk.XIIIをもとに、エンジンを低空性能が優れた大馬力の1段2速過給器付ロールスロイス・マーリン25（1,640hp）としたバリエーション。機首の"ブルノーズ"レドームにはアメリカ製SCR720／730空中邀撃レーダー、またはイギリス製A.I.Mk.X空中邀撃レーダーを搭載した。1944年4月から1945年9月まで280機が作られ、後期生産分の機体にはパドルブレードタイプのプロペラが装備されている。N.F.Mk.XIXを最初に受領したのはNo.85Sqn。また、No.68Sqnでは本機の高速性を生かして、V-1飛行爆弾迎撃やイントルーダー任務に使用している。燃料搭載量はN.F.Mk.XIIIと同じ。

Mosquito N.F.Mk.XIX

Derived from the N.F.Mk.XIII, this model was equipped with the powerful 1,640hp Rolls-Royce Merlin 25 one-stage supercharger engine, giving it a performance advantage at low levels. The N.F.Mk.XIX had either the American-built SCR720/729 airborne interception radar or the British-built A.I.Mk.X installed in a "bull nose" radome as soon as the units were available. 280 Mosquito N.F.Mk.XIXs were completed from April 1944 to September 1945. Late production aircraft were fitted with paddle blade propellers. No.85 Squadron was the first unit to receive this advanced fighter variant, which was more of a night fighter than a specialized aircraft for bomber support operations. No.68 Squadron operated the speedy N.F.Mk.XIXs for the interception of V-1 flying bombs and counter-intruder operations. The fuel capacity was the same as for the N.F.Mk.XIII.

N.F.Mk.XIX Nose left side
機首左側面

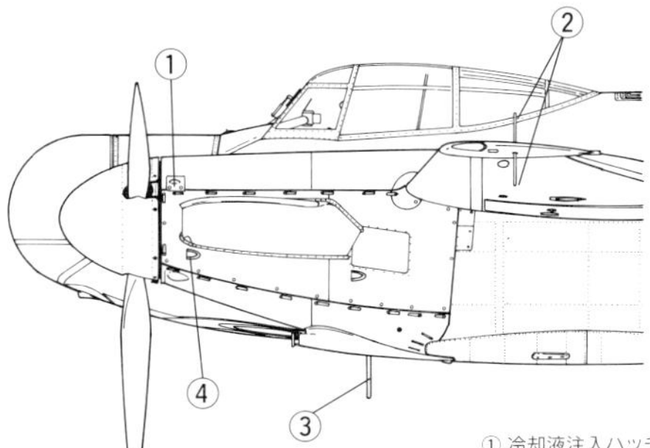

① 冷却液注入ハッチのバルジ
② 方位受信レーダーアンテナ
③ レーダーシステム用アンテナ
④ 冷却空気取入口

① Coolant fill hatch bulge
② Radar receiving azimuth antenna
③ Antenna for radar system
④ Cooling air intake

N.F.Mk.XV Left view
左側面図

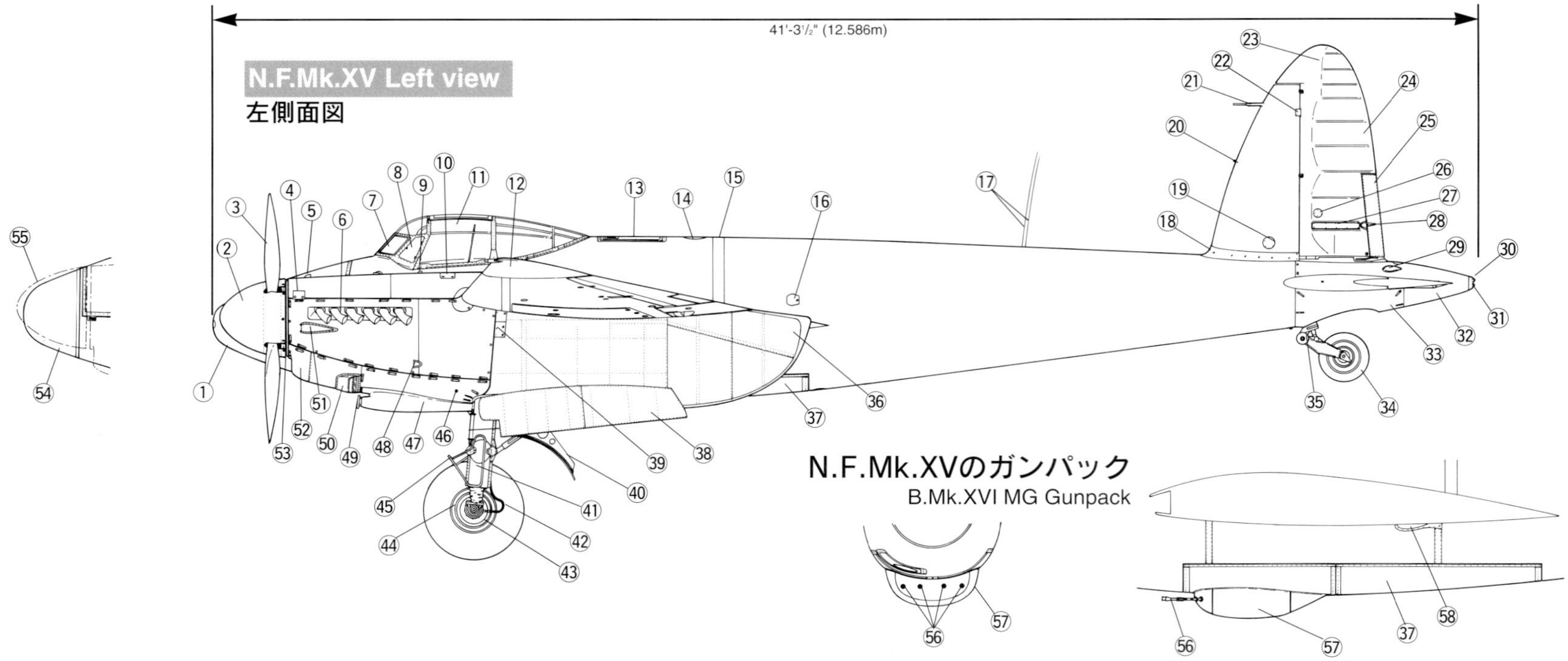

N.F.Mk.XVのガンパック
B.Mk.XVI MG Gunpack

① 延長タイプの"シンブルノーズ"
② スピナー
③ デ・ハヴィランド・油圧式4翅プロペラ
④ 冷却液注入ハッチ
⑤ 冷却液タンク・ベント
⑥ 改修された推力式単排気管
⑦ 後期型の中空合わせガラス式V字タイプ前部風防
⑧ 内側開閉式の前部風防側面窓
⑨ 中間冷却器タンク・ベント
⑩ 中間冷却器タンク冷却液注入ハッチ
⑪ 与圧コクピット・キャノピー
⑫ 延長された主翼端
⑬ 救命ボート収納部パネル（投棄式）
⑭ ベークライト製アンテナ保護板
⑮ 胴体補強用帯金
⑯ 地上スターターソケットハッチ
⑰ ウィップアンテナ
⑱ 垂直尾翼基部フェアリング
⑲ 垂直尾翼基部点検ハッチ
⑳ アンテナ空中線接続部(アンテナ空中線はなし)
㉑ ピトー管
㉒ 外部電源接続部パネル
㉓ 方向舵ホーンバランス
㉔ 方向舵（羽布張り、作動角26°-1°+2°）
㉕ 方向舵トリムタブ（作動角16°±2°30"）
㉖ 方向舵操舵装置点検ハッチ
㉗ 方向舵トリムタブ操作ロッド・アクセスパネル
㉘ 方向舵トリムタブ操作ロッド・フェアリング
㉙ 昇降舵操舵リンケージ・フェアリング
㉚ 航法灯
㉛ 編隊灯
㉜ 初期型のテイルコーン・フェアリング（取り外し式）
㉝ 引込式尾脚アクセスパネル
㉞ 後期型のシミーダンパー付"マーストランド"タイヤの引込式尾脚
㉟ 後期型の尾輪ストラット
㊱ ナセル後部フェアリング
㊲ 胴体下面コンポーネント・ドア
㊳ 初期型の主脚収納ドア
㊴ ジャッキ用アクセスハッチ(両ナセル外側のみ)
㊵ マッドガード
㊶ ショック・アブソーバー・ストラット
㊷ ブレーキパイプ
㊸ 直径が小さくなった主車輪
㊹ 衝撃吸収用ラバーパッド
㊺ 初期型の主脚収納ドア・ガイド
㊻ 燃料ポンプ・ドレイン（両ナセル左側のみ）
㊼ 気化器空気取入口フェアリング
㊽ 燃料ポンプ冷却空気取入口

① Thimble nose radome (longer-type)
② Spinner
③ De Havilland Hydromatic four-blade propeller
④ Coolant fill hatch
⑤ Coolant header tank vent
⑥ Stub exhaust pipes (modified)
⑦ "Dry air" sandwich type Perspex V-windscreen
⑧ Direct view panel (opens inwards)
⑨ Intercooler header tank vent
⑩ Intercooler header tank fill hatch
⑪ Pressurized cockpit canopy
⑫ Extended wingtip
⑬ Dinghy access panel (jettisonable)
⑭ Antenna bakelite securing plate
⑮ Reinforcing strip
⑯ Ground starter socket hatch
⑰ Whip antenna
⑱ Vertical stabilizer root fairing
⑲ Root fitting inspection hatch
⑳ Antenna wire mount(antenna wire not fitted)
㉑ Pitot tube
㉒ Electrical connection access hatch
㉓ Rudder horn balance
㉔ Rudder (fabric-covered, 26°-1°+2°arc of movement)
㉕ Rudder trimtab (16°±2°30"arc of movement)
㉖ Rudder controls inspection hatch
㉗ Rudder trim tab control rod access panel
㉘ Rudder trim tab control rod fairing
㉙ Elevator operating linkage fairing
㉚ Tail navigation light
㉛ Tail formation light
㉜ Detachable tail cone fairing (early-type)
㉝ Retracting gear access panel
㉞ Rearward-retracting tail wheel with Marstrand two-track anti-shimmy tire (late-type)
㉟ Tail wheel strut construction (late-type)
㊱ Nacelle tail fairing
㊲ Ventral compartment door
㊳ Main wheel door (early-type)
㊴ Access hatch for jacking (outboard side only of both nacelles)
㊵ Mudguard
㊶ Shock-absorber strut
㊷ Brake pipe
㊸ Smaller diameter plan-type main wheel
㊹ Rubber shock pads
㊺ Main wheel door guide (early-type)

モスキートN.F.Mk.XV

N.F.Mk.XVはイギリス上空に侵入するルフトヴァッフェのユンカースJu86P高高度偵察機を迎撃するために開発された機体。1942年9月15日に初飛行した時の原型機は、機体上面を戦闘機型の基本塗装であるダークグリーン／オーシャングレイ、機体下面とスピナーをPRUブルーで塗られていた。この原型機は与圧キャビン付爆撃機型の原型機（MP469）を改造して作られ、機首ははじめブローニング .303 in 機関銃を装備したMk.IIタイプであった。機関銃はのちに胴体下面の武装パックに移動され、"シンブルノーズ"レドームに改められた機首にA.I.Mk.VIIIレーダーを搭載。機体全面はPRUブルー、レドームは黒で塗装された。 原型機につづいてB.Mk.IV（DZ366、DZ385、DZ409、DZ417）をベースに4機のN.F.Mk.XVが作られ、最初に完成した機体は1943年1月3日に初飛行。これら5機のN.F.Mk.XVは1943年3月に、No.85スコードロンの特別"C"飛行隊へ配属され、全面ディープスカイブルー１色で塗られた。さまざまなテストを行った結果、翼長を62フィート7インチ (19.075m) に延長。機首の"シンブルノーズ"レドームも、長さのあるタイプに改められた結果、全長は41フィート3½インチ（12.586m）になっている。延長主翼の翼端灯は前方に突き出す小型なものであった。エンジンは左右とも1,565hpのマーリン61か、または1,680hpのマーリン72／73か、もしくはニードルブレードタイプの4翅プロペラを装備した2段過給器付き1,710hpのマーリン76／77を搭載。機体は余分な防弾装備や重量物をすべて取り外して、軽量化をはかっていた。軽量化にあたっては、着陸灯を撤去、主車輪に直径の小さなものを採用、胴体下面コンポーネント・ドアを軽量なものにし、通常型のアンテナ支柱を撤去して、ウィップアンテナに変更するなどの措置がとられた。また、前部キャノピー前面と上面、バルジ型の航法士側面窓に中空合わせ式ガラスが採用された。N.F.Mk.XVの塗装は1943年3月の通達に従って一時的に全面スムーズナイトDTD308となったが、1943年8月には全面ディープスカイブルーに再塗装された。最終的にこの高々度型N.F.Mk.XVは戦闘の機会がおとずれないまま、テストベッド用の特別機となっている。燃料搭載量は24ガロン（109.1リッター）入落下タンク2個使用時で335ガロン（1,523リッター）であった。

Mosquito N.F.Mk.XV

This model was expressly developed in order to fight the Luftwaffe Ju86P high altitude reconnaissance air raids over the Britain. The prototype flew for the first time on September 15, 1942, camouflaged with the upper surfaces in the standard day fighter finish of Dark Green and Ocean Grey with undersurfaces and spinners in PRU Blue. It was a modified bomber prototype with a pressurized cabin, serialled MP469, which was initially fitted with four Browning .303in machine guns in a Mk.II-type nose, later moved to an underbelly pack for the fitting of the A.I.Mk.VIII radar in a thimble nose type radome. At that time, the aircraft was finished in PRU Blue overall with Black radome. The following four production aircraft built were transformations of B.Mk.IV Series II airframes serialled DZ366, DZ385, DZ409 and DZ417. The first production aircraft flight took place on January 3, 1943, and the five N.F.Mk.XVs were delivered in March 1943 to the Special "C" Flight of No.85 Squadron sporting a finish of Deep Sky Blue overall. After various tests, all aircraft had their wingspan extended to 62ft-7in (19.075m). The fitting of a longer-type thimble nose radome extended the fuselage length to 41ft-3½in (12.586m). The new wingtip model had one navigation light fitted on a small projection. The aircraft were powered either by the 1,565hp Merlin 61 or 1,680hp Merlin 72/73 or 1,710hp Merlin 76/77 two-stage supercharger engines driving needle-type four-blade propellers. Most of the surplus armor fittings and other weight were removed. This included removal of the underwing landing lights, fitting the landing gear with smaller diameter wheels, installing lighter ventral compartment doors and replacing the standard antenna mast with two whip antennas. The canopy direct view panels, the upper front panel and the bulged side panels were of the "dry air" sandwich type. For a short period, the N.F.Mk.XVs were marked in overall Smooth Night DTD308 following a directive issued in March 1943 but by August 1943 they were once again painted in overall Deep Sky Blue. Ultimately, these high-flying Mosquitoes were never engaged in combat and were operated as special testbed aircraft. Fuel capacity was 335gal (1,523 liters) including two 24gal (109.1-liter) underwing tanks.

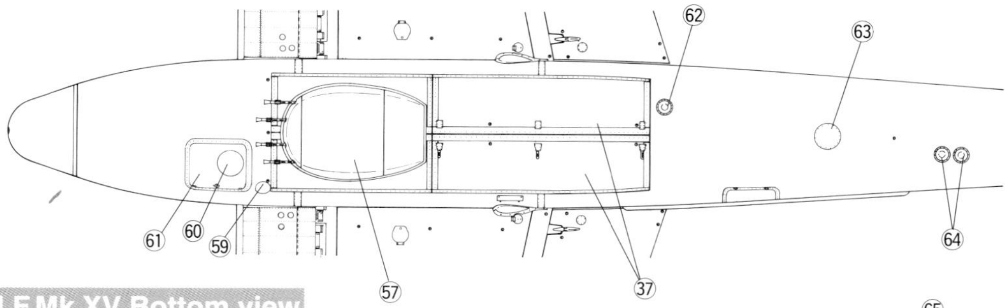

N.F.Mk.XV Bottom view
下面図

㊾ 空気取入口保護用防氷カバー
㊿ 中間冷却器用空気排出口
51 磁石発電機冷却用空気取入口
52 中間冷却器用空気取入口
53 スピナー後部とエンジン・ナセル先端に段差あり
54 長さが短いタイプの"シンブルノーズ"
55 点線はスピナーをあらわす
56 消焔器付ブローニングMk.II .303in(7.7mm)機関銃（装弾数各500発、最大750発）
57 機関銃用ガンパック
58 燃料パイプ・フェアリング
59 曳航式アンテナ用フェアリング
60 下方確認窓
61 搭乗員乗降ハッチ

46 Fuel pump drain (left side only of both nacelles)
47 Carburetor air intake fairing
48 Fuel pump cooling air intake
49 Air intake guard
50 Intercooler duct exhaust
51 Magneto cooling air intake
52 Intercooler radiator air intake
53 Gap between spinner backplate and front engine nacelle
54 Thimble nose radome (shorter-type)
55 Spinner (dashed)
56 Browning Mk.II .303in MG with flash hider (500 rounds each, provision for 750)
57 Four-MG gunpack
58 Fuel pipe fairing

62 地上用敵味方識別灯（赤）
63 カバーパネル
64 地上用敵味方識別灯（前方が緑、後方がアンバー）
65 航法灯（左翼が赤、右翼が緑）
66 点線は標準型の翼端を示す
67 燃料注入ハッチ
68 ベントパイプ・アクセスハッチ
69 補助翼中央ヒンジ・フェアリング
70 延長主翼端下面
71 航法灯点検パネル

59 Trailing antenna fairlead fairing covering panel
60 Drift sight window
61 Crew entry door
62 Downward identification code light (red)
63 Cover panel
64 Downward identification code lights (forward: green, aft: amber)
65 Navigation light (left: red, right: green)
66 Standard wingtip (dashed)
67 Fuel fill hatch
68 Vent pipe access hatch
69 Aileron center hinge fairing
70 Wingtip underside
71 Navigation light inspection hatch

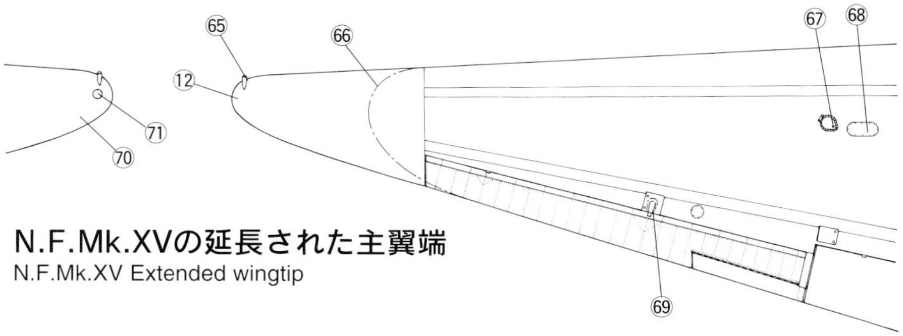

N.F.Mk.XVの延長された主翼端
N.F.Mk.XV Extended wingtip

モスキートN.F.Mk.30

N.F.Mk.30は1段過給器付N.F.Mk.XIXからの派生型で、第二次大戦後半の夜間戦闘機型モスキートのなかで、もっとも重要なバリエーションである。1944年3月に初飛行した原型機（MM686）は、パドルブレードタイプのプロペラを装備した2段過給器付き1,680hpマーリン72エンジンを搭載。量産機のうち最初の70機は原型機と同じマーリン72エンジンを搭載して完成したが、その後は主に1,710hpマーリン76（1,710hp）エンジンを搭載し、さらに後期生産分はマーリン113エンジンに換装された。総生産機数は1944年6月から1945年6月までのあいだに530機であった。N.F.Mk.30が最初に実戦配備された部隊はNo.219Sqnで、このほかに23のSqnもN.F.Mk.30を受領した。N.F.Mk.30のキャノピーは航法士右側の窓がバルジ状に膨らんだタイプを採用、コクピットは与圧式ではなかった。機首の"ブルノーズ"レドーム内にはA.I.Mk.Xレーダーを装備。ごく一部の機体のみSCR720／729レーダーを装備し、機関砲フェアリングと胴体下面コンパートメント・ドアのあいだに、レーダ一用アンテナが設置されている。また、両翼端の上下にはそれぞれ2本ずつ、計8本のダイポールアンテナが設置されており、これは全機に共通する特徴であった。N.F.Mk.30は実戦配備の初期において、従来型排気管シュラウドと単排気管に関する問題からトラブルが続出。テストをくりかえした結果、従来型排気管シュラウドのかわりに、ルーバー型の新排気管シュラウドを採用することでトラブルを克服している。なお、このルーバー型の新排気管シュラウドはナセルの左右で形が異なる（105頁、ディテール・イラスト参照）。N.F.Mk.30のテイルコーン・フェアリングは、ほとんどモスキート後期型のものであるが、一部の機体は尾部に"モニカ"後方警戒レーダーと"パーフェクトス"敵味方識別装置を搭載していた。燃料最大搭載量は716ガロン（3,255リッター）、通常の作戦時は453ガロン（2,059リッター）であった。

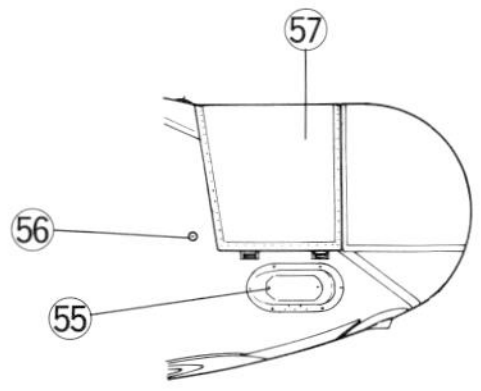

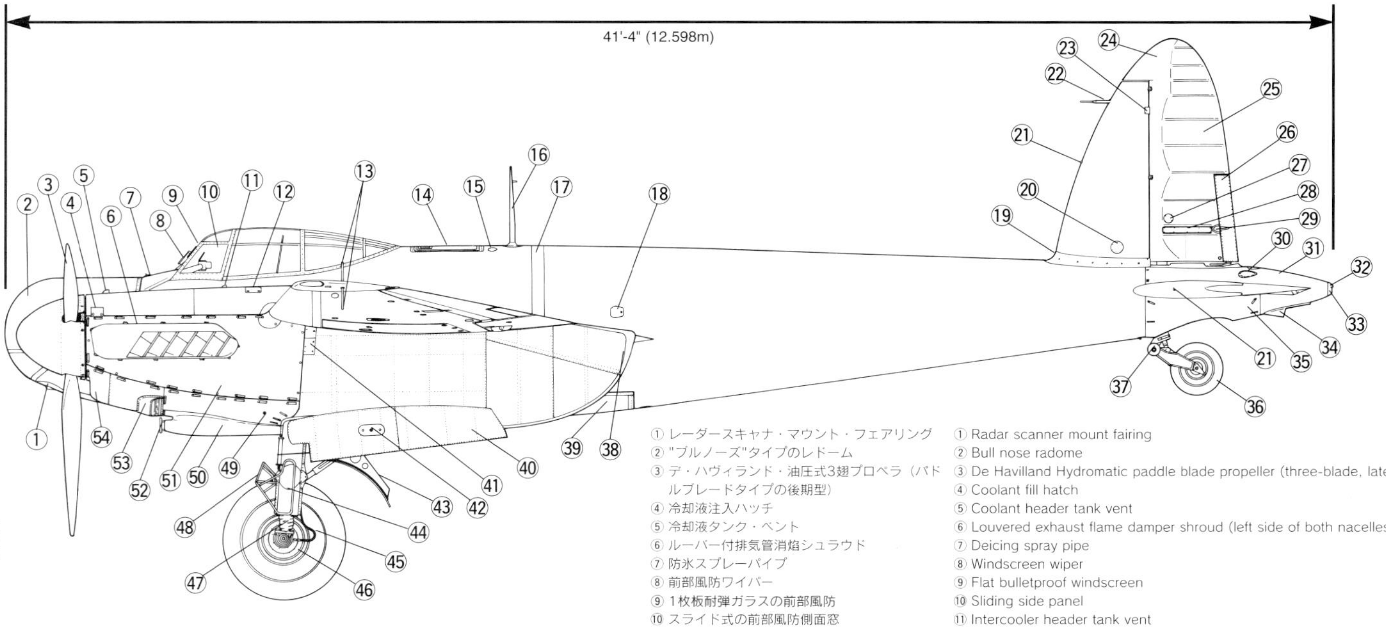

N.F.Mk.30 Left side
左側面

① レーダースキャナ・マウント・フェアリング
② "ブルノーズ"タイプのレドーム
③ デ・ハヴィランド・油圧式3翅プロペラ（パドルブレードタイプの後期型）
④ 冷却液注入ハッチ
⑤ 冷却液タンク・ベント
⑥ ルーバー付排気管消焔シュラウド
⑦ 防氷スプレーパイプ
⑧ 前部風防ワイパー
⑨ 1枚板耐弾ガラスの前部風防
⑩ スライド式の前部風防側面窓

① Radar scanner mount fairing
② Bull nose radome
③ De Havilland Hydromatic paddle blade propeller (three-blade, late-type)
④ Coolant fill hatch
⑤ Coolant header tank vent
⑥ Louvered exhaust flame damper shroud (left side of both nacelles, latetype)
⑦ Deicing spray pipe
⑧ Windscreen wiper
⑨ Flat bulletproof windscreen
⑩ Sliding side panel
⑪ Intercooler header tank vent

N.F.Mk.30 Front view
正面図

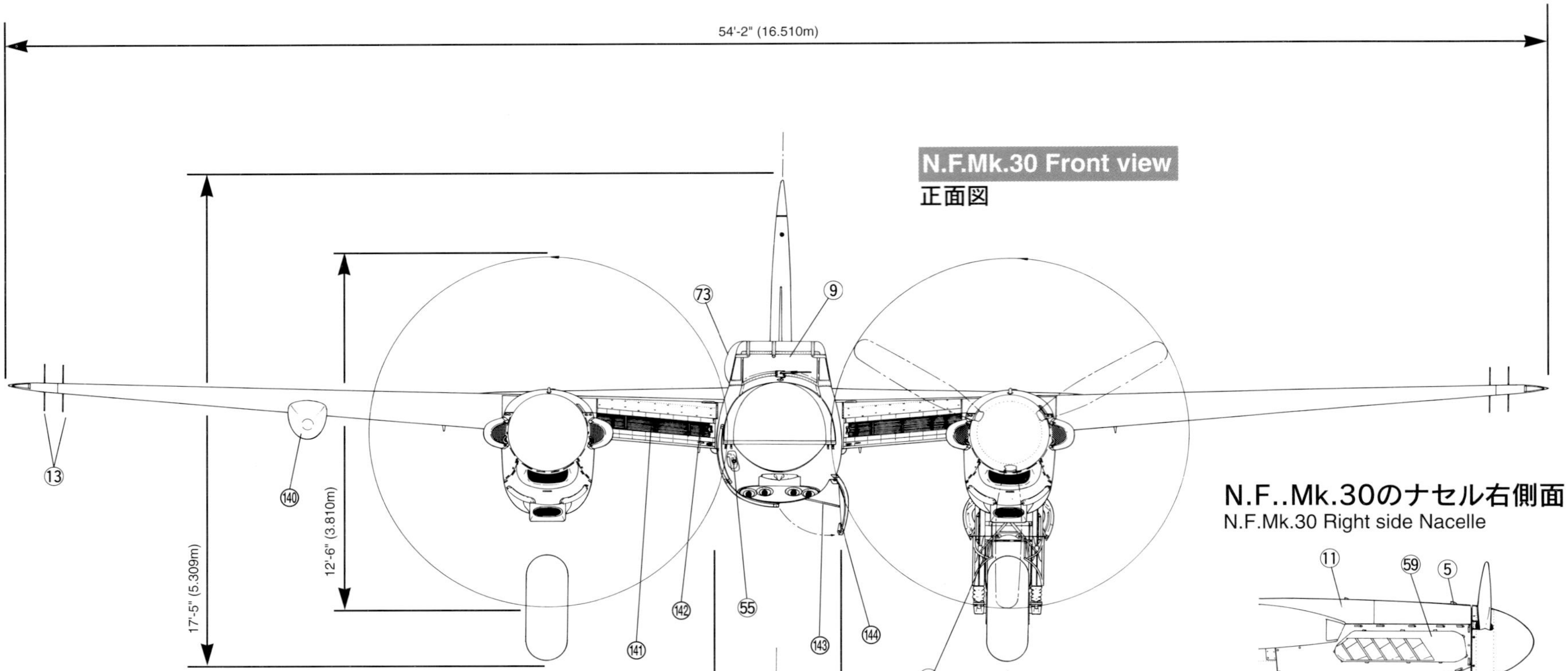

N.F..Mk.30のナセル右側面
N.F.Mk.30 Right side Nacelle

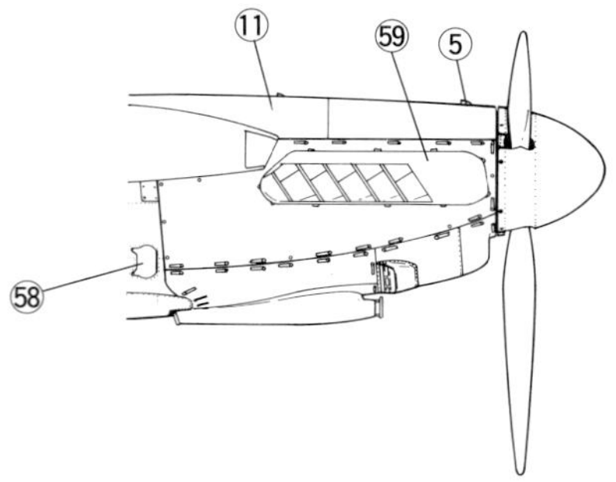

N.F.Mk.30のナセル左側面
N.F.Mk.30 Left side Nacelle

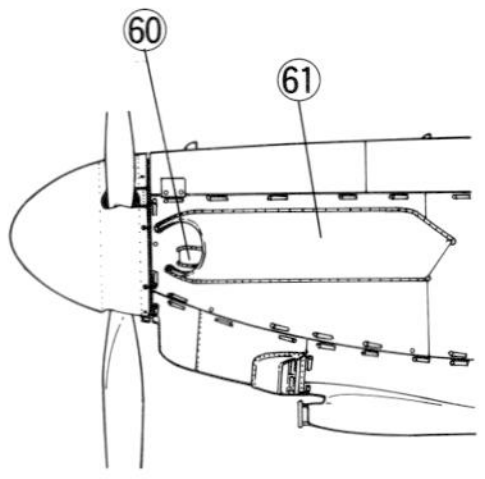

⑪ 中間冷却器タンク・ベント
⑫ 中間冷却器タンク冷却液注入ハッチ
⑬ 方位受信レーダーアンテナ
⑭ 救命ボート収納部パネル（投棄式）
⑮ 燃料注入ハッチ
⑯ アンテナ支柱（アンテナ空中線はなし）
⑰ 胴体補強用帯金
⑱ 地上スターターソケットハッチ
⑲ 垂直尾翼基部フェアリング
⑳ 垂直尾翼基部点検ハッチ
㉑ アンテナ空中線接続部（アンテナ空中線はなし）
㉒ ピトー管
㉓ 外部電源接続部パネル
㉔ 方向舵ホーンバランス
㉕ 方向舵（羽布張り、作動角26°-1°+2°）
㉖ 方向舵トリムタブ（作動角16°±2°30"）
㉗ 方向舵操舵装置点検ハッチ
㉘ 方向舵トリムタブ操作ロッド・アクセスパネル
㉙ 方向舵トリムタブ操作ロッド・フェアリング
㉚ 昇降舵操舵リンケージ・フェアリング
㉛ 後期型のテイルコーン・フェアリング
㉜ 航法灯
㉝ 編隊灯
㉞ 後方警戒レーダーのマウント・フェアリング
㉟ 引込式尾脚アクセスパネル
㊱ 後期型のシミーダンパー付"マーストランド"タイヤの引込式尾脚
㊲ 後期型の尾輪ストラット
㊳ 後期型のナセル後部フェアリング
㊴ 胴体下面コンパートメント・ドア
㊵ 後期型の主脚収納ドア
㊶ ジャッキ用アクセスハッチ(両ナセル外側のみ)
㊷ 主脚収容ドアロック用ラッチ・アクセスハッチ
㊸ マッドガード
㊹ ショック・アブゾーバー・ストラット
㊺ ブレーキパイプ
㊻ 後期型の主車輪
㊼ 衝撃吸収用ラバーパッド
㊽ 後期型の主脚収納ドア
㊾ 燃料ポンプ・ドレイン（両ナセル左側のみ）
㊿ 気化器空気取入口フェアリング
51 図は燃料ポンプ冷却空気取入口のないモデル
52 空気取入口保護用防氷カバー
53 中間冷却器用空気排出口
54 中間冷却器用空気取入口
55 射撃カメラ用フェアリング
56 ピトー管静圧ヘッド（右側のみ）
57 レーダー点検パネル
58 エンジン始動用カートリッジスターター用点検ハッチ（両ナセル右側のみ）
59 後期型のナセル右側消焔排気管シュラウド（両ナセル右側共通）
60 磁石発電機冷却用空気取入口
61 初期型消焔排気管シュラウド
62 航法灯（左翼が赤、右翼が緑）
63 燃料注入ハッチ

⑫ Intercooler header tank fill hatch
⑬ Radar receiving azimuth antenna
⑭ Dinghy access panel (jettisonable)
⑮ Fuel fill hatch
⑯ Antenna mast (antenna wire not fitted)
⑰ Reinforcing strip
⑱ Ground starter socket hatch
⑲ Vertical stabilizer root fairing
⑳ Root fitting inspection hatch
㉑ Antenna wire mount (antenna wire notfitted)
㉒ Pitot tube
㉓ Electrical connection access hatch
㉔ Rudder horn balance
㉕ Rudder (fabric-covered, 26°-1°+2°arc of movement)
㉖ Rudder trim tab (16°±2°30" arc of movement)
㉗ Rudder controls inspection hatch
㉘ Rudder trim tab control rod access panel
㉙ Rudder trim tab control rod fairing
㉚ Elevator operating linkage fairing
㉛ Detachable tail cone fairing (late-type)
㉜ Tail navigation light
㉝ Spare tail-socket fairing
㉞ Tail warning radar mount fairing
㉟ Retracting gear access panel
㊱ Rearward-retracting tail wheel with Marstrand two-track anti-shimmy tire (late-type)
㊲ Tail wheel strut construction (late-type)
㊳ Nacelle tail fairing
㊴ Ventral compartment door (fitted with door locks)
㊵ Main wheel door (late-type)
㊶ Access hatch for jacking
㊷ Door locking latch access hatch
㊸ Mudguard
㊹ Shock-absorber strut
㊺ Brake pipe

Mosquito N.F.Mk.30

This model, derived from the one-stage supercharger engined N.F.Mk.XIX, was the most important cannon-equipped night fighter variant operated by RAF in late WWII. The prototype, serialled MM686, made its maiden flight in March 1944, powered by the two-stage 1,680hp Merlin 72 driving paddle blade propellers. The first 70 production aircraft had the same engine model as the prototype, and the following were primarily equipped with the 1,710hp Merlin 76s. Later airframes had Merlin 113s. A total of 530 aircraft were completed and delivered from June 1944 to June 1945. The first unit to operate this model was No.219 Squadron, followed by twenty-three other RAF Squadrons. The N.F.Mk.30 did not have a pressurized cabin, and the cockpit canopy had an observer/radar operator's bulged window in the right side panel. They were fitted with the bull nose type radomes, most of which housed the A.I.Mk.X radar, since only a few had the SCR720/729 radar with an antenna located under the belly between the 20mm cannon under fairing and the ventral compartment doors. All had two dipole antennas extending above and below each wingtip. During early operations, the aircraft suffered a series of problems due to the standard flame damper shrouds fitted to the engine stub exhaust pipes, but this was solved when, after numerous tests, the standard shrouds were replaced by louvered-type shrouds. Those fitted on the right side of each engine nacelle had different shape from those fitted on the left side. Most of the N.F.Mk.30s had the late-type tail cone fairing and some were also equipped with Monica tail warning radar and Perfectos "Identification Foe or Friend" apparatus to interrogate enemy I.F.F. equipment. The max. fuel capacity was 716gal (3,255 liters) and the usual service load was 453gal (2,059 liters).

N.F.Mk.30 Top view
上面図

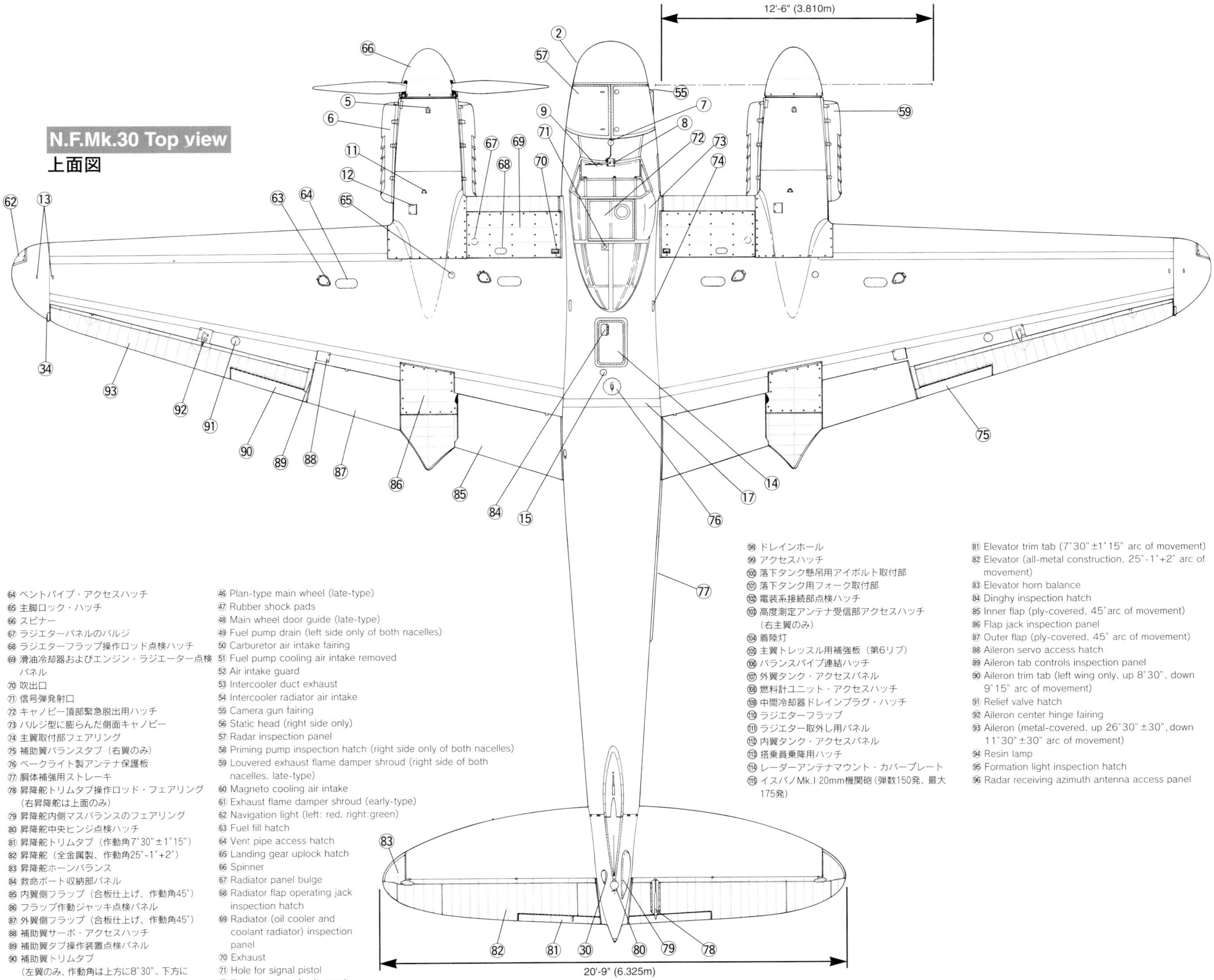

64 ベントパイプ・アクセスハッチ
65 主脚ロック・ハッチ
66 スピナー
67 ラジエターパネルのバルジ
68 ラジエターフラップ操作ロッド点検ハッチ
69 滑油冷却器およびエンジン・ラジエーター点検パネル
70 吹出口
71 信号弾発射口
72 キャノピー頂部緊急脱出用ハッチ
73 バルジ型に膨らんだ側面キャノピー
74 主翼取付部フェアリング
75 補助翼バランスタブ（右翼のみ）
76 ベークライト製アンテナ保護板
77 胴体補強用ストレーキ
78 昇降舵トリムタブ操作ロッド・フェアリング（右昇降舵は上面のみ）
79 昇降舵内側マスバランスのフェアリング
80 昇降舵中央ヒンジ点検ハッチ
81 昇降舵トリムタブ（作動角7°30"±1°15"）
82 昇降舵（全金属製、作動角25°-1°+2°）
83 昇降舵ホーンバランス
84 救命ボート収納部パネル
85 内翼側フラップ（合板仕上げ、作動角45°）
86 フラップ作動ジャッキ点検パネル
87 外翼側フラップ（合板仕上げ、作動角45°）
88 補助翼サーボ・アクセスハッチ
89 補助翼タブ操作装置点検パネル
90 補助翼トリムタブ（左翼のみ、作動角は上方に8°30"、下方に9°15"）
91 リリーフバルブ・ハッチ
92 補助翼中央ヒンジ・フェアリング
93 補助翼（金属外皮、作動角は上に26°30"±30"、下に11°30"±30"）
94 レジンランプ
95 航法灯点検ハッチ
96 方位受信レーダーアンテナ・アクセスパネル
97 補助翼ヒンジ・アクセスハッチ

46 Plan-type main wheel (late-type)
47 Rubber shock pads
48 Main wheel door guide (late-type)
49 Fuel pump drain (left side only of both nacelles)
50 Carburetor air intake fairing
51 Fuel pump cooling air intake removed
52 Air intake guard
53 Intercooler duct exhaust
54 Intercooler radiator air intake
55 Camera gun fairing
56 Static head (right side only)
57 Radar inspection panel
58 Priming pump inspection hatch (right side only of both nacelles)
59 Louvered exhaust flame damper shroud (right side of both nacelles, late-type)
60 Magneto cooling air intake
61 Exhaust flame damper shroud (early-type)
62 Navigation light (left: red, right:green)
63 Fuel fill hatch
64 Vent pipe access hatch
65 Landing gear uplock hatch
66 Spinner
67 Radiator panel bulge
68 Radiator flap operating jack inspection hatch
69 Radiator (oil cooler and coolant radiator) inspection panel
70 Exhaust
71 Hole for signal pistol
72 Emergency roof exit panel
73 Bulged window panel
74 Wing fix attachment fairing
75 Aileron balance tab (right wing only)
76 Antenna bakelite securing plate
77 Reinforcing strake
78 Elevator trim tab control rod fairing (right upper surface only)
79 Elevator static balance fairing
80 Elevator center hinge inspection hatch

98 ドレインホール
99 アクセスハッチ
100 落下タンク懸吊用アイボルト取付部
101 落下タンク用フォーク取付部
102 電装系接続部点検ハッチ
103 高度測定アンテナ受信部アクセスハッチ（右主翼のみ）
104 着陸灯
105 主翼トレッスル用補強板（第6リブ）
106 バランスパイプ連結ハッチ
107 外翼タンク・アクセスパネル
108 燃料計ユニット・アクセスハッチ
109 中間冷却器ドレインプラグ・ハッチ
110 ラジエターフラップ
111 ラジエター取外し用パネル
112 内翼タンク・アクセスパネル
113 搭乗員乗降用ハッチ
114 レーダーアンテナマウント・カバープレート
115 イスパノMk.I 20mm機関砲（弾数150発、最大175発）

81 Elevator trim tab (7°30"±1°15" arc of movement)
82 Elevator (all-metal construction, 25°-1°+2° arc of movement)
83 Elevator horn balance
84 Dinghy inspection hatch
85 Inner flap (ply-covered, 45°arc of movement)
86 Flap jack inspection panel
87 Outer flap (ply-covered, 45° arc of movement)
88 Aileron servo access hatch
89 Aileron tab controls inspection panel
90 Aileron trim tab (left wing only, up 8°30", down 9°15" arc of movement)
91 Relief valve hatch
92 Aileron center hinge fairing
93 Aileron (metal-covered, up 26°30"±30", down 11°30"±30" arc of movement)
94 Resin lamp
95 Formation light inspection hatch
96 Radar receiving azimuth antenna access panel

N.F.Mk.30 Bottom view

下面図

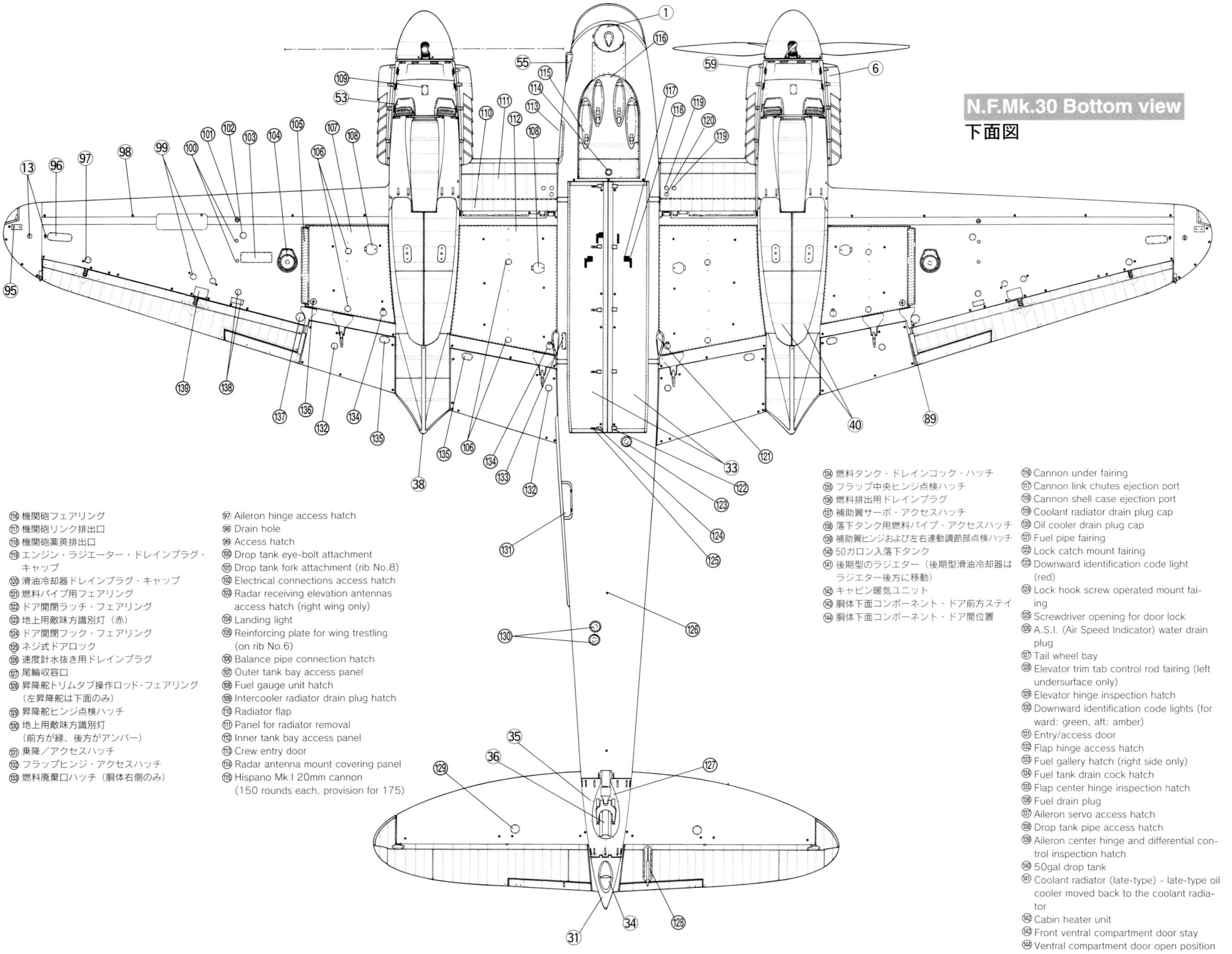

⑯ 機関砲フェアリング
⑰ 機関砲リンク排出口
⑱ 機関砲薬莢排出口
⑲ エンジン・ラジエーター・ドレインプラグ・キャップ
⑳ 滑油冷却器ドレインプラグ・キャップ
㉑ 燃料パイプ用フェアリング
㉒ ドア開閉ラッチ・フェアリング
㉓ 地上用敵味方識別灯（赤）
㉔ ドア開閉フック・フェアリング
㉕ ネジ式ドアロック
㉖ 速度計水抜き用ドレインプラグ
㉗ 尾輪収容口
㉘ 昇降舵トリムタブ操作ロッド・フェアリング（左昇降舵は下面のみ）
㉙ 昇降舵ヒンジ点検ハッチ
㉚ 地上用敵味方識別灯（前方が緑、後方がアンバー）
㉛ 乗降／アクセスハッチ
㉜ フラップヒンジ・アクセスハッチ
㉝ 燃料廃棄口ハッチ（胴体右側のみ）
㉞ 燃料タンク・ドレインコック・ハッチ
㉟ フラップ中央ヒンジ点検ハッチ
㊱ 燃料排出用ドレインプラグ
㊲ 補助翼サーボ・アクセスハッチ
㊳ 落下タンク用燃料パイプ・アクセスハッチ
㊴ 補助翼ヒンジおよび左右連動調節部点検ハッチ
㊵ 50ガロン入落下タンク
㊶ 後期型のラジエター（後期型滑油冷却器はラジエター後方に移動）
㊷ キャビン暖気ユニット
㊸ 胴体下面コンポーネント・ドア前方ステイ
㊹ 胴体下面コンポーネント・ドア開位置

97 Aileron hinge access hatch
98 Drain hole
99 Access hatch
100 Drop tank eye-bolt attachment
101 Drop tank fork attachment (rib No.8)
102 Electrical connections access hatch
103 Radar receiving elevation antennas access hatch (right wing only)
104 Landing light
105 Reinforcing plate for wing trestling (on rib No.6)
106 Balance pipe connection hatch
107 Outer tank bay access panel
108 Fuel gauge unit hatch
109 Intercooler radiator drain plug hatch
110 Radiator flap
111 Panel for radiator removal
112 Inner tank bay access panel
113 Crew entry door
114 Radar antenna mount covering panel
115 Hispano Mk.I 20mm cannon (150 rounds each, provision for 175)
116 Cannon under fairing
117 Cannon link chutes ejection port
118 Cannon shell case ejection port
119 Coolant radiator drain plug cap
120 Oil cooler drain plug cap
121 Fuel pipe fairing
122 Lock catch mount fairing
123 Downward identification code light (red)
124 Lock hook screw operated mount faiing
125 Screwdriver opening for door lock
126 A.S.I. (Air Speed Indicator) water drain plug
127 Tail wheel bay
128 Elevator trim tab control rod fairing (left undersurface only)
129 Elevator hinge inspection hatch
130 Downward identification code lights (forward: green, aft: amber)
131 Entry/access door
132 Flap hinge access hatch
133 Fuel gallery hatch (right side only)
134 Fuel tank drain cock hatch
135 Flap center hinge inspection hatch
136 Fuel drain plug
137 Aileron servo access hatch
138 Drop tank pipe access hatch
139 Aileron center hinge and differential control inspection hatch
140 50gal drop tank
141 Coolant radiator (late-type) - late-type oil cooler moved back to the coolant radiator
142 Cabin heater unit
143 Front ventral compartment door stay
144 Ventral compartment door open position

F.B.Mk.VI Series I Left side

左側面図

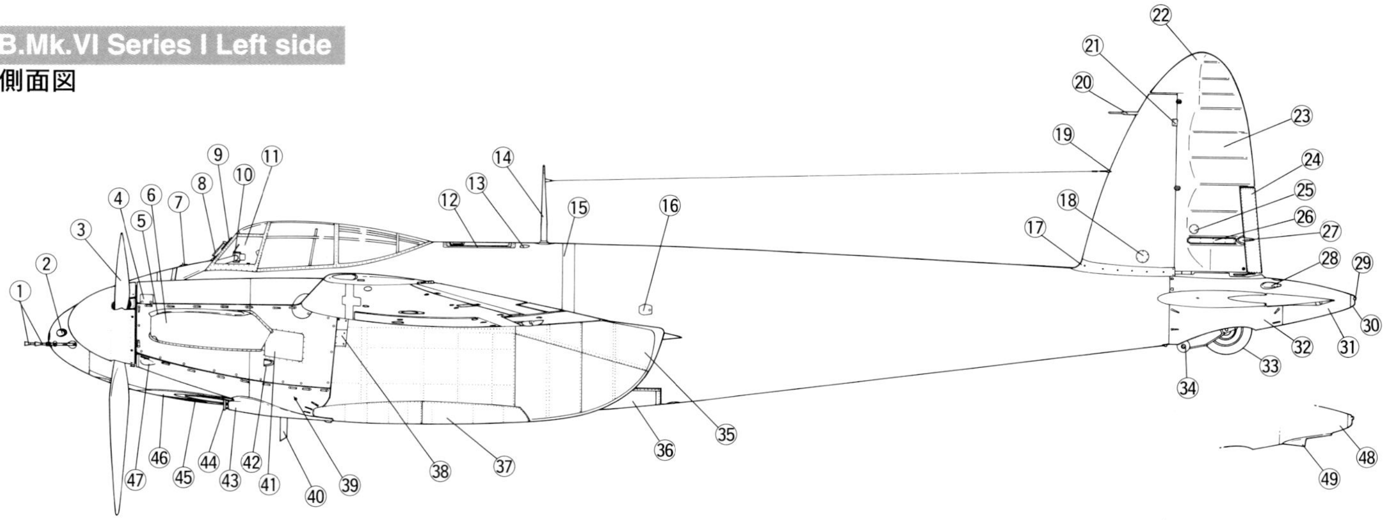

① 消焔器付ブローニングMk.II .303 in (7.7mm)機関銃（弾数各500発、最大750発）
② 射撃カメラ取付穴
③ デ・ハヴィランド・油圧式3翅プロペラ（ニードルブレードタイプの初期型）
④ 冷却液注入ハッチ
⑤ 排気マニフォールド・フランジ冷却空気ダクト
⑥ 排気管消焔シュラウド
⑦ 防氷スプレーパイプ
⑧ 前部風防ワイパー
⑨ 1枚板耐弾ガラスの前部風防
⑩ 光像式射撃照準器
⑪ スライド式側面窓
⑫ 救命ボート収納部パネル（投棄式）
⑬ 燃料注入ハッチ
⑭ アンテナ支柱
⑮ 胴体補強用帯金
⑯ 地上スターターソケットハッチ
⑰ 垂直尾翼基部フェアリング
⑱ 垂直尾翼基部点検ハッチ
⑲ アンテナ空中線接続部
⑳ ピトー管
㉑ 外部電源接続部パネル
㉒ 方向舵ホーンバランス
㉓ 方向舵（羽布張り、作動角26°-1°+2°）
㉔ 方向舵トリムタブ（作動角16°±2°30"）
㉕ 方向舵操舵装置点検ハッチ
㉖ 方向舵トリムタブ操作ロッド・アクセスパネル
㉗ 方向舵トリムタブ操作ロッド・フェアリング
㉘ 昇降舵操舵リンケージ・フェアリング
㉙ 航法灯
㉚ 尾部スペアソケット・フェアリング
㉛ 初期型のテイルコーン・フェアリング（取り外し式）
㉜ 引込式尾脚アクセスパネル
㉝ 後期型のシミーダンパー付"マーストランド"タイヤの引込式尾脚
㉞ 後期型の尾輪ストラット
㉟ 後期型のナセル後部フェアリング
㊱ 爆弾倉ドア
㊲ 初期型の主脚収納ドア
㊳ ジャッキ用アクセスハッチ(両ナセル外側のみ)
㊴ 燃料ポンプ・ドレイン（両ナセル左側のみ）
㊵ 曳航式アンテナ用フェアリング
㊶ 排気用耐熱パネル
㊷ 燃料ポンプ冷却空気取入口
㊸ 気化器空気取入口フェアリング
㊹ 空気取入口保護用防氷カバー
㊺ イスパノMk.I 20mm機関砲(装弾数150発、最大175発)
㊻ 機関砲フェアリング
㊼ 冷却液パイプのバルジ（両ナセル左側のみ）
㊽ 後期型のテイルコーン・フェアリング（取り外し式）
㊾ 後方警戒レーダーのマウント・フェアリング

① Browning Mk.II .303in MG with flash hider (500 rounds each, provision for 750)
② Camera gun port
③ De Havilland Hydromatic needle blade propeller (three-blade, early-type)
④ Coolant fill hatch
⑤ Cooling spouts air duct fairing for exhaust manifold flanges
⑥ Exhaust flame damper shroud
⑦ Deicing spray pipe
⑧ Windscreen wiper
⑨ Flat bulletproof windscreen
⑩ Reflecting gunsight
⑪ Sliding side panel
⑫ Dinghy access panel (jettisonable)
⑬ Fuel fill hatch
⑭ Antenna mast
⑮ Reinforcing strip
⑯ Ground starter socket hatch
⑰ Vertical stabilizer root fairing
⑱ Root fitting inspection hatch
⑲ Antenna wire mount
⑳ Pitot tube
㉑ Electrical connection access hatch
㉒ Rudder horn balance
㉓ Rudder (fabric-covered, 26° -1° +2° arc of movement)
㉔ Rudder trim tab (16° ±2°30" arc of movement)
㉕ Rudder controls inspection hatch
㉖ Rudder trim tab control rod access panel
㉗ Rudder trim tab control rod fairing
㉘ Elevator operating linkage fairing
㉙ Tail navigation light
㉚ Spare tail-socket fairing
㉛ Detachable tail cone fairing (early-type)
㉜ Retracting gear access panel
㉝ Rearward-retracting tail wheel with Marstrand two-track anti-shimmy tire (late-type)
㉞ Tail wheel strut construction (late-type)
㉟ Nacelle tail fairing (late-type)
㊱ Bomb bay door (fitted with hydraulic jacks)
㊲ Main wheel door (early-type)
㊳ Access hatch for jacking (outboard side only of both nacelles)
㊴ Fuel pump drain (left side only of both nacelles)
㊵ Trailing antenna fairlead fairing
㊶ Heat resistant panel
㊷ Fuel pump cooling air intake
㊸ Carburetor air intake fairing
㊹ Air intake guard
㊺ Hispano Mk.I 20mm cannon (150 rounds each, provision for 175)
㊻ Cannon under fairing
㊼ Coolant pipe bulge (left side only of both nacelles)
㊽ Detachable tail cone fairing (late-type)
㊾ Tail warning radar mount fairing

モスキートF.B.Mk.VIシリーズ I & II

F.B.Mk.VIはF.Mk.IIから発展した戦闘爆撃機型で、その活躍と生産機数の点で、もっとも重要なモスキートのバリエーションである。F.Mk.IIと同様に機首にはブローニング .303 in 機関銃4挺、胴体下面にイスパノ20mm機関砲4門を装備。さらに機関砲収容部の後方に爆弾倉を設け、主翼下面のラックにも爆弾2発を懸吊できた。外観上の特徴は、胴体下面のドアが、前方の機関砲点検扉と後方の爆弾倉ドアにわけられた点である。原型機はB.MkIVシリーズII（DZ434）をベースに製作され、のちにシリアルナンバーHJ662／Gを与えられて、1942年6月1日に初飛行した。生産型1号機の完成は1943年2月。1,480hpのマーリン21または23エンジンを搭載した最初の113機が、F.B.Mk.VIシリーズ I として1943年5月から同年末にかけて配備開始された。これにつづいて生産され、より高出力のマーリン25（1,640hp）を装備する2000機あまりの機体は、F.B.Mk.VIシリーズ II として1943年末から1946年夏にかけて配備された。F.B.Mk.VIを最初に装備した部隊は、No.418Sqn（RCAF）で、1943年5月11日のことである。F.B.Mk.VIによる最初のソーティは1943年10月3日に行われた。F.B.Mk.VIシリーズ I は250ポンド爆弾4発、または151ガロン（686.5リッター）入増設燃料タンクを搭載でき、主翼下面へ50-100ガロン増加タンクを懸吊できた。F.B.Mk.VIシリーズ IIでは胴体爆弾倉内に500ポンド爆弾2発を搭載可能になり、さらに主翼下面に500ポンド爆弾2発を懸吊するか、25／60ポンド（11.3／27.2kg）ロケット弾8発を搭載することができた。ロケット弾発射用レールは主翼下面にそれぞれ4本ずつ、計8本装備され、また、8本のシングルレールのうち外側の2本ずつを残して内側の4本を外した場合には、落下タンクを発射用レールの内翼側に懸吊することもできた。モスキートが実戦でロケット弾をはじめて使用したのは1944年10月26日であった。F.B.Mk.VIの主翼は"ユニバーサルウイング"、前部風防には耐弾ガラスを採用し、胴体下面には曳航式アンテナ用フェアリングがある。主車輪は配備の初期から左側にスロットのないモスキート後期型のものが用いられ、生産の途中で主脚収容ドア・ガイドと尾輪ストラットも、モスキート後期型のものになった。また、多くのF.Mk.VIでは曳航式アンテナ用フェアリングとアンテナ支柱は撤去され、コクピット頂部と胴体上面にウィップアンテナを装備するようになった。ほとんどの機体はモスキート後期型テイルコーン・フェアリングを取り付けていた。大戦中、実戦に参加したF.B.Mk.VIの多くは排気管シュラウドを取り外して行動し、またマーリン25エンジンを搭載したF.B.Mk.VIの一部は、パドルブレードタイプのプロペラを装備していた。極東に送られたF.B.Mk.VIのほとんどは、大型化した気化器空気取入口フェアリングに、改修型の空気取入口保護カバーを取り付けた機体である。後期生産分の機体には主翼前縁にP.R.MkXVIと同様の着陸／滑走灯が装備されていた。F.B.Mk.VIシリーズ IIの一部には、Gee航法装置、"モニカ"後方警戒用レーダー、A.I.Mk.IV、Mk.V と Mk.XV ASHレーダーを装備した機体もあった。燃料搭載量は通常453ガロン（2,059リッター）、最大716ガロン（3,225リッター）。F.B.Mk.VIは昼間／夜間戦闘爆撃機、イントルーダー／レンジャー機、長距離戦闘機としてヨーロッパ、中東、極東と広い範囲で活躍した。ビルマでは日本軍に対して地上攻撃を行ない、日本軍機と交戦している。

F.B.Mk.VI Series II Left view

左側面図

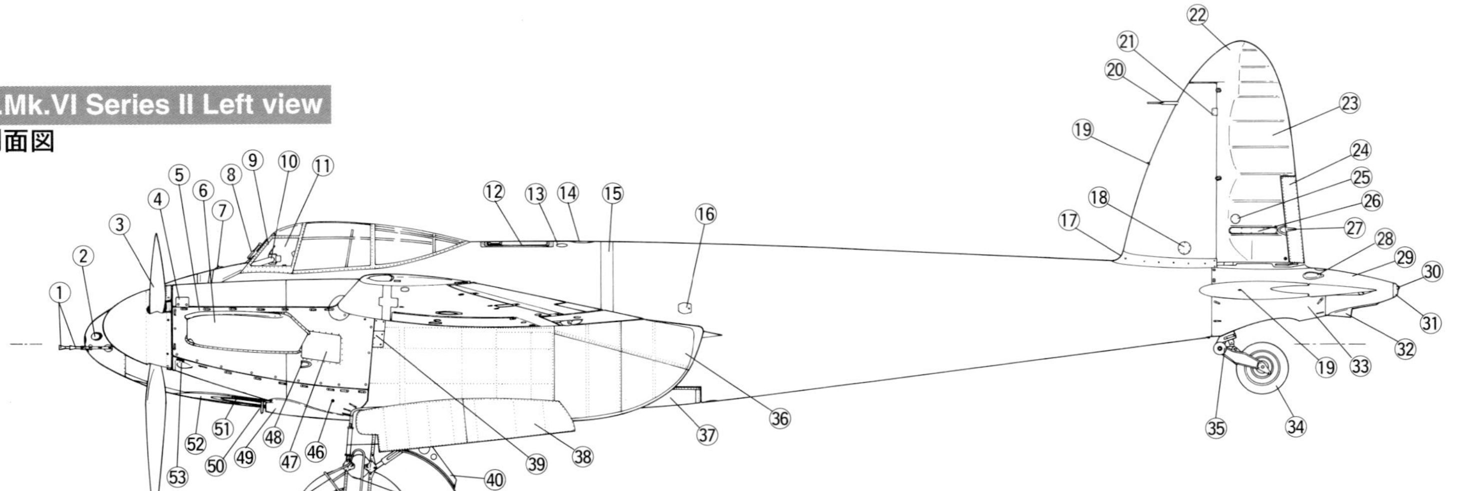

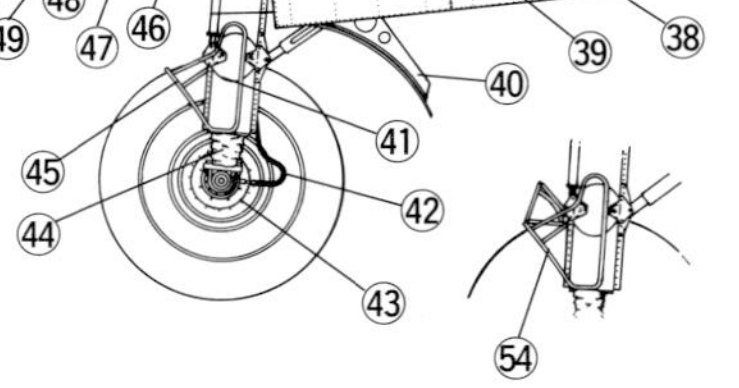

①消焔器付ブローニングMk.II.303 in (7.7mm) 機関銃(装弾数各500発、最大750発)
②射撃カメラ取付穴
③デ・ハヴィランド・油圧式3翅プロペラ（ニードルブレードタイプの初期型）
④冷却液注入ハッチ
⑤排気マニフォールド・フランジ冷却空気ダクト
⑥排気管消焔シュラウド
⑦防氷スプレーパイプ
⑧前部風防ワイパー
⑨1枚板耐弾ガラスの前部風防
⑩光像式射撃照準器
⑪スライド式側面窓
⑫救命ボート収納部パネル（投棄式）
⑬燃料注入ハッチ
⑭ベークライト製アンテナ保護板
⑮胴体補強用帯金
⑯地上スターターソケットハッチ

① Browning Mk.II .303in MG with flash hider (500 rounds each, provision for 750)
② Camera gun port
③ De Havilland Hydromatic needle blade propeller (three-blade,early-type)
④ Coolant fill hatch
⑤ Cooling spouts air duct fairing for exhaust manifold flanges
⑥ Exhaust flame damper shroud
⑦ Deicing spray pipe
⑧ Windscreen wiper
⑨ Flat bulletproof windscreen
⑩ Reflecting gunsight
⑪ Sliding side panel
⑫ Dinghy access panel (jettisonable)
⑬ Fuel fill hatch
⑭ Antenna bakelite securing plate

F.B.Mk.VI Series II Fuselage right side

胴体右側面

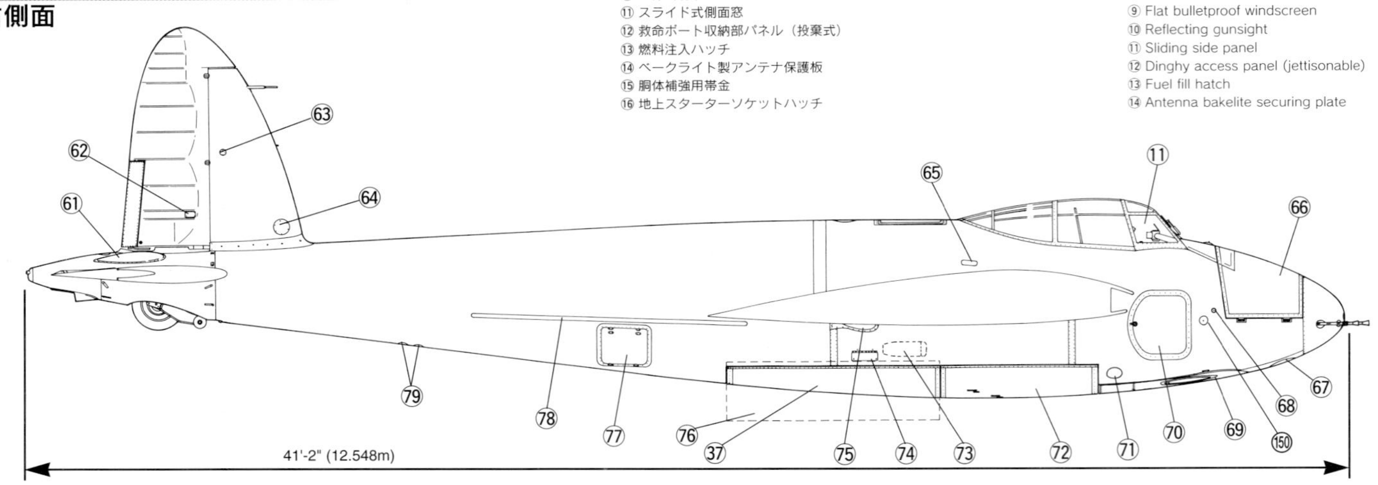

Mosquito F.B.Mk.VI Series I & II

most important Mosquito variant produced. Similar to the F.Mk.II, with a standard armament of four Browning .303in machine guns and four Hispano 20mm it also had the capability to carry two bombs inside a narrow bomb bay located behind the cannon bay, and two bombs on underwing racks. The major distinguishing feature of this model was the two standard belly bay doors split into rear bomb bay doors and forward cannon bay doors. The first prototype, a modified B.Mk.IV Series II with serial DZ434 (later re-serialled as HJ662/G) made its maiden flight on June 1, 1942. The first production aircraft rolled out in February 1943. The first 113 examples, powered with either the 1,480hp Merlin 21 or 23 engines, were designated as F.B.Mk.VI Series I, and delivered from May to late 1943. The over 2,000 following production aircraft, equipped with the more powerful 1,640hp Merlin 25 engines, were designated as F.B.Mk.VI Series II and delivered between late 1943 and the summer of 1946. No.418 Squadron (RCAF) was the first unit equipped with the F.B.Mk.VI on May 11, 1943. The first operational sortie was mounted on October 3, 1943. The F.B.Mk.VI Series I could carry either four 250-lb bombs or two belly-mounted 151gal (686.5-liter) fuel tanks and two 50-100gal underwing drop tanks in the place of the bomb load. The F.B.Mk.VI Series II increased its bomb load to two 500-lb bombs mounted inside the belly with the addition of either two 500-lb bombs or eight 25/60-lb (11.3/27.2kg) rocket projectiles on underwing racks. When the aircraft was equipped with two double-tiered rails under each wing instead of four single rails, it was capable of mounting underwing fuel tanks as well, located inboard of the rocket rails. Mosquitoes used rockets in action for the first time on October 26, 1944. The aircraft had the "universal wing", the underbelly trailing antenna fairlead fairing and the bulletproof windscreen. From the beginning of the first deliveries, these Mosquitoes were equipped with the late-type main wheels without slots on the left side. In the course of production they were fitted with late-type main wheel door guides and the late-type tail wheel strut construction, and most aircraft were fitted with late-type tail cone fairings as well. They also had both the trailing antenna fairlead fairing and the antenna mast removed. A number of F.B.Mk.VIs were equipped with different antenna fittings, among these whip antennas located on the roof of the cockpit canopy and dorsal fuselage. During war operations, many aircraft had the exhaust flame damper shrouds removed, especially the Series II variant equipped with the Merlin 25s, some of which drove paddle blade propellers as well. Most examples detached to Far East had an enlarged carburetor air intake fairing with revised intake guard. Later production aircraft had landing/taxi lights on the right wing leading edge like the P.R.Mk. XVI. Some F.B.Mk.VI Series IIs were fitted with Gee long-range navigation equipment, the tail cone-mounted Monica warning radar and the nose-mounted A.I.Mk.IV, Mk.V and Mk.XV ASH radar devices. The normal fuel capacity was 453gal (2,059 liters) and the maximum load was 716gal (3,255 liters). Operated as a day and night fighter-bomber, intruder/ranger and long-range fighter, the F.B.Mk.VIs saw action over Europe, the Middle East and the Far East, where they were engaged in the destruction of Japanese ground targets throughout Burma, meeting opposition by the Japanese Air Force.

F.B.Mk.VI Series II Top view
上面図

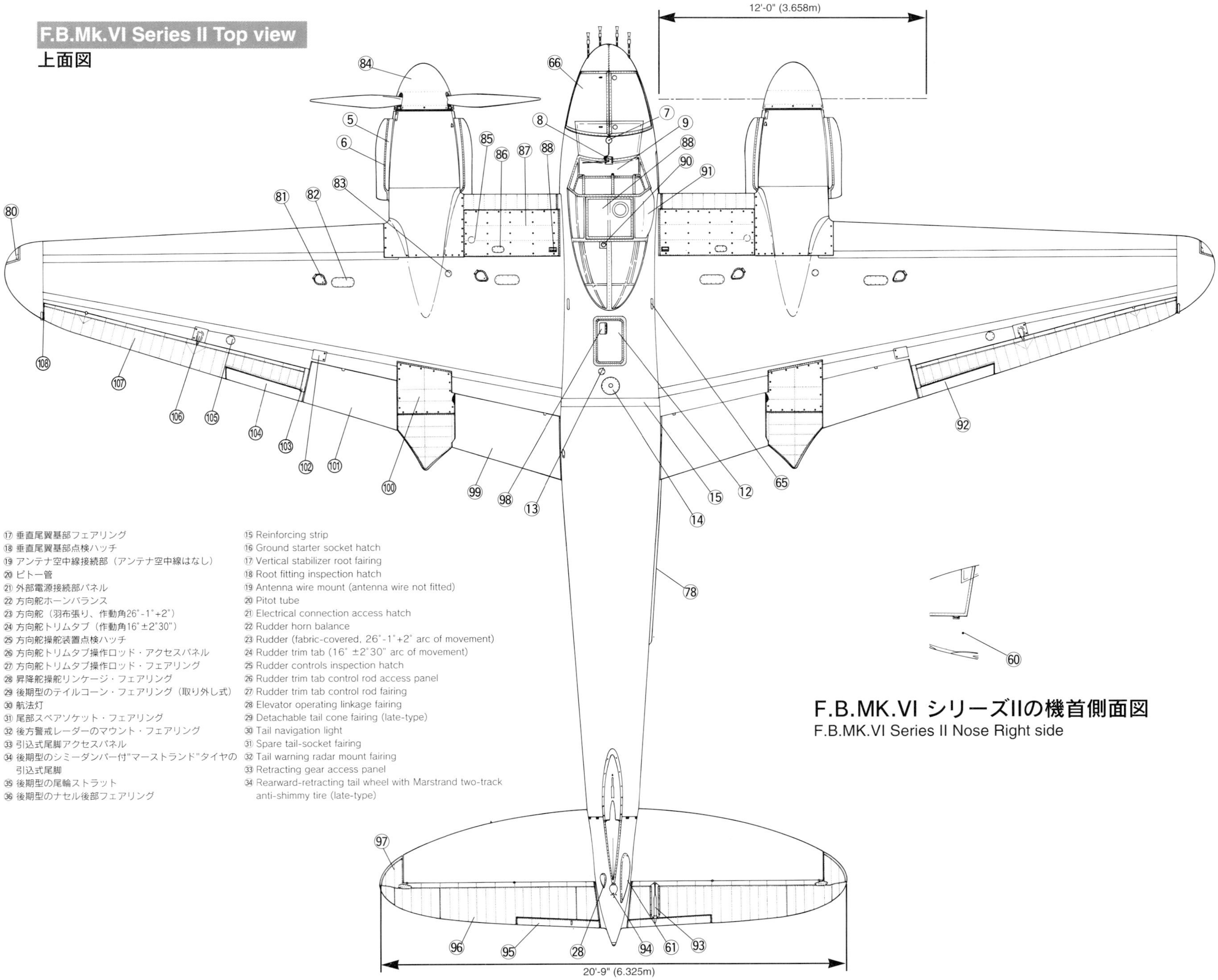

⑰ 垂直尾翼基部フェアリング
⑱ 垂直尾翼基部点検ハッチ
⑲ アンテナ空中線接続部（アンテナ空中線はなし）
⑳ ピトー管
㉑ 外部電源接続部パネル
㉒ 方向舵ホーンバランス
㉓ 方向舵（羽布張り、作動角26°-1°+2°）
㉔ 方向舵トリムタブ（作動角16°±2°30"）
㉕ 方向舵操舵装置点検ハッチ
㉖ 方向舵トリムタブ操作ロッド・アクセスパネル
㉗ 方向舵トリムタブ操作ロッド・フェアリング
㉘ 昇降舵操舵リンケージ・フェアリング
㉙ 後期型のテイルコーン・フェアリング（取り外し式）
㉚ 航法灯
㉛ 尾部スペアソケット・フェアリング
㉜ 後方警戒レーダーのマウント・フェアリング
㉝ 引込式尾脚アクセスパネル
㉞ 後期型のシミーダンパー付"マーストランド"タイヤの引込式尾脚
㉟ 後期型の尾輪ストラット
㊱ 後期型のナセル後部フェアリング

⑮ Reinforcing strip
⑯ Ground starter socket hatch
⑰ Vertical stabilizer root fairing
⑱ Root fitting inspection hatch
⑲ Antenna wire mount (antenna wire not fitted)
⑳ Pitot tube
㉑ Electrical connection access hatch
㉒ Rudder horn balance
㉓ Rudder (fabric-covered, 26°-1°+2° arc of movement)
㉔ Rudder trim tab (16° ±2°30" arc of movement)
㉕ Rudder controls inspection hatch
㉖ Rudder trim tab control rod access panel
㉗ Rudder trim tab control rod fairing
㉘ Elevator operating linkage fairing
㉙ Detachable tail cone fairing (late-type)
㉚ Tail navigation light
㉛ Spare tail-socket fairing
㉜ Tail warning radar mount fairing
㉝ Retracting gear access panel
㉞ Rearward-retracting tail wheel with Marstrand two-track anti-shimmy tire (late-type)

F.B.MK.VI シリーズIIの機首側面図
F.B.MK.VI Series II Nose Right side

F.B.Mk.VI Series II Bottom view

下面図

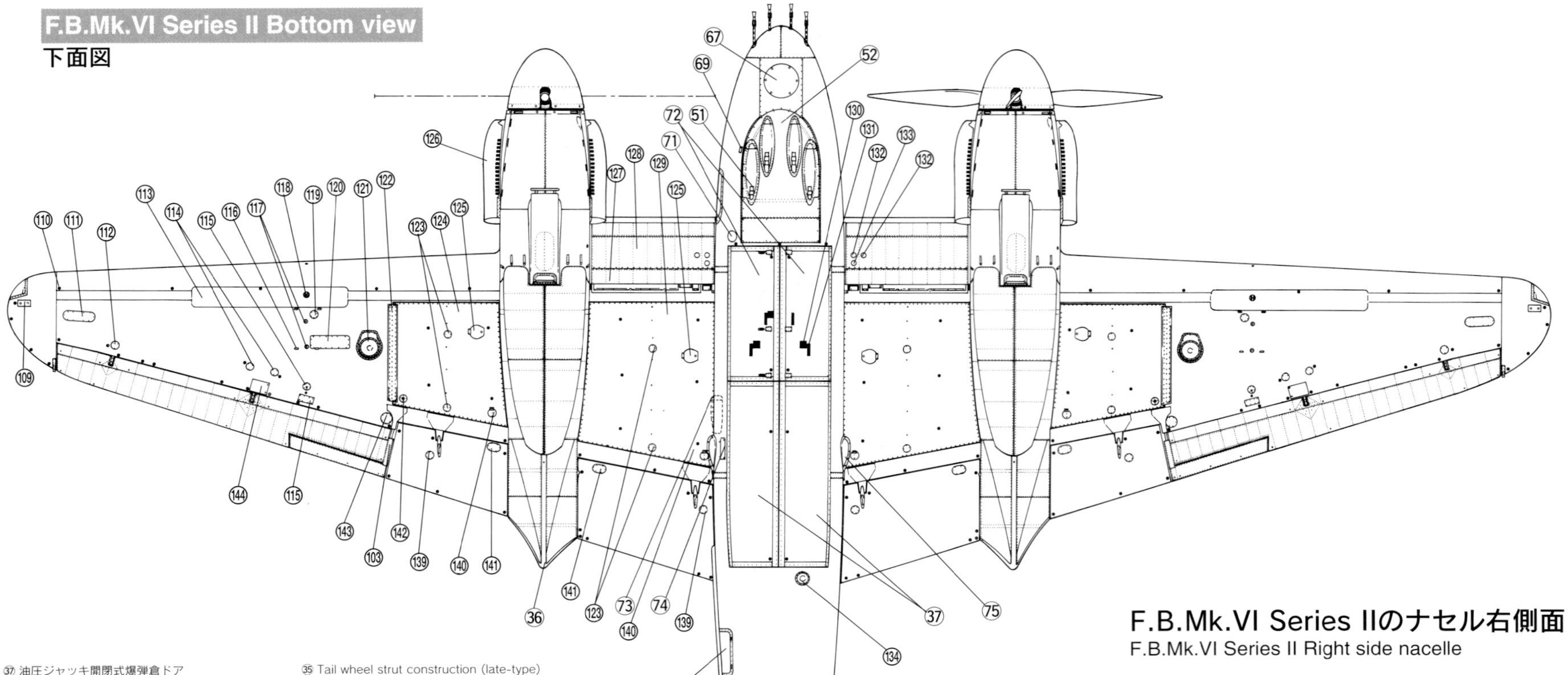

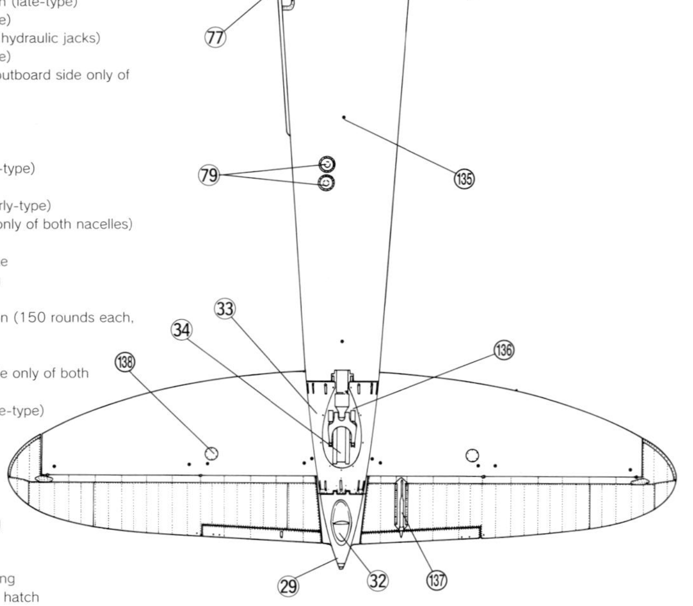

F.B.Mk.VI Series IIのナセル右側面

F.B.Mk.VI Series II Right side nacelle

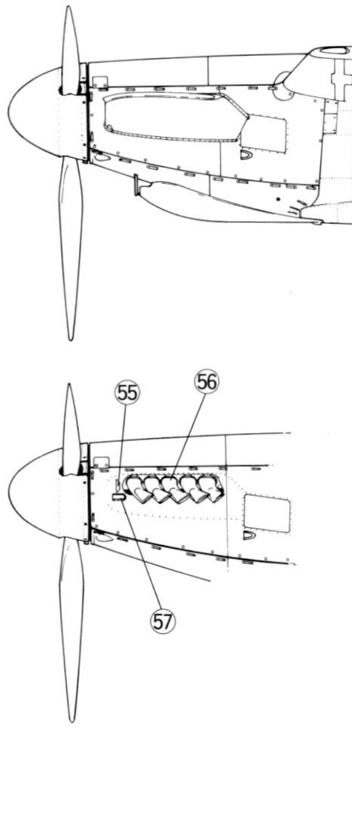

37 油圧ジャッキ開閉式爆弾倉ドア
38 初期型の主脚収容ドア
39 ジャッキ用アクセスハッチ（両ナセル外側のみ）
40 マッドガード
41 ショック・アブゾーバー・ストラット
42 ブレーキパイプ
43 後期型の主車輪
44 衝撃吸収用ラバーパッド
45 初期型の主脚収納ドア・ガイド
46 燃料ポンプ・ドレイン（両ナセル左側のみ）
47 排気用耐熱パネル
48 燃料ポンプ冷却空気取入口
49 気化器空気取入口フェアリング
50 空気取入口保護用防氷カバー
51 イスパノMk.I 20mm機関砲（装弾数150発、最大175発）
52 機関砲フェアリング
53 冷却液パイプのバルジ（両ナセル左側のみ）
54 後期型の主脚収納ドア・ガイド
55 排気管冷却用空気取入口
56 推力式単排気管
57 点火プラグ冷却空気取入口
58 改修された空気取入れ口保護用防氷カバー
59 大型化された気化器空気取入口フェアリング
60 外部ソケット（左側のみ）
61 昇降舵内側マスバランスのフェアリング
62 左側と位置が異なる方向舵操舵装置点検ハッチ
63 方向舵中央ヒンジ点検ハッチ
64 左側と位置が異なる垂直尾翼基部点検ハッチ
65 主翼取付部フェアリング
66 機関銃点検パネル
67 機関銃弾リンク／薬莢空回収用アクセスハッチ
68 ピトー管静圧ヘッド（右側のみ）
69 外気温度計
70 搭乗員乗降用ハッチ
71 曳航式アンテナ用開口部カバーパネル
72 機関砲収容部ドア閉状態
73 F.B.Mk.VIの一部にみられる燃料冷却器
74 燃料廃棄口ハッチ（胴体右側のみ）
75 燃料パイプ用フェアリング
76 爆弾倉ドア開状態

35 Tail wheel strut construction (late-type)
36 Nacelle tail fairing (late-type)
37 Bomb bay door (fitted with hydraulic jacks)
38 Main wheel door (early-type)
39 Access hatch for jacking (outboard side only of both nacelles)
40 Mudguard
41 Shock-absorber strut
42 Brake pipe
43 Plan-type main wheel (late-type)
44 Rubber shock pads
45 Main wheel door guide (early-type)
46 Fuel pump drain (left side only of both nacelles)
47 Heat resistant panel
48 Fuel pump cooling air intake
49 Carburetor air intake fairing
50 Air intake guard
51 Hispano Mk.I 20mm cannon (150 rounds each, provision for 175)
52 Cannon under fairing
53 Coolant pipe bulge (left side only of both nacelles)
54 Main wheel door guide (late-type)
55 Exhaust air intake
56 Stub exhaust pipes
57 Sparking plug cooling air intake
58 Air intake guard (modified)
59 Carburetor air intake fairing (enlarged)
60 Outlet (left side only)
61 Elevator static balance fairing
62 Rudder controls inspection hatch (different position this side)
63 Rudder center hinge inspection hatch
64 Root fitting inspection hatch (different position on this side)
65 Wing fix attachment fairing
66 MG inspection panel
67 MG empty shell and link access hatch

F.B.Mk.VI Series II Front view

正面図

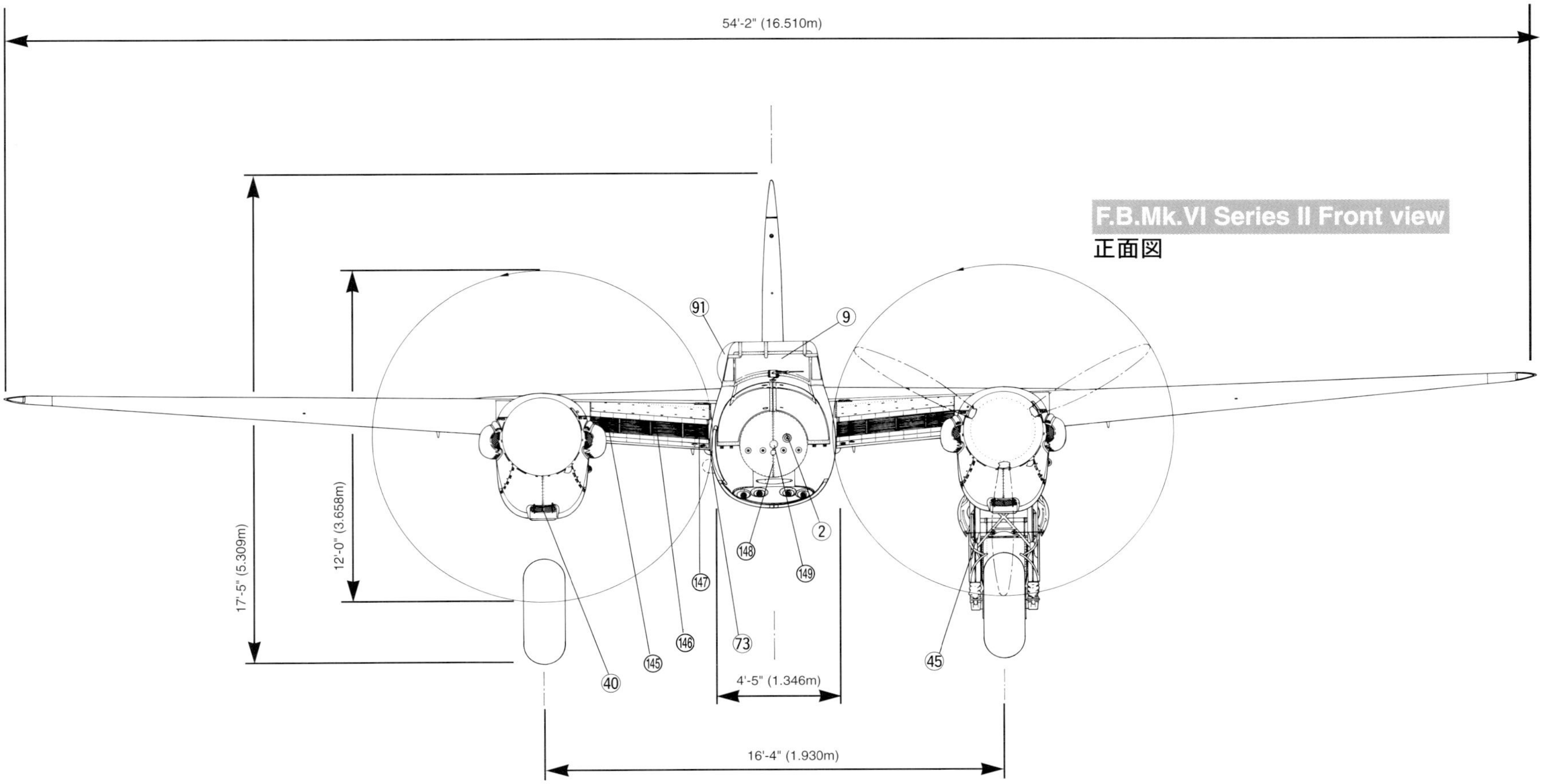

77 乗降／アクセスハッチ
78 胴体補強用ストレーキ
79 地上用敵味方識別灯（前方が緑、後方がアンバー）
80 航法灯（左翼が赤、右翼が緑）
81 燃料注入ハッチ
82 ベントパイプ・アクセスハッチ
83 主脚ロック・ハッチ
84 スピナー
85 ラジエターパネルのバルジ
86 ラジエターフラップ操作ロッド点検ハッチ
87 油冷却器およびエンジン・ラジエーター点検パネル
88 吹出口
89 キャノピー頂部緊急脱出用ハッチ
90 信号弾発射口
91 バルジ型に膨らんだ側面キャノピー
92 補助翼バランスタブ（右翼のみ）
93 昇降舵トリムタブ操作ロッド・フェアリング（右昇降舵は上面のみ）
94 昇降舵中央ヒンジ点検ハッチ
95 昇降舵トリムタブ（作動角7°30" ±1°15"）
96 昇降舵（全金属製、作動角25°-1°+2°）
97 昇降舵ホーンバランス
98 救命ボート収納部パネル
99 内翼側フラップ(合板仕上げ、作動角45°)
100 フラップ作動ジャッキ点検パネル
101 外翼側フラップ(合板仕上げ、作動角45°)
102 補助翼サーボ・アクセスハッチ
103 補助翼タブ操作装置点検パネル
104 補助翼トリムタブ（左翼のみ、作動角は上方に8°30"、下方に9°15"）
105 リリーフバルブ・ハッチ
106 補助翼中央ヒンジ・フェアリング
107 補助翼(金属外皮、作動角は上に26°30'±30'、下に11°30" ±30")
108 レジンランプ
109 航法灯点検ハッチ
110 ドレインホール
111 方位受信レーダーアンテナ・アクセスパネル
112 補助翼ヒンジ・アクセスハッチ
113 ロケット弾発射レイル用補強板
114 アクセスハッチ
115 落下タンク用燃料パイプ・アクセスハッチ
116 爆弾懸吊ポイント（4カ所）
117 落下タンク懸吊用アイボルト取付部
118 落下タンク用フォーク取付部（第8リブ）
119 電装系接続部点検ハッチ
120 高度測定アンテナ受信部アクセスハッチ（右主翼のみ）
121 着陸灯
122 主翼トレッスル用補強板（第6リブ）
123 バランスパイプ連結ハッチ
124 外翼タンク・アクセスパネル
125 燃料計ユニット・アクセスハッチ
126 冷却用ルーバー
127 ラジエターフラップ
128 ラジエター着脱用パネル
129 内翼タンク・アクセスパネル
130 機関砲リンク排出口
131 機関砲薬莢排出口
132 エンジン・ラジエーター・ドレインプラグ・キャップ
133 滑油冷却器ドレインプラグ・キャップ
134 地上用敵味方識別灯（赤）
135 速度計水抜き用ドレインプラグ
136 尾輪収容口
137 昇降舵トリムタブ操作ロッド・フェアリング（左昇降舵は下面のみ）
138 昇降舵ヒンジ点検ハッチ
139 フラップヒンジ・アクセスハッチ
140 燃料タンク・ドレインコック・ハッチ
141 フラップ中央ヒンジ点検ハッチ
142 燃料排出用ドレインプラグ
143 補助翼サーボ・アクセスハッチ
144 補助翼ヒンジおよび左右連動調節部点検ハッチ
145 初期型の滑油冷却器
146 初期型のラジエター
147 キャビン暖気ユニット
148 機首先端航法灯取付穴カバープレート
149 レーダー受信アンテナ取付穴カバープレート
150 ドリフトレコーダー・ドア

68 Static head (right side only)
69 Air thermometer
70 Crew entry door
71 Trailing antenna fairlead fairing mount covering panel
72 Cannon bay door (fitted with door locks)
73 Fuel cooler (non-standard equipment, seen on a nuber of F.B.Mk.VIs)
74 Fuel gallery hatch (right side only)
75 Fuel pipe fairing
76 Bomb bay door opening position
77 Entry/access door
78 Reinforcing strake
79 Downward identification code lights (forward: green, aft: amber)
80 Navigation light (left: red, right: green)
81 Fuel fill hatch
82 Vent pipe access hatch
83 Landing gear uplock hatch
84 Spinner
85 Radiator panel bulge
86 Radiator flap operating jack inspection hatch
87 Radiator (oil cooler and coolant radiator) inspection panel
88 Exhaust
89 Emergency roof exit panel
90 Hole for signal pistol
91 Bulged window panel
92 Aileron balance tab (right wing only)
93 Elevator trim tab control rod fairing (right upper surface only)
94 Elevator center hinge inspection hatch
95 Elevator trim tab (7°30"±1°15" arc of movement)
96 Elevator (all-metal construction, 25°-1° +2° arc of movement)
97 Elevator horn balance
98 Dinghy inspection hatch
99 Inner flap (ply-covered, 45° arc of movement)
100 Flap jack inspection panel
101 Outer flap (ply-covered, 45° arc of movement)
102 Aileron servo access hatch
103 Aileron tab controls inspection panel
104 Aileron trim tab (left wing only, up 8°30", down 9°15" arc of movement)
105 Relief valve hatch
106 Aileron center hinge fairing
107 Aileron (metal-covered, up 26°30"±30", down 11°30" ±30" arc of movement)
108 Resin lamp
109 Navigation light inspection hatch
110 Drain hole
111 Radar receiving azimuth antenna access panel
112 Aileron hinge access hatch
113 RP rails mount reinforcing plate
114 Access hatch
115 Drop tank pipe access hatch
116 Bomb carrier pick-up point (four)
117 Drop tank eye-bolt attachment
118 Drop tank fork attachment (on rib No.8)
119 Electrical connections access hatch
120 Radar receiving elevation antennas access hatch (right wing only)
121 Landing light
122 Reinforcing plate for wing trestling (on rib No.6)
123 Balance pipe connection hatch
124 Outer tank bays access panel
125 Fuel gauge unit hatch
126 Cooling louvers
127 Radiator flap
128 Panel for radiator removing
129 Inner tank bays access panel
130 Cannon link chutes ejection port
131 Cannon shell case ejection port
132 Coolant radiator drain plug cap
133 Oil cooler drain plug cap
134 Downward identification code light (red)
135 A.S.I. (Air Speed Indicator) water drain plug
136 Tail wheel bay
137 Elevator trim tab control rod fairing (left undersurface only)
138 Elevator hinge inspection hatch
139 Flap hinge access hatch
140 Fuel tank drain cock hatch
141 Flap center hinge inspection hatch
142 Fuel drain plug
143 Aileron servo access hatch
144 Aileron center hinge and differential control inspection hatch
145 Oil cooler (early-type)
146 Coolant radiator (early-type)
147 Cabin heater unit
148 Forward navigation light port covering plate
149 Radar transmitting antenna port covering plate
150 Drift recorder door

F.B.Mk.VI Series II equipped with A.I.Mk.IV radar
F.B.Mk.VIシリーズII A.I.Mk.IVレーダー装備機

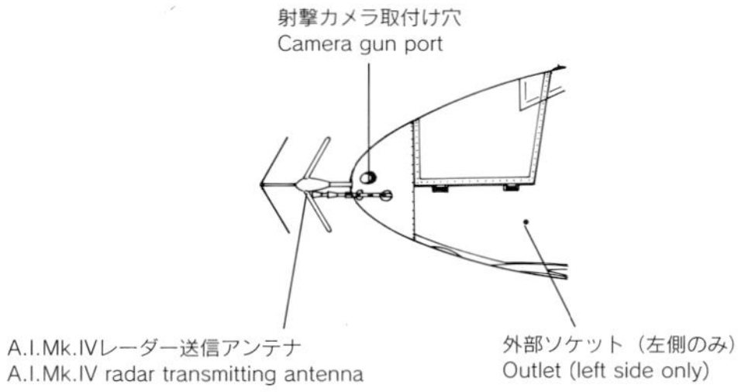

F.B.Mk.VI Series II equipped with A.I.Mk.XV ASH radar
F.B.Mk.VIシリーズII A.I.Mk.XV ASH radar装備機

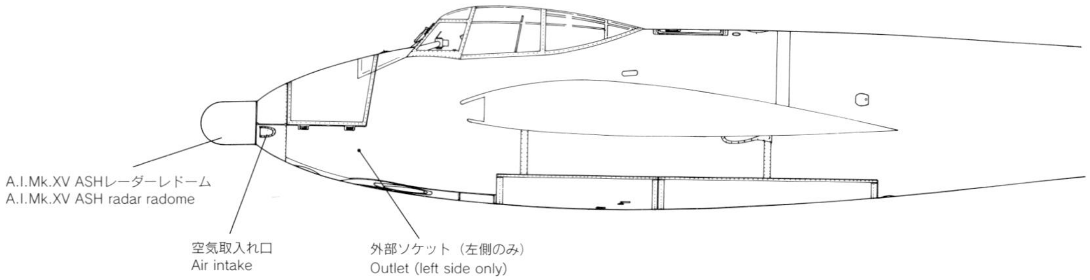

F.B.Mk.VI Series II later production model - mainly operated by Coastal Command
F.B.Mk.VIシリーズII後期生産機（主に沿岸航空軍団で使用された機体）

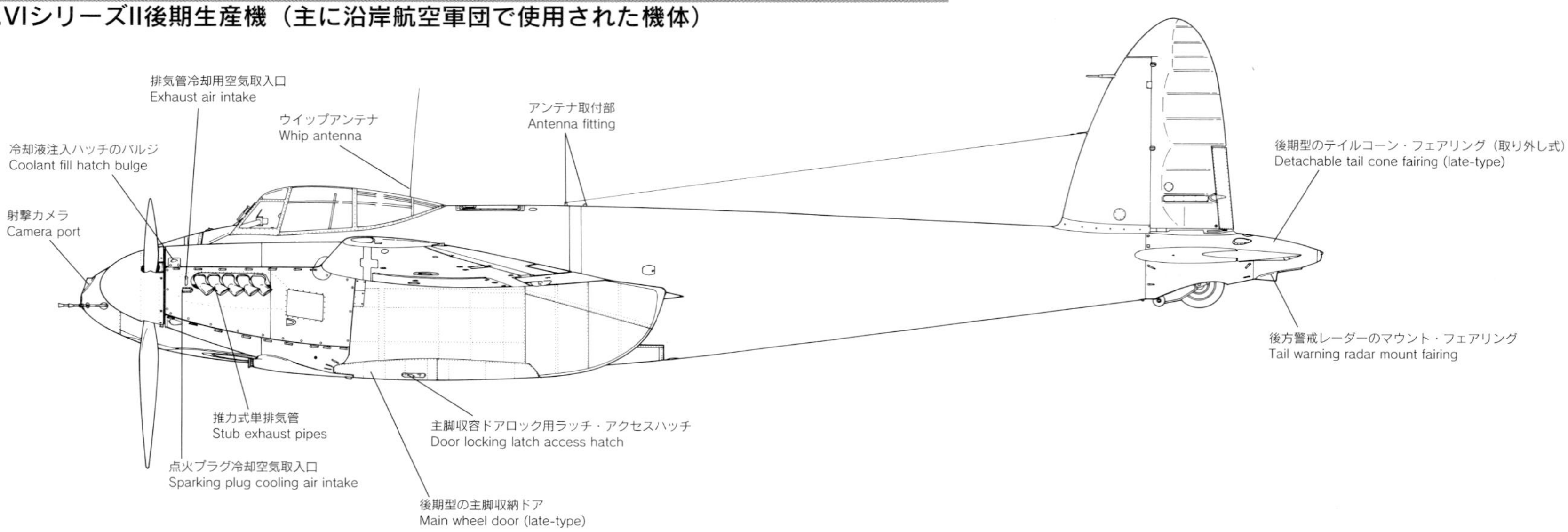

F.B.Mk.XVIII Front view
正面図

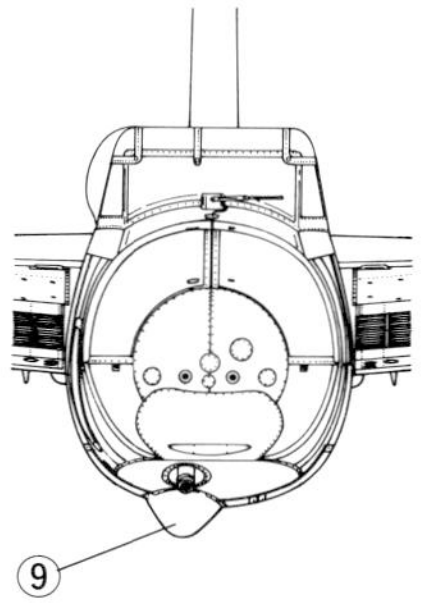

① 機首下面補強板
② 機銃発射口カバープレート
③ 消焔器付ブローニングMk.II .303 in (7.7mm)機関銃（装弾数各500発、最大750発）
④ 射撃カメラ取付穴カバープレート
⑤ デ・ハヴィランド・油圧式3翅プロペラ（後期型）
⑥ 冷却液注入ハッチのバルジ
⑦ 後期型のテイルコーン・フェアリング（取り外し式）
⑧ 後方警戒レーダーのマウント・フェアリング

① Reinforcing panel
② MG port covering plate
③ Browning Mk.II .303in MG with flash hider (500 rounds each, provision for 750)
④ Camera gun covering plate
⑤ De Havilland Hydromatic paddle blade propeller (three-blade, late-type)
⑥ Coolant fill hatch bulge
⑦ Detachable tail cone fairing (late-type)
⑧ Tail warning radar mount fairing
⑨ Bulged cannon fairing

F.B.Mk.XVIII Left side
左側面図

⑨ バルジタイプの57mm砲フェアリング
⑩ 57mmモリンズ砲（装弾数25発）
⑪ 消焔器
⑫ 機関銃点検ハッチ
⑬ 防弾鋼板
⑭ 改修された機関砲格納部ドア
⑮ 機関砲薬莢排出口
⑯ 機関砲マウント・フェアリング
⑰ ピトー管静圧ヘッド（右側のみ）
⑱ 補強された内翼側フラップ
⑲ ドリフト・レコーダー・ドア

⑩ Molins 57mm cannon (25 rounds)
⑪ Flash hider
⑫ MG inspection panel
⑬ Armor plates
⑭ Cannon bay doors (modified)
⑮ Cannon shell case ejection port
⑯ Cannon mount fairing
⑰ Static head (right side only)
⑱ Inner flap (reinforced)
⑲ Drift recorder door

F.B.Mk.XVIII Bottom view
下面図

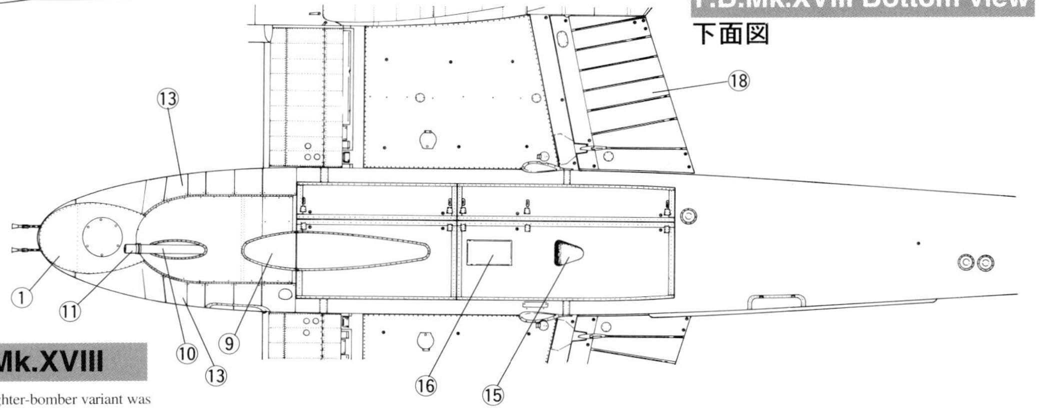

モスキートF.B.Mk.XVIII

F.B.Mk.XVIIIは6ポンド対戦車砲を航空機用に改修した大口径のモリンズ57mm(2¹/₄インチ)砲を機首に装備する、地上攻撃用／対艦船攻撃用モスキート。F.B.Mk.VIシリーズI（HJ732／G）を改修し、マーリン23エンジンを搭載した原型機は、1943年6月8日に初飛行している。1943年10月から1945年1月まで月産1機のペースで合計16機を生産。すべての生産機にはマーリン25エンジンが搭載された。1943年10月22日にはNo.248Sqnが最初の2機（HX902、HX903）を受領している。"ツェツェ蝿"（アフリカに生息するハエの一種、眠り病を媒介する）と呼ばれたモスキートF.B.Mk.XVIIIは、胴体下面の20mm機関砲4門を撤去、主翼下面にはロケット弾8発を装備するか、または500ポンド爆弾2発をラックに懸吊した。自動装填式の57mm砲は胴体中心線から4インチ（101.6mm）右、機体の水平方向に対して3°45"下方に向けて取り付けられ、装弾数は25発であった。左右対称の胴体下面ドアは右側が大きなものに改められ、胴体下面前方にも57mm砲をクリアするためのバルジが設けられている。機首の武装は最初 .303 in 機関銃4挺であったが、57mm砲発射の衝撃で機首に亀裂がはいるという問題を解決するために、外側の2挺を撤去。機関銃のマウントと機首のあいだに補強用ロッドが取り付けられた。なお、.303 in 機関銃は基本的に57mm砲の照準のために用いられていた。内翼側のフラップは厚い外板で覆われ、補強用のリブを追加。機首の搭乗員区画とエンジン・ナセルにも重装甲が施されている。F.B.Mk.XVIIIは主にコースタル・コマンド（沿岸航空軍団）におけるドイツ軍Uボートと艦船に対する攻撃で活躍した。最大燃料搭載量は668ガロン(3,037リッター)、作戦時には403ガロン(1,832リッター)。ごく一部の機体が胴体に65ガロン(295.5リッター)入長距離飛行用増加燃料タンクを装備していた。

Mosquito F.B.Mk.XVIII

This ground-attack and anti-shipping fighter-bomber variant was equipped with the Molins 57mm (2¹/₄in) cannon, officially designated as "Airborne Six-Pounder Class M Gun". The prototype, a modified F.B.Mk.VI Series I serialled HJ732/G and powered by Merlin 23s, first flew on June 8, 1943, and the following 16 production aircraft completed were delivered from October 1943 to January 1945, at the rate of one per month. All were equipped with Merlin 25s. No.248 Squadron received the first two aircraft (HX902 and HX903) on October 22, 1943. The Mosquito F.B.Mk.XVIII, nicknamed "TseTse", had the four 20mm cannons removed, retaining the full load of two 50-100gal drop tanks and either eight rocket projectiles or two 500-lb bombs on underwing racks. The 57mm quick fairing cannon was installed 4in (101.6mm) to the right side of the fuselage center line, and aligned to fire at a slightly downward angle of 3°45' to the aircraft's horizontal datum, carrying an ammunition load of 25 rounds. The symmetric arrangement of the ventral compartment doors was modified and a bulged cannon fairing was fitted under the forward belly. Initially the four .303in machine guns were retained, but later, to solve the problem of the 57mm cannon's blast causing cracks in the nose, the outer pair of MGs was removed and the area was reinforced with rods fitted between the MG mounts and the nose. The .303in MGs were mainly engaged for sighting the 57mm cannon. Also the inner wing flaps were fitted with heavier skin and additional strengthened ribs. Crew and engine nacelles were heavily armor-plated. Mainly operated by Coastal Command against German U-boats and shipping, the F.B.Mk.XVIII had a maximum fuel capacity of 668gal (3,037 liters) and an operational load of 403 gal (1,832 liters). A very few examples had a 65gal (295.5-liter) long-range fuel tank fitted in the fuselage.

F.B.Mk.XVIII Fuselage right side
胴体右側面

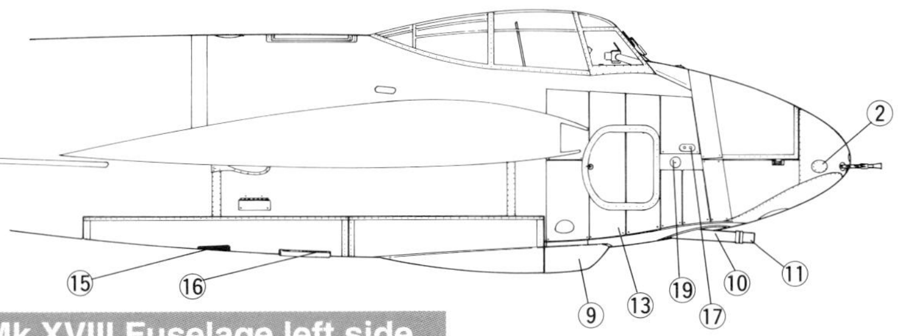

F.B.Mk.XVIII Fuselage left side
胴体左側面

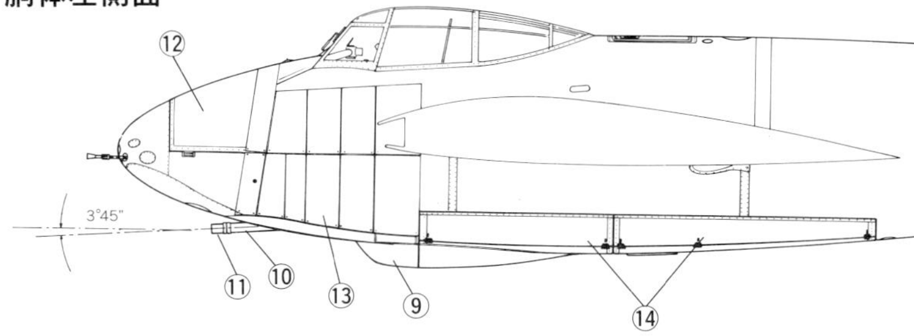

Aero Detail 23 Production Team

DE HAVILLAND MOSQUITO

PLANNING, COORDINATION & RESEARCH
カルミネ・ディ・ナポリ　Carmine Di Napoli
ラファエレ・マンチーニ　Raffaele Mancini

PHOTOGRAPHS
ラファエレ・マンチーニ (exept photos 11, 76,-79, 134 and photo page 7)　Raffaele Mancini
エリク・デスルー (photo 11)　Eric Dessouroux / B.A.M.R.S.
ベルギー王立軍事博物館 (photos 76-79)　Collections Royal Army Museum - Brussels
アメリカ空軍博物館 (photo 134 and page 7)　U.S. Air Force Museum

ILLUSTRATIONS
カルミネ・ディ・ナポリ　Carmine Di Napoli
ラファエレ・マンチーニ　Raffaele Mancini

ORIGINAL TEXT
ラファエレ・マンチーニ　Raffaele Mancini
ステファニア・グイドリッツィ　Stefania Guidorizzi

JAPANESE TEXT
山田智夫　Tomoo Yamada

ENGLISH TEXT
スコット・ハーズ　Scott T. Hards

COVER DESIGN
丹羽和夫　Kazuo Niwa

LAYOUT
佐藤 理　Osamu Sato

EDITOR
田中理人　Masato Tanaka
佐藤 理　Osamu Sato

SPECIAL THANKS TO :
ブリティッシュ・エアロスペース社　British Aerospace Aerostructures Limited
モスキート・エアクラフト・ミュージアム　Mosquito Aircraft Museum
モスキート搭乗員協会　Mosquito Aircrew Association
ベルギー王立軍事博物館　Musée Royal de l'Armée et d'Histoire Militaire
イギリス空軍博物館　Royal Air Force Museum
デ・ハヴィランド・エアクラフト・ミュージアム・トラスト　The de Havilland Aircraft Museum Trust Limited
アメリカ空軍博物館　U.S. Air Force Museum
IPMS UK　IPMS UK
ニック・アレン　Nick Allen
フィリップ・バートルズ　Philip Birtles
エドガー・ブルックス　Edgar Brooks
レット・バーンズ　Rhett Burns
エリク・デスルー　Eric Dessouroux
ロン・アイル　Ron Eyre
ピーター・ジェイムズ　Peter James
アルフォンソ・ミノ　Alfonso Mino
デレク・パーチェス　Derek Purchese
エテイアンヌ・レユニ　Etienne Reunis
ヘンリー・ロルフ　Henry Rolfe
グレアム・トーマス　Graham Thomas
アンドリュー・シンプソン　Andrew Simpson
トニー・ウィルソン　Tony Wilson

In memory of Kevin Moorhouse and Steve Watson, the crew members of the British Aerospace-operated Mosquito T.Mk.III RR299 which crashed during an airshow on July 21, 1996.

1996年のエアショーでブリティッシュ・エアロスペースのモスキートT.Mk.III RR299に搭乗し、墜落事故で命を失ったふたりのクルー、ケヴィン・ムアハウス氏とスティーヴ・ワトソン氏に捧げる。

〈編集後記〉

●英国とは不思議な国である。仕事の関係で何回も訪れているところだが、英国人のフレンドリーな態度とマナーはいつも素晴らしいと思う。そしてビール好きの私にとっては、正に天国のようなところ。とにかく生で飲んだギネスビールは最高（余談だが、最近流行っているあるメーカーの「スーパー薄い」ビールや発泡酒など美味しいと思って飲んでいる日本人は、とんでもない勘違いをしていると思う）。しかし、料理が出てくると、ことは一転して、「何でこの国の飯はこんなにマズイだろう」と思うばかりだ。とにかくポテトや豆の山に味のない肉料理。あれほどのビールが作れるんだから、完全に英国人は味音痴じゃないはずだが……どうしてだろう。
　彼らが半世紀前に作っていた航空機も同じだと思う。翻訳しながらこの本の研究を読んで、モスキートはおそらく「第二次世界大戦最高の双発機」であることを確信しました。そして同時に英国が作っていたあのスピットファイアも「第二次世界大戦最高の単発機」に近い傑作機である（当然トップは米海軍のコルセアだが）。ですので、英国がものすごい航空機設計技術をもっているのはあきらかです。ところが、同じ国がフェアリー・バトルやアヴロ・マンチェスター、ボールトン・ポール・デファイアントとかホーカー・タイフーン（初期型）などの失敗作で真面目に戦いにいどんだのですから、不思議でしかたがありません。とくに艦上機では、英国は自国での開発がまったくダメで、米国製に頼らなければならなかった。なぜでしょうね。
　このところ、タミヤさんは結構英国機に力を入れているので、ファンにはうれしい日々が続いているはずです。機会があればまたADシリーズの英国機のラインアップを増やしたいと思います。

スコット・ハーズ

●スコットさんも書いていますけど、モスキートってほんとにスゴイ機体だったんですね。モスキートのライバルといわれる（？）He219ウーフーが、エンジンの出力不足に悩まされつづけた爆撃機なみの大型戦闘機だったことを考えると、小型の機体に優秀な大馬力エンジンを搭載したモスキートの勝ち、という気がします。デ・ハヴィランド社がこれだけの傑作機を生み出した背景には、やはり、かなりの技術的蓄積があったのだなと、あらためて思いました。

編（T）

●モスキートほど（作戦的に成功しているという意味で）特別な任務に使われた機体はないと思います。そのひとつひとつがドラマ的で、とても印象深い。だれか「モスキート爆撃隊」みたいな映画をつくってくれないかな。

編（S）

In North America, the *Aero Detail* series, *Achtung Panzer* series, and other Model Graphix books are available from :

RZM IMPORTS
PO Box 995 Southbury, CT 06488 U. S. A.
Tel: 203-264-0774
Fax: 203-264-4967
Email: RZMIMPORT@aol.com

●The authors wish to express their sincere thanks to Art Box and all the people around the world who allowed them to start a reliable historical revision concerning the Mosquito's technical data and features. The most persistant instance of data misinterpretation, regrettably perpetuated by many sources for too many years, regards the overall length of some two-stage engine-equipped Mosquito marks, including the B.Mk.XVI and P.R.Mk.XVI. These have been reported to be 44ft-6in (13.564m), a full 3ft-6in (1.067m) longer than the actual measurement! We earnestly hope that some of the previously overlooked and unpublished information provided here will be useful for enthusiasts to build and accurately paint models of one of the most beautiful and fascinating airplanes ever created.

Carmine Di Napoli & Raffaele Mancini
Website: http://www.aeroslides.com
E-mail: inform@aeroslides.com

●You have in your hands the largest and most information-packed volume in the Aero Detail series to date. We are extremely proud of it, and hope that you won't regret having parted with so much of your hard-earned money to acquire a copy. Messrs. De Napoli and Mancini do superb work, and we are grateful to have them as contributors to our series. Upcoming volumes in the Aero Detail series will cover the Japanese Hayate, F4U Corsair and more. As always, we welcome your comments. You can reach us by mail at Art Box to the address shown on this page, or by e-mail to the address below.

Scott T. Hards
HobbyLink Japan　www.hlj.com
E-mail: scott@hlj.com

エアロ・ディテール23
デ・ハヴィランド・モスキート
発行日　1999年1月　初版第1刷

発行所　株式会社大日本絵画
〒101-0054　東京都千代田区神田錦町1丁目7番地
03-3294-7861（代表）
編集　株式会社アートボックス
Edited by ART BOX Co., Ltd.
No.7, 1 Chome Kanda-Nishiki-cho Chiyoda-ku Tokyo
〒101-0054 Japan
印刷／製本　錦印刷株式会社
Published January 1999 by Dai Nippon Kaiga Co., Ltd.
No.7, 1-chome Kanda-Nishiki-cho, Chiyoda-ku Tokyo
〒101-0054 Japan